Waterlogging Signalling and Tolerance in Plants

Stefano Mancuso • Sergey Shabala
Editors

Waterlogging Signalling and Tolerance in Plants

Editors
Prof. Dr. Stefano Mancuso
Polo Scientifico - University of Florence
Dpt. Plant, Soil and Environmental
Science
LINV International Laboratory of Plant
Neurobiology
Viale delle idee, 30
50019 Sesto Fiorentino (FI), Italy
e-mail: stefano.mancuso@unifi.it

Prof. Dr. Sergey Shabala
University of Tasmania
School of Agricultural Science
Private Bag 54
Hobart, Tas, 7001, Australia
e-mail: sergey.shabala@utas.edu.au

ISBN 978-3-642-10304-9   e-ISBN 978-3-642-10305-6
DOI 10.1007/978-3-642-10305-6
Springer Heidelberg Dordrecht London New York

Library of Congress Control Number: 2010920284

© Springer-Verlag Berlin Heidelberg 2010
This work is subject to copyright. All rights are reserved, whether the whole or part of the material is concerned, specifically the rights of translation, reprinting, reuse of illustrations, recitation, broadcasting, reproduction on microfilm or in any other way, and storage in data banks. Duplication of this publication or parts thereof is permitted only under the provisions of the German Copyright Law of September 9, 1965, in its current version, and permission for use must always be obtained from Springer. Violations are liable to prosecution under the German Copyright Law.
The use of general descriptive names, registered names, trademarks, etc. in this publication does not imply, even in the absence of a specific statement, that such names are exempt from the relevant protective laws and regulations and therefore free for general use.

Cover design: WMXDesign GmbH, Heidelberg, Germany, from a sketch of S. Mancuso

Printed on acid-free paper

Springer is part of Springer Science+Business Media (www.springer.com)

Preface

In the last half century, because of the raising world population and because of the many environmental issues posed by the industrialization, the amount of arable land per person has declined from 0.32 ha in 1961–1963 to 0.21 ha in 1997–1999 and is expected to drop further to 0.16 ha by 2030 and therefore is a severe menace to food security (FAO 2006). At the same time, about 12 million ha of irrigated land in the developing world has lost its productivity due to waterlogging and salinity.

Waterlogging is a major problem for plant cultivation in many regions of the world. The reasons are in part due to climatic change that leads to the increased number of precipitations of great intensity, in part to land degradation. Considering India alone, the total area suffering from waterlogging is estimated to be about 3.3 million ha (Bhattacharya 1992), the major causes of waterlogging include superfluous irrigation supplies, seepage losses from canal, impeded sub-surface drainage, and lack of proper land development. In addition, many irrigated areas are subjected to yield decline because of waterlogging due to inadequate drainage systems. Worldwide, it has been estimated that at least one-tenth of the irrigated cropland suffers from waterlogging.

Higher plants are aerobic organisms needing a continuous delivering of oxygen to support respiration and oxidation reactions. When air spaces in the soil become saturated with water (e.g. flooding or water logging) plants can experience oxygen shortage because of the low solubility and diffusion rate of oxygen in water. The oxygen deficiency under such conditions is even aggravated by soil microorganisms that consume the residual oxygen and may also reduce the supply of nitrate (Gibbs and Greenway 2003). In addition, toxic microelements, CO_2, and ethylene may also accumulate in the rhizosphere (Armstrong and Drew 2002).

Subterranean organs of higher plants are normally exposed to large changes of the oxygen partial pressure (pO_2) in their local environment. These fluctuations can range from value close to 21 KPa, which is the value of the pO_2 of the air, to value close to zero in flooded soils. Thus, even in their normal habitat, most plants frequently experience hypoxic periods that result in many unfavorable morphological and physiological changes, including increased root mortality and reduced root metabolism. Thereby, the comprehension of the consequences of oxygen deficiency on plant growth is of crucial importance.

During anoxia, the oxidative phosphorylation by the respiratory chain stops and glycolysis becomes the only available route for ATP production. Moreover, the chemical oxidizing power (e.g. NAD^+) must be generated via pathways that use molecules other than oxygen as acceptor of the reductant. It seems that fermentation processes such as ethanolic or lactate fermentation are the principal means of supporting the reoxidation and recycling of NADH during anoxic events (Armstrong 1979).

Although the mechanisms of response to hypoxic or anoxic conditions are relatively well understood as a result of the work over a period of half a century, a number of unexplained issues still remain. Among others, three most critical questions are (1) what are the mechanisms of perception and signalling of oxygen deficiency, (2) which pathways consuming ATP are down-regulated, and (3) how membrane electrochemical gradients are stabilized under hypoxic conditions. Accordingly, in an attempt to give an updated representation of the state of the research, the book starts with a comprehensive biophysical examination of oxygen transport in plants and analysing the impact of waterlogging stress on a range of key biophysical and physiological parameters. Then the mechanism of intracellular signalling, mainly pH, oxygen, and nitrogen reactive species (ROS and RNS), the role of specific membrane transporters such as acquaporins, and the nature of putative "oxygen sensors" are discussed. Finally, in the last chapters, agronomical and genetics aspects of waterlogging tolerance and prospects of plant breeding are considered.

The volume consists of 13 chapters grouped into four main parts, namely,

1. Whole-plant regulation
2. Intracellular signalling
3. Membrane transporters in waterlogging tolerance
4. Agronomical and environmental aspects

collected to answer the needs of the largest audience of science students and plant researchers. We hope that the large range of subjects and the variety of approaches to the study of waterlogging signalling and tolerance in plants could raise the interest of even the scientists who are not directly involved in the study of such a subject but are interested to learn about this fascinating theme.

At the end, the editors gratefully acknowledge all the contributors who have made this book possible. Stefano Mancuso is grateful for the financial support given to his lab by the *Fondazione Ente Cassa di Risparmio di Firenze*; the waterlogging research in Sergey Shabala's laboratory was supported by funding from Grain Research and Development Corporation and by the Australian Research Council. Last but not least, support of the Springer's staff and specifically of Dr. Andrea Schlitzberger is greatly acknowledged.

January 2010
Stefano Mancuso
Sergey Shabala

References

Armstrong W (1979) Aeration in higher plants. Adv Bot Res 7:225–232

Armstrong W, Drew M (2002) Root growth and metabolism under oxygen deficiency. In: Waisel Y, Eshel A, Kafkafi U (eds) Plant roots the hidden half. Marcell Dekker, New York, USA, pp 729–761

Bhattacharya AK (1992) Status paper on waterlogged/drainage affected areas. Proceedings on Drainage Congestion in Northern Bihar, vols. I and II, Water and Land Management Institute Patna, Bihar, India

FAO (2006) World agriculture: towards 2030/2050 – Interim report, Rome

Gibbs J, Greenway H (2003) Mechanisms of anoxia tolerance in plants. I. Growth, survival and anaerobic catabolism. Funct Plant Biol 30:1–47

Contents

Part I Whole-Plant Regulation

1 Oxygen Transport in Waterlogged Plants 3
Lars H. Wegner
1.1 Introduction 4
1.2 O_2 Transport in Plants: Some Basic Physics, and Modelling
of O_2 Diffusion 5
1.3 A Survey of Methods to Study O_2 Transport and Related
Parameters in Higher Plants 7
1.4 Anatomical Adaptations to Flooding Stress: Barriers
to Radial Oxygen Loss 10
1.5 Anatomical Adaptations to Flooding Stress:
Formation of Aerenchyma 11
1.6 Mechanisms of O_2 Transport in Plants 13
1.7 O_2 Transport in Plants: Ecological Implications 18
1.8 Open Questions and Directions of Further Research 18
References 19

2 Waterlogging and Plant Nutrient Uptake 23
J. Theo M. Elzenga and Hans van Veen
2.1 Introduction 23
2.2 Effects of Hypoxia on Nutrient Uptake 26
2.2.1 Physiological Effects of Hypoxia Change Root Elongation
Rate, k, and Maximal Nutrient Uptake Rate, I_{max} 26
2.2.2 Waterlogging Leads to Changes in the Availability, C_{li},
and the Effective Diffusion Coefficient, D_e, of Some of the
Nutrients in the Soil 28
2.2.3 In Waterlogged Conditions, Some Plant Species Show
More Root Hair Development, Longer and Thinner

ix

Roots and Increased Levels of Infection With Mycorhizal
Fungi – Effectively Increasing k 29

2.2.4 Waterlogging Decrease Evaporation and Bulk
Water Flow, V_o 30

2.2.5 In Response to Waterlogging the Kinetics of Root
Transport Systems, K_m and I_{max}, Can be Modified 31

2.3 Summary and Concluding Remarks 31

References 32

**3 Strategies for Adaptation to Waterlogging and Hypoxia
in Nitrogen Fixing Nodules of Legumes** 37

Daniel M. Roberts, Won Gyu Choi, and Jin Ha Hwang

3.1 Introduction: The Oxygen Diffusion Barrier in Nodules 38

3.1.1 Nodule Morphology and the Gas Diffusion Barrier 38

3.1.2 Modulation of the Gas Diffusion Barrier 40

3.1.3 Control of the Gas Diffusion Barrier in Response
to Sub-Ambient O_2 and Flooding 40

3.1.4 Mechanism of Regulation of the Gas Diffusion Barrier
in Response to pO_2 41

3.2 Developmental and Morphological Adaptations of
Nitrogen-Fixing Nodules to Low Oxygen Stress 43

3.2.1 Secondary Aerenchyma Formation 43

3.2.2 The Inner Cortex and Infected Zone 44

3.2.3 Influence of Adaptive Changes on Nitrogen Fixation
Under Altered Rhizosphere pO_2 Conditions 45

3.3 Strategies of Adaptation: Flood-Tolerant Legumes
and Oxygen Diffusion 46

3.3.1 Tropical Wetland Legumes 46

3.3.2 *Lotus uliginosus*: A Temperate Wetland Legume 49

3.4 Strategies of Adaptation: Alternate Nodulation Pathways
for Flooding Tolerant Legumes 50

3.4.1 Intercellular-Based Mechanism of Nodulation:
The Lateral Root Boundary Pathway 50

3.4.2 *Sesbania rostrata*: A Model Legume for
Aquatic Nodulation 51

3.5 Summary and Concluding Remarks 53

References 55

4 Oxygen Transport in the Sapwood of Trees 61

Sergio Mugnai and Stefano Mancuso

4.1 Brief Anatomy of a Woody Stem 62

4.2 Atmosphere Inside a Stem: Gas Composition
and its Effects on Respiration 63

4.3 Gas Transport and Diffusion 66

Contents   xi

4.4 Radial and Axial Oxygen Transport to Sapwood 68
4.5 Sapwood Respiration 70
References 73

Part II Intracellular Signalling

5 **pH Signaling During Anoxia** 79
Hubert H. Felle
5.1 Introduction 79
5.2 pH, Signal and Regulator 81
     5.2.1 pH as Systemic Signal 82
     5.2.2 The Nature of pH Transmission 83
     5.2.3 What is the Information? 83
5.3 Anoxic Energy Crisis and pH Regulation 85
     5.3.1 The Davis-Roberts-Hypothesis: Aspects of pH Signaling 85
     5.3.2 Cytoplasmic Acidification, ATP and Membrane
            Potential 86
     5.3.3 Cytoplasmic pH (Change), An Error Signal? 87
5.4 pH Interactions Between the (Major) Compartments
     During Anoxia 88
     5.4.1 The pH Trans-Tonoplast pH Gradient 88
     5.4.2 Cytoplasm and Apoplast 90
     5.4.3 The Apoplast Under Anoxia 90
5.5 Anoxia Tolerance and pH 91
     5.5.1 pH as a Stress Signal – Avoidance of Cytoplasmic
            Acidosis 92
5.6 pH as Signal for Gene Activation 93
5.7 pH Signaling and Oxygen Sensing 94
5.8 Conclusions 94
References 95

6 **Programmed Cell Death and Aerenchyma Formation Under**
   Hypoxia 99
Kurt V. Fagerstedt
6.1 Introduction 100
6.2 Description of Aerenchyma Formation: Induced and
     Constitutive 102
6.3 Evidence for PCD During Lysigenous Aerenchyma Formation 103
6.4 Description of the sequence of events leading to induced
     lysigenous aerenchyma formation 104
     6.4.1 Stimuli for Lysigenous Aerenchyma Development
            (Low Oxygen, Cytosolic Free Calcium, Ethylene,
            P, N, and S Starvation, and Mechanical Impedance) 105
     6.4.2 PCD and the Clearing of the Cell Debris 110

6.4.3 What Determines the Architecture of
Aerenchyma? – Targeting and Restricting PCD 112
6.5 Future Prospects 113
References 113

7 Oxygen Deprivation, Metabolic Adaptations and Oxidative Stress 119
Olga Blokhina and Kurt V. Fagerstedt
7.1 Introduction 120
7.2 Anoxia: Metabolic Events Relevant for ROS Formation 121
7.2.1 "Classic" Metabolic Changes Under Oxygen Deprivation Related to ROS Formation 121
7.2.2 Changes in Lipid Composition and Role of Free Fatty Acids Under Stress 124
7.2.3 Modification of Lipids: LP 125
7.3 ROS and RNS Chemistry Overview and Sources of Formation Under Lack of Oxygen 126
7.3.1 Reactive Oxygen Species 126
7.3.2 Reactive Nitrogen Species 127
7.3.3 Plant Mitochondria as ROS Producers: Relevance for Oxygen Deprivation Stress 129
7.4 O_2 Fluxes in Tissues and Factors Affecting O_2 Concentration In Vivo 131
7.5 Microarray Experiments in the Study of Hypoxia-Associated Oxidative Stress 132
7.6 Update on Antioxidant Protection 133
7.6.1 Low Molecular Weight Antioxidants 134
7.6.2 Enzymes Participating in Quenching ROS 136
7.7 Concluding Remarks 138
References 139

Part III   Membrane Transporters in Waterlogging Tolerance

8 Root Water Transport Under Waterlogged Conditions and the Roles of Aquaporins 151
Helen Bramley and Steve Tyerman
8.1 Introduction 151
8.2 Variable Root Hydraulic Conductance (L_r) 152
8.3 Changes in Root Morphology and Anatomy 153
8.3.1 Root Death and Adventitious Roots 153
8.3.2 Barriers to Radial Flow 154
8.3.3 Varying the Root or Root Region Involved in Water Uptake 157
8.4 Volatile and Toxic Compounds in Anaerobic Soils 158

8.5 Water Permeability of Root Cells and Aquaporins 158
    8.5.1 Plant Aquaporins 159
    8.5.2 Responses at the Cell Level Affecting Water
        Permeability and Potential Mechanisms 161
    8.5.3 Other Changes Under Oxygen Deficiency that Could
        Affect Water Transport 169
    8.5.4 Transport of Other Molecules Besides Water
        Through MIPs Relevant to Flooding 170
8.6 Signalling 171
8.7 Conclusion and Future Perspectives 172
References 173

9 Root Oxygen Deprivation and Leaf Biochemistry in Trees 181
Laura Arru and Silvia Fornaciari
9.1 Introduction 182
9.2 Root O_2 Deprivation 183
    9.2.1 Root O_2 Deprivation: Effects on Leaves 185
9.3 The Role of ADH 185
9.4 Carbon Recovery 186
9.5 Differential mRNA Translation 188
9.6 Effects on Cell Metabolism 189
9.7 Conclusions 191
References 192

10 Membrane Transporters and Waterlogging Tolerance 197
Jiayin Pang and Sergey Shabala
10.1 Introduction 198
10.2 Waterlogging and Plant Nutrient Acquisition 198
    10.2.1 Root Ion Uptake 198
    10.2.2 Transport Between Roots and Shoots 199
    10.2.3 Ionic Mechanisms Mediating Xylem Loading 200
    10.2.4 Control of Xylem Ion Loading Under Hypoxia 201
10.3 Oxygen Sensing in Mammalian Systems 201
    10.3.1 Diversity and Functions of Ion Channels as
         Oxygen Sensors 201
    10.3.2 Mechanisms of Hypoxic Channel Inhibition 203
    10.3.3 The Molecular Mechanisms of Oxygen Sensing in
         Plant Systems Remain Elusive 203
10.4 Impact of Anoxia and Hypoxia on Membrane Transport
    Activity in Plant Cells 204
    10.4.1 Oxygen Deficiency and Cell Energy Balance 204
    10.4.2 H^+ and Ca^{2+} Pumps 204
    10.4.3 Ca^{2+}-Permeable Channels 205
    10.4.4 K^+-Permeable Channels 206

10.5 Secondary Metabolites Toxicity and Membrane
Transport Activity in Plant Cells 206
10.5.1 Waterlogging and Production of Secondary
Metabolites 206
10.5.2 Secondary Metabolite Production and Plant
Nutrient Acquisition 207
10.6 Secondary Metabolites and Activity of Key Membrane
Transporters 208
10.6.1 Pumps 208
10.6.2 Carriers 209
10.6.3 Channels 209
10.7 Breeding for Waterlogging Tolerance by Targeting Key
Membrane Transporters 211
10.7.1 General Trends in Breeding Plants for
Waterlogging Tolerance 211
10.7.2 Improving Membrane Transporters Efficiency
Under Hypoxic Conditions 211
10.7.3 Reducing Sensitivity to Toxic Secondary Metabolites 212
References 213

11 Ion Transport in Aquatic Plants 221
Olga Babourina and Zed Rengel
11.1 Introduction 221
11.2 Morphological and Physiological Adaptations of
Aquatic Plants 222
11.3 Ion Transport 224
11.3.1 Cation Transport Systems 228
11.3.2 Anion Transport Systems 230
11.4 Root Versus Leaf Uptake 230
11.5 Molecular Characterisation of Transporter Genes 232
11.6 The Relevance of Aquatic Plants to Terrestrial Plants
in Regards to Waterlogging and Inundation Stresses 233
11.7 Conclusions 233
References 234

Part IV Agronomical and Environmental Aspects

**12 Genetic Variability and Determinism of Adaptation
of Plants to Soil Waterlogging** 241
Julien Parelle, Erwin Dreyer, and Oliver Brendel
12.1 Introduction 242
12.2 Diversity Among Populations: Adaptation to
Water-Logged Soils? 246
12.3 Genetic Control of Traits Related to Hypoxia Tolerance 249

12.4	Genetic Determinism of Tolerance to Waterlogging and Identification of the Involved Genome Regions	250
	12.4.1 Methodology of the Detection of QTL for Hypoxia Tolerance: Caution and Strategies	251
	12.4.2 Major Loci Detected for Hypoxia Tolerance	256
12.5	Conclusions	260
References		260

13 Improvement of Plant Waterlogging Tolerance 267
Meixue Zhou

13.1	Introduction	267
13.2	Genetic Resources of the Tolerance	268
13.3	Selection Criteria	271
13.4	Genetic Studies on Waterlogging Tolerance	273
13.5	Marker-Assisted Selection	275
	13.5.1 QTL Controlling Waterlogging Tolerance	275
	13.5.2 Accurate Phenotyping is Crucial in Identifying QTLs for Waterlogging Tolerance	278
References		281

Index 287

Contributors

Laura Arru Department of Agricultural and Food Sciences, Besta University of Modena and Reggio Emilia, via Amendola 2 –pad., 42100 Reggio Emilia, Italy, laura.arru@unimore.it

Olga Babourina School of Earth and Environment, University of Western Australia, 35 Stirling Hwy, Crawley, WA 6009, Australia, Olga.Babourina@uwa.edu.au

Olga Blokhina Division of Plant Biology, Department of Biological and Environmental Sciences, University of Helsinki, Viikki Biocenter 3, POB 65, 00014 Helsinki, Finland, olga.blokhina@helsinki.fi

Helen Bramley Institute of Agriculture, The University of Western Australia, 35 Stirling Highway, Crawley, Western Australia 6009, Australia

Oliver Brendel INRA, UMR1137 "Ecologie et Ecophysiologie Forestières", F 54280 Champenoux, France; UMR1137 "Ecologie et Ecophysiologie Forestières", Faculté des Sciences, Nancy-Université, 54500 Vandoeuvre, France, brendel@nancy.inra.fr

Won Gyu Choi Department of Biochemistry and Cellular and Molecular Biology, The University of Tennessee, Knoxville, TN 37996, USA

Erwin Dreyer INRA, UMR1137 "Ecologie et Ecophysiologie Forestières", F 54280 Champenoux, France; UMR1137 "Ecologie et Ecophysiologie Forestières", Faculté des Sciences, Nancy-Université, 54500 Vandoeuvre, France, dreyer@nancy.inra.fr

J. Theo M. Elzenga Laboratory of Plant Physiology, Center for Ecological and Evolutionary Studies, University of Groningen, P.O. Box 14, Haren NL-9750AA, The Netherlands, j.t.m.elzenga@rug.nl

Kurt V. Fagerstedt Division of Plant Biology, Department of Biological and Environmental Sciences, University of Helsinki, Viikki Biocenter 3, POB 65, 00014 Helsinki, Finland, kurt.fagerstedt@helsinki.fi

Hubert H. Felle Botanisches Institut I, Justus-Liebig-Universität Gießen, Senckenbergstr 17, 35390 Gießen, Germany, Hubert.Felle@bio.uni-giessen.de

Silvia Fornaciari Department of Agricultural and Food Sciences, Besta University of Modena and Reggio Emilia, via Amendola 2 –pad., 42100 Reggio Emilia, Italy

Jin Ha Hwang Program in Genome Science and Technology, The University of Tennessee, Knoxville, TN 37996, USA

Stefano Mancuso Dpt. Plant, Soil and Environmental Science, University of Florence, viale delle Idee 30, 50019 Sesto Fiorentino, Italy, stefano.mancuso@unifi.it

Sergio Mugnai Dpt. Plant, Soil and Environmental Science, University of Florence, viale delle Idee 30, 50019 Sesto Fiorentino, Italy, sergio.mugnai@unifi.it

Jiayin Pang School of Plant Biology, The University of Western Australia, 35 Stirling Highway, Crawley, WA 6009, Australia

Julien Parelle University of Franche-Comté, UMR UFC/CNRS 6249 USC INRA "Chrono-Environnement", Place Leclerc, 25030 Besancon, France, julien. parelle@univ-fcomte.fr

Zed Rengel School of Earth and Environment, University of Western Australia, 35 Stirling Hwy, Crawley, WA 6009, Australia

Daniel M. Roberts Department of Biochemistry and Cellular and Molecular Biology, The University of Tennessee, Knoxville, TN 37996, USA; Program in Genome Science and Technology, The University of Tennessee, Knoxville, TN 37996, USA, drobert2@utk.edu

Sergey Shabala School of Agricultural Science, University of Tasmania, Private Bag 54, Hobart, Tas 7001, Australia, sergey.shabala@utas.edu.au

Steve Tyerman Wine and Horticulture, School of Agriculture, Food and Wine, The University of Adelaide (Waite Campus), Plant Research Centre, PMB 1, Glen Osmond SA 5064, Australia, stephen.tyerman@adelaide.edu.au

Hans van Veen Laboratory of Plant Physiology, Center for Ecological and Evolutionary Studies, University of Groningen, P.O. Box 14, Haren NL-9750AA, The Netherlands

Lars H. Wegner Plant Bioelectrics Group, Karlsruhe Institute of Technology, Institute of Botany I and Institute of Pulsed Power and Microwave Technology, Hermann-von-Helmholtz Platz 1 76344, Eggenstein-Leopoldshafen, Germany, Lars.Wegner@ihm.fzk.de

Meixue Zhou Tasmanian Institute of Agricultural Research, University of Tasmania, P.O. Box 46, Kings Meadows, TAS 7249, Australia, mzhou@utas.edu.au

Part I
Whole-Plant Regulation

Chapter 1
Oxygen Transport in Waterlogged Plants

Lars H. Wegner

Abstract In flooded soils, roots are exposed to a reducing environment with low oxygen availability. In order to supply roots with sufficient oxygen for respiration, most wetland plants form extended gas-filled cavities known as aerenchyma (in the root) or lacunae (in the shoot). Oxygen transport can be either diffusive or convective; according to general belief, the latter is restricted to the existence of separate "inlet" and "outlet" structures, e.g. in extended rhizomes. Mainly three separate mechanisms of convective oxygen transport have been identified so far: (1) Humidity-induced pressurization/convection that is driven by a difference in vapour pressure between the water-saturated gas phase of the leaf intercellulars and the micro-environment. When a significant overpressure in the gas phase of the leaf is maintained, gas flow to a distant "outlet", being at atmospheric pressure, is generated. (2) Thermal osmosis, a mass flow of gas into the leaf driven by heat transfer in the opposite direction and (3) Venturi-induced convection, a gas flow initiated by an underpressure in the gas phase of broken culms (e.g. of *Phragmites communis*) that is generated by wind passing over them. Species with extended pathways for gas transport (with formation frequently being induced by hypoxia or anoxia) dominate in wetland communities and can even provide the rhizosphere of flooded soils with oxygen.

Abbreviations

ATP   Adenosine triphosphate
c_{O_2} Oxygen concentration

L.H. Wegner
Plant Bioelectrics Group, Karlsruhe Institute of Technology, Institute of Botany I and Institute of Pulsed Power and Microwave Technology, Hermann-von-Helmholtz Platz 1, 76344 Eggenstein-Leopoldshafen, Germany
e-mail: Lars.Wegner@ihm.fzk.de

S. Mancuso and S. Shabala (eds.), *Waterlogging Signalling and Tolerance in Plants*,
DOI 10.1007/978-3-642-10305-6_1, © Springer-Verlag Berlin Heidelberg 2010

C_0	Initial Oxygen concentration
COPR	Critical oxygen pressure for respiration
D	Diffusion coefficient
k_H	Henry's gas constant for O_2 in water
J_{O_2}	Oxygen flux
l	Root length
M	Respiratory demand
$P(O_2)$	Oxygen partial pressure
Q	Oxygen consumption by respiration
R_0	Root radius
ROL	Radial oxygen loss
R_{SH}	Diffusional resistance between root surface and oxygen electrode
R_T	Complex resistance of the root tissue (includes root porosity, consumption of O_2 along the pathway by respiration and radial loss, and the diffusional resistance for radial O_2 transport)
ϵ	Fractional porosity of the root
T	Temperature
τ	Tortuosity of a diffusion pathway
ρ	Air density
v	Wind speed
∇^2	Laplacian operator

1.1 Introduction

Submerged plants and plants rooting in flooded soils have to cope with a lack of oxygen, or anaerobiosis. Root growth and maintenance of root tissue mainly rely on respiration (oxidative phosphorylation) as the most efficient source of energy. Respiration, however, depends on the supply of oxygen; when the O_2 concentration drops below a certain level that varies somewhat among species ("Pasteur point"; ca. 0.2%), respiration is completely inhibited. Below the critical oxygen pressure for respiration (COPR; lowest concentration to support maximum respiration rate; Armstrong et al. 2009), respiration is still active but its rate is strongly reduced.

When respiration is impaired at low O_2 supply (hypoxia) or in the absence of O_2 (anoxia), root cells are able to switch to fermentation for ATP synthesis (Thomson and Greenway 1991). However, fermentation is thermodynamically much less efficient than respiration: While a total number of 36 molecules ATP is synthesized per mol glucose during respiration, it is only about three molecules during fermentation (depending on the process of fermentation). Fermentation in roots usually results in the accumulation of ethanol as the end product that can reach a toxic concentration level when the tissue suffers from prolonged oxygen deprivation. Moreover, toxic by-products are produced under anaerobic conditions not only by the plant tissue (e.g. acetaldehyde), but are also released by bacteria that prevail under these conditions, an adverse factor that is still frequently undervalued. As a consequence, growth and vitality are negatively affected in plant tissues that suffer from oxygen deprivation, e.g. in roots growing in waterlogged soils, and these conditions may even be lethal under extreme conditions. Interestingly, there is no evidence that O_2 is used more efficiently in plants that are adapted to submergence or to waterlogged soils (Lambers et al. 1978), even though the tolerance towards this factor shapes the composition of plant communities on soils with temporal or permanent flooding, and to some extent also in communities under partial or complete submergence (although factors for competition are more complex under these conditions). When species of the same genus differing in flooding tolerance where compared at the tissue level with respect to the dependence of the respiration rate on O_2 partial pressure, or with respect to the minimum partial pressure required for respiration, no major difference between species or treatments was observed (Lambers et al. 1978; Lambers and Steingröver 1978; Lambers and Smakman 1978).

The alternative is that cells are supplied more efficiently with oxygen in those plants that can cope better with anoxic conditions, and this is indeed the case: Tolerant plants develop anatomical structures known as lacunae (in the shoot) and/ or aerenchyma (in the root) that serve as internal pathways for effective oxygen supply to organs that suffer from O_2 deprivation. Oxygen transport in these structures can be diffusive, but especially in shoots and extended rhizomes, convective mechanisms of oxygen delivery have also been identified, that are much more efficient when long distances have to be traversed. Both diffusive and convective mechanisms will be treated in detail below. Beforehand, some basic physics related to oxygen transport in plants will be discussed.

1.2 O_2 Transport in Plants: Some Basic Physics, and Modelling of O_2 Diffusion

Oxygen that is consumed in plant tissues by respiration is eventually replenished by the oxygen reservoir provided by the atmosphere and by air-filled cavities in the soil. A phase transition of O_2 from the gas phase to the fluid phase (or vice versa), usually occurring in soil or at sites within plant organs (sub-stomatal cavities, gas-filled intercellular spaces, lacunae, aerenchyma) is governed by Henry's law, provided that the two phases are approximately at equilibrium:

$$P(O_2) = k_H c_{O_2} \quad \text{(for } T = \text{const)} \tag{1.1}$$

stating that the oxygen concentration, c_{O_2}, in solution is proportional to the partial pressure in the adjacent gas phase, $P(O_2)$; k_H is Henry's gas constant for O_2 in water. The solubility of O_2 in water is relatively high; at the usual partial pressure of O_2 in air (0.2 bar), a concentration of 8 mg O_2 l^{-1}, corresponding to 0.25 mM, is calculated at $T = 298$ K. This means that (1.1) does not predict any shortage of oxygen in plants. However, equilibrium thermodynamics is of very limited value when dealing with living organisms. A more adequate mathematical description is obtained using equations that derive local O_2 concentration from equations of oxygen transport J_{O_2} and consumption by respiration (Q). The conservation of mass requires the local O_2 concentration anywhere in the plant to be

$$dc_{O_2}/dt = J_{O_2} - Q \qquad (1.2)$$

Oxygen transport can either be diffusive or convective. Diffusive transport is described by Fick's law:

$$J_{O_2,\text{diff}} = D\nabla^2 c_{O_2} \qquad (1.3)$$

With D being the diffusion constant for O_2 and ∇^2 being the Laplacian operator that is defined as $\partial^2/\partial x^2 + \partial^2/\partial y^2 + \partial^2/\partial z^2$. When a (local) steady state is established, the transport rate equals the rate of respiration.

$$J_{O_2,\text{diff}} = Q \qquad (1.4)$$

The diffusion constant for O_2 in water is quite low; O_2 diffusion is about 10^4 times slower in water than in air. The reason for this is that movement of O_2 in water is extremely retarded by the interaction of O_2 and H_2O molecules via hydrogen bonds.

In order to predict steady-state concentration profiles of oxygen across a tissue based on these equations, it is assumed that at a hypothetical, initial state, the oxygen concentration is uniform across the tissue fulfilling Henry's law. While this value will prevail locally at the air/water boundary, the O_2 concentration in the tissue will drop with increasing distance from this boundary, since O_2 consumption initially exceeds delivery until O_2 gradients are established within the tissue that accelerate diffusion; under steady-state conditions, O_2 re-supply, matches respiratory demand. This general model can be adapted to specific scenarios taking the anatomical structures into account. Among the most obvious applications is the calculation of radial O_2 profiles in roots when the root surface is in contact with air, or is covered by a water film. This situation was analysed in detail by De Willingen and van Noordwijk (1984). For the root surface being exposed to air, which is the simplest situation, they arrived at the following steady-state solution for the oxygen concentration c_{O_2} normalized to the initial value C_0:

$$c_{O_2} = 1 - \frac{q}{4}(1 - r^2) \qquad (1.5)$$

r is the distance from the root centre divided by the root radius R_0, and q is a factor defined as $q = QR_0^2/DC_0$ that includes Q, the rate of respiration. Equation 1.5 describes a hyperbolic decrease of the O_2 concentration with respect to C_0 at the surface when moving towards the root centre. The magnitude of the drop strongly depends on R_0 that is raised to the second power, indicating that oxygen depletion at the root centre is more severe in thick roots. A linear relationship between c and r^2 was indeed confirmed when experimental data of Bowling (1973) obtained by radial insertion of an O_2-sensitive microelectrode (see below) were plotted in this way (Van Noordwijk and De Willingen 1984). These authors also considered more complicated boundary conditions with a water film covering the root surface, and with oxygen supply to the root surface being (locally) impeded by solid soil particles. These models were based on the assumption that respiration was constant throughout the tissue and was not dependent on the O_2 concentration unless it was virtually zero.

Under waterlogging conditions, when the root is exposed to anoxic conditions, no radial oxygen supply is provided, and O_2 supply to the root tip is restricted to longitudinal diffusion from aerial parts of the plant via the root tissue. Assuming that no radial efflux of O_2 occurs, that respiration and other root properties are constant along the pathway, the oxygen concentration profile with root length, $c_{O_2}(l)$, is calculated according to (Greenwood 1967; Armstrong et al. 1983):

$$c_{O_2}(l) = C_0 - \frac{M}{2D\varepsilon\tau}l^2 \tag{1.6}$$

M is the respiratory demand (gram oxygen per cubic centi-metre per second), ϵ is the fractional porosity of the root, i.e. the gas space normalized to the total volume, τ is the tortuosity of the pathway and l is the length. A more sophisticated version of this model was presented by Armstrong and Beckett (1987), taking different rates of O_2 consumption in cortex and stele into account. The diffusion pathway is much more extended under these conditions compared to radial O_2 uptake, and it is obvious that the apical part of the root will soon suffer from severe anoxia in the absence of any anatomical structures that support longitudinal O_2 transport (Fig. 1.1); gas-filled spaces facilitate long-distance O_2 transport and thus serve this purpose most efficiently. Therefore, it is no surprise that species forming extended, continuous gas-filled spaces (aerenchyma) in roots or rhizomes under waterlogging conditions dominate in wetland vegetation (Blom 1999; Jackson and Armstrong 1999). Movement of O_2 in aerenchyma can either be diffusive or convective; mechanisms are outlined in more detail below. Before, a brief overview of the methods used to measure O_2 transport in plants is given, and relevant aspects of plant anatomy are discussed.

1.3 A Survey of Methods to Study O_2 Transport and Related Parameters in Higher Plants

Standard methods for measuring O_2 concentrations and fluxes use polarographic techniques. W. Armstrong was the first to introduce this approach to measure O_2 transport in plants (Armstrong 1964, 1979; Armstrong and Wright 1975). He employed cylindrical platinum electrodes that ensleeve the root closely behind

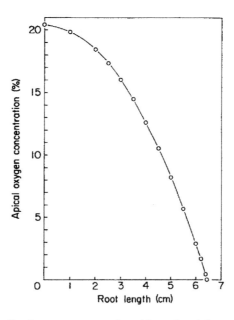

Fig. 1.1 Calculated profile of oxygen concentration with root length in pea using (1.6). Values for c_0, ϵ and τ were taken from experimental data obtained on pea seedlings. From Armstrong et al. 1983, with permission

the apex. At the surface of the polarized platinum electrode, O_2 is electrolytically reduced to two OH^- requiring four electrons that are drawn from the reference electrode, giving rise to an electric current. It turned out that the rate of O_2 reduction being proportional to the recorded current is not related to the exact electrode potential, but rather to the rate of diffusion of oxygen towards the electrode, which, in turn, is a function of the gradient in O_2 concentration between the root surface and the surface of the platinum electrode (where the concentration is effectively zero; Armstrong 1979). When the root is bathed in a de-oxygenated medium, radial oxygen loss from the root (ROL) is related to O_2 transport in the plant according to:

$$\text{ROL} = \frac{\Delta c_{O_2}}{R_\text{T} - R_\text{SH}} \quad (1.7)$$

Δc_{O_2} is the concentration difference between the root base and the electrode surface and R_SH is the diffusional resistance between root surface and electrode. The parameter R_T is a complex resistance of the root tissue that includes root porosity, consumption of O_2 along the pathway by respiration and radial loss, and the diffusional resistance for radial O_2 transport into the medium at the site of the electrode. When, radial oxygen loss from the root outside the electrode is minimized and respiration is inhibited (e.g. by lowering the temperature), this experimental approach can be used to estimate the effective longitudinal resistance for oxygen transport in the root. Moreover, the internal oxygen concentration in the root can be inferred (Armstrong and Wright 1975), at least in relative terms when comparing ROL values at different physiological situations or at different sites of the root.

For measuring oxygen fluxes, an interesting alternative to root sleeving electrodes is provided by oxygen microelectrodes that are used to measure directly the O_2 concentration gradient at the root surface that drives oxygen flux. With this technique, fluxes can be measured at a high spacial resolution. Mancuso et al. (2000) described a vibrating O_2 selective probe for this purpose. A similar system was used by Pang et al. (2006) to characterize O_2 fluxes in barley cultivars differing in waterlogging tolerance. Hypoxic stress resulted in a marked decrease in oxygen uptake in the mature root zone; this decrease was more pronounced with the waterlogging-sensitive cultivar than with the insensitive one.

By recording oxygen fluxes, only indirect information on processes in the plant can be obtained. A bulk O_2 partial pressure *in* the root tissue can be measured locally by attaching an O_2 electrode of the Clark type, similar to the one described above, e.g. to the stump of a cut lateral root (Skelton and Allaway 1996). While these "macroscopic" electrodes render overall values, more detailed information on local O_2 partial pressures on tissue level and the location of diffusion barriers is provided by the insertion of oxygen microelectrodes. Radial O_2 gradients in roots were first studied by Bowling (1973) using platinum-tipped microelectrodes. He observed a flat O_2 gradient in maize roots. The approach was criticized for methodological reasons, since artifacts could be introduced by the insertion of the electrode into the tissue, but later Armstrong and coworkers (Armstrong et al. 2000), using miniaturized, gold-tipped Clark electrodes, could show that O_2 profiles across the whole root were identical during penetration and subsequent microelectrode retraction. The microelectrodes fabricated by Armstrong et al. (2000) had a rapid response time of less than 1 s. Tip diameter ranged between 12 and 18 µm; electrodes with narrow tips proved to be unstable. Therefore, the technique was restricted to thick roots like those of *Phragmites australis* and *Zea mays* in order to minimize the damage. Recently, Colmer and Pedersen (2008) used commercially available O_2 microelectrodes to measure oxygen partial pressures in rice to study, among other things, the dynamics of oxygen during complete submergence. According to the authors, microelectrodes were inserted into lacunae of the midrib of leaf sheaths, and into root aerenchyma; unfortunately it remains unclear how the exact position of the electrode tip was verified.

The use of microelectrodes for measuring oxygen concentrations and fluxes in plants requires micromanipulation of the electrodes and, in case of electrode insertion into the tissue, can be quite cumbersome. A range of alternative techniques are available to study gas transport in a more convenient way, where this is feasible. Qualitative data on oxygen concentrations at the root surface can be obtained by using O_2-sensitive dyes, e.g. methylene blue (Armstrong et al. 1992). An obvious approach is to expose the shoot to an artificial atmosphere, e.g. to 100% ethene (Raskin and Kende 1983; Grosse et al. 1992), and to measure the change in the gas composition at some point "downstream" in a small perspex chamber (so-called head space technique; Brix 1989; Brix et al. 1996), or in a gas flask that contains the root system, with the shoot protruding through a tight rubber seal (Grosse et al. 1992). An interesting modification is to expose whole stands of plants to an artificial atmosphere; this is a most elegant approach when plants like *Phragmites communis* are studied that form extensive rhizomes. However, results obtained with the head space technique and related techniques have to be interpreted with care, since artificial gradients in gas concentration are introduced and the microclimate experienced by the shoot is changed. Brix et al. (1996) advised to apply this method to study diffusional transport only, but the artificial experimental situation may per se give rise to some convective transport. An alternative is to extract gas samples from lacunae and aerenchyma using syringes to analyse (natural) gradients of O_2 partial pressure within the plant, but this is again difficult from a technical viewpoint and restricted to a few species only. In order to quantify mass flow of gases, it is sometimes possible to attach parts of a plant like excised culms of *Phragmites* to a flow meter; more elegant is the by-pass technique that leaves the plant largely undisturbed (Brix et al. 1996). The internal gas transport in the plant is interrupted by local flooding of a gas space at an internodium with water or another liquid; this obstruction is bypassed by a tube with integrated flow meter. Samples can be taken regularly from the tube to analyse the composition of the gas phase. The technique has been developed for *Phragmites*, and it is not known whether it can be transferred to other species. Mass flow of gas is usually believed to be driven by gradients in gas pressure (although this is not necessarily so under all conditions, see below); pressure can be measured locally by inserting small manometers into air spaces, e.g. pith cavities of *Phragmites* culms (Afreen et al. 2007). Care has to be taken not to affect native pressure gradients by this procedure.

A very important parameter that connects physical aspects of gas flow with anatomy is the fractional porosity of the tissue, i.e. the fractional volume that is taken by gas space. This can be done conveniently by measuring the density of a tissue block using a pycnometer (Armstrong 1979), or alternatively the change in weight upon an infiltration of the tissue with water (Michael et al. 1999). Recently, a non-invasive method to determine the local porosity of root tissue has been suggested by Kaufmann and co-authors using ^{1}H MR microscopy (Kaufmann et al. 2010). It is important to note, though, that porosity is just a rough indicator for the impedance of a tissue to gas transport; in order to understand mechanisms of oxygen transfer in more detail, an anatomical investigation into the pathways for gas transport and the location of diffusion barriers is required. These aspects will be discussed in the following sections.

1.4 Anatomical Adaptations to Flooding Stress: Barriers to Radial Oxygen Loss

When a root is situated in a reducing environment, and oxygen supply can only be provided by downward transport from those parts of the plant that are in contact with the atmosphere, it will be crucial to minimize radial oxygen loss from the root to provide the root apex with sufficient oxygen. ROL is prevented by tissues serving as a barrier to radial oxygen loss. In some species such as rice (Colmer et al. 1998), *Hordeum marinum* (Garthwaite et al. 2008) and *Caltha palustris* (Visser et al. 2000), these barriers are inducible by flooding stress. This aspect is only briefly discussed here; for a more thorough treatise of this topic, the reader is referred to Colmer (2003). Several anatomical features seem to contribute to the formation of "tight" barriers to ROL: Densely packed rows of parenchymatous cells in the outer cortex, suberin deposits in the cell wall of a cell layer (hypodermis), sclerenchymatous fibres that develop further into an exodermis. These structures will also provide additional mechanical support when large aerenchyma is formed in the cortex in response to flooding stress.

1.5 Anatomical Adaptations to Flooding Stress: Formation of Aerenchyma

Here, I will focus on structural aspects of aerenchyma in rhizomes and roots with reference to their function, that is, to facilitate long-distance gas transport. Biochemical and molecular aspects of aerenchyma formation upon flooding that have been unravelled in detail in recent years will be covered by a separate contribution to this book (chapter 8, p. 151–180). Evolutionary aspects are discussed in a recent review by Jackson et al. (2009). The summary given here is mainly based on the comprehensive study by Justin and Armstrong (1987) and more recent ones by Visser et al. (2000) and Seago et al. (2005) on this topic.

Originally, two basic types of aerenchyma were identified derived from the mechanism of their formation (Arber 1920; Fig. 1.2). *Schizogenous aerenchyma* is formed by a local disassembly of the cell wall and growth patterns that favour the formation of a gas phase of increasing size with distance from the apex. In *Lysigenous* aerenchyma, by contrast, gas-filled cavities are generated by programmed cell death of single cells or groups of cells. Seago et al. (2005) introduced the term "expansigeny" to denote a separate type of aerenchyma that is formed from intercellulars by programmed patterns of cell division and growth only, frequently giving rise to "honeycomb structures" in traverse sections of the root, e.g. in *Rumex* and *Hydrocotyle* (Fig. 1.2e). No information is provided to what extent this type of aerenchyma (that is hard to distinguish from the schizogenuous type) is inducible by O_2 deficiency. A more detailed typology of aerenchyma was developed by Justin and Armstrong (1987) based on anatomical investigation of 91 species. Their classification is based on anatomical criteria, rather than on ontogenesis. Generally, the structure of aerenchyma is the result of a trade-off between gas conductance and mechanical stability, since a mechanical collapse of the aerenchyma would counteract its function. It is frequently observed that radial sectors of cells disintegrate, sometimes with cell walls remaining like in the gramineean type; in other cases, tangential rows of cells decay for aerenchyma formation. Justin and Armstrong

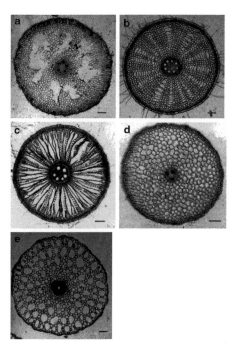

Fig. 1.2 Anatomical adaptation of roots to waterlogging conditions. Fresh cross-sections at 5 cm behind the apex of adventitious roots of five wetland species. (**a**) *Caltha palustris*, (**b**) *Carex acuta*, (**c**) *Juncus effusus*, (**d**) *Ranunculus sceleratus*, (**e**) *Rumex palustris*. Plants were grown in stagnant deoxygenated nutrient solution containing 0.1% agar for at least 2 weeks prior to taking the sections. Scale bars: 0.1 mm. Lysogenuous aerenchyma in **a** and **c**, schizogenuous/expansigenous aerenchyma in **b** and honeycomb structure in **e**. No aerenchyma formation in **d**. From Visser et al. 2000, with permission

(1987) stresssed that cubic packaging of cells (cells adjusted radially in consecutive rows) favoured both porosity of the tissue and aerenchyma formation compared to a hexagonal arrangement (alternating position of cells in consecutive rows). At a given contact surface with neighbouring cells, intercellulars were much more extended with the former than with the latter. Moreover, aerenchyma formation under flooding stress was almost always associated with cubic, but not with hexagonal arrangement of cells. Aerenchyma was also absent in most secondary tissues (with the exception of phellem); accordingly, formation of secondary tissue is frequently suppressed when wetland dicots are exposed to flooding stress.

Coupling these observations to function, Justin and Armstrong (1987) found that among those species they investigated, aerenchyma formation was favoured by flooding stress (based on measurements of fractional root porosity) in >50% of the typical wetland plants, but only in few "non-wetland" or "intermediate" plants. When porosity of wetland plants remained unaffected by flooding, aerenchyma formation was independent of flooding stress, or these species were shallow-rooting.

Both depth of root penetration and root length were positively correlated with aerenchyma formation/increase in porosity in flooded soil; a curvilinear relationship was found that could be explained by O_2 profiles generated by diffusion from aerial parts into the root. Wetland-plants clearly out-competed non-wetland plants. These results were later confirmed by Visser et al. (2000); these authors also reported an increase in root diameter induced by soil flooding in three out of five wetland plants investigated in that study.

It should be noted that aerenchyma are not restricted to roots, but are of equal importance in rhizomes. Several macrophytic wetland species, among them *Typha* and *Phragmites* develop extended rhizomes that are buried in anoxic soil. Oxygen supply under these conditions is provided by interconnected gas spaces that form a low-impedance network for O_2 transport.

These studies clearly demonstrated the beneficial effect of aerenchyma in waterlogged soils. A detailed account of mechanisms that favour efficient O_2 conduction in plants is given in the following section.

1.6 Mechanisms of O_2 Transport in Plants

Oxygen transport in gas-filled spaces is often diffusive, but also various mechanisms for convective mass flow (ventilation) have been advanced. Both theoretical aspects and experimental evidence raised in favour of the different mechanisms will be critically discussed here.

Diffusion. Diffusion is supposed to be the only mechanism for O_2 transport in plants towards the root apex in the absence of any gas circulation or separate inlet- and outlet for a throughflow (but see Raskin and Kende 1983). In waterlogged soil, longitudinal gradients for O_2 diffusion are established along the pathway by respiratory oxygen consumption of tissues (see above), and by radial loss to the soil that forms an oxygen sink under these conditions. In most plants adapted to flooding, gradients of O_2 concentration are flattened by aerenchyma formation, thus decreasing the longitudinal resistance for O_2 diffusion, while local O_2 demand is decreased when programmed cell death is involved. For pea, a plant with generally low fractional root porosity (0.038; Armstrong et al. 1982) that is not affected by waterlogging, evidence was obtained that flooding induced a down-regulation of respiration in apical parts of the root. This assured sufficient O_2 supply to the root apex for sustained growth (Armstrong et al. 1983), up to a root length of 7–8 cm. Radial loss of oxygen to the rhizosphere can be minimized by diffusion barriers (see above). Nonetheless, it has to be acknowledged that the efficiency of a diffusive supply with oxygen is limited and could not maintain sufficient oxygen transport over the distance of metres, as required for macrophytic species with long rhizomes such as *Phragmites* and *Typha*. From a theoretical viewpoint, mechanisms that involve mass flow should be much more efficient.

Convection. Several convective mechanisms of gas transport in plants have been advanced and confirmed by experiments. Although each mechanism operates on a distinct physical basis, it is frequently difficult to separate them experimentally, especially under field conditions.

- *Humidity-induced pressurization/convection.* In the intercellular gas phase of a leaf, an overpressure is generated when a significant vapour pressure deficit exists in the microenvironment of the leaf, while the gas phase in the leaf remains saturated with water. The partial pressures of the other gases will be at equilibrium in both phases. A prerequisite for the maintenance of a pressure differential is the existence of small pores in the barrier separating the gas phases from the atmosphere that do support diffusion, but not mass flow between compartments. Stomatal pores are generally believed to fulfil this property. In order to support convective transport by Poiseuille-flow, there has to be an "outlet" at some distance from the leaf (e.g. a dead, neighbouring culm; Fig. 1.3a) wherein, gas flow can escape to the atmosphere. Overpressure in leaf intercellulars will support gas flow through the plant as long as the diffusion rate of gas molecules from the atmosphere into the leaf is sufficient to prevent a breakdown of the overpressure, and provided that the difference in vapour pressure between the compartments is maintained. A mathematical model on humidity-induced convection has been presented by Armstrong et al. (1996b), While this "operative" model may describe the effect adequately under range of conditions, it suffers from obvious shortcomings: The model predicts a decrease in convective flow with "venting resistance" (the resistance for mass flow from the leaf to the outlet where pressure is at the atmospheric level; Fig. 2 in

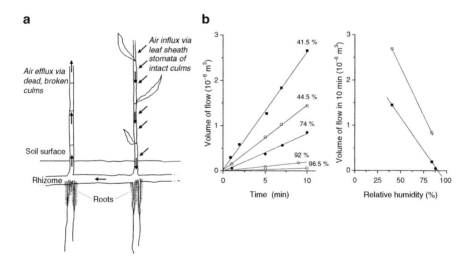

Fig. 1.3 Convective gas flow driven by humidity-induced pressure in *Phragmites*. The ventilation pathway is shown schematically in **a**. As indicated by the data shown in **b**, mass flow is strongly affected by temperature (open symbols: 24°C; closed symbols: 17°C) and by relative humidiy (see numbers at traces), underpinning the relevance of this mechanism for convective gas flow. Adopted from Armstrong et al. 1996a, with modifications

Armstrong et al. 1996b), but it should be an optimum curve, since no flow is supported when the resistance is infinitely low, and no pressure differential is generated under these conditions

- *Thermal osmosis.* When the temperature of the internal gas phase in a leaf exceeds the temperature in the atmosphere, this will induce heat flow across the barrier separating the compartments. This heat flow can drive a mass flow (uptake of gas molecules by the leaf) under isobaric conditions. An influx of gas molecules, in turn, will lead to the build-up of an overpressure in leaf inter-cellulars. Coupling of flows implies very low pore size (<1 μm) supporting this so-called "Knudsen diffusion". The effect was derived theoretically and has been demonstrated experimentally (Grosse 1996). It should not be mixed up with a temperature-induced increase in vapour pressure when the gas phase in the leaf remains saturated with air (Colmer 2003). For thermal osmosis driving convective flow, again a distant outlet that supports mass flow is required.
- *Venturi-induced convection.* Wind passing over broken culms e.g. of *P. australis* will generate a sub-atmospheric pressure in the stationary gas phase within the culm (so-called Venturi effect). The pressure difference to the ambient, atmospheric pressure, ΔP, is described by the following equation:

$$\Delta P = -\tfrac{1}{2}\rho * v^2 \tag{1.8}$$

with ρ and v being the air density and the wind speed, respectively.

When other broken culms are protected from the wind, e.g. due to their location within a stand or in case they are broken just above the water level, a pressure differential will be generated between these dead culms that will drive ventilation through the rhizome connecting them (Fig. 1.4a). This mechanism was first identified by Armstrong and co-workers (1992, 1996a).

Theoretical concepts were verified by performing model experiments (Armstrong et al. 1996b; Dedes and Woermann 1996). Convective flow was shown to depend strongly on the pore size; for humidity-induced convection, an optimum at about 0.2–0.4 μm was found (Armstrong et al. 1996b). Moreover, contribution of these mechanisms to convective flow in a range of aquatic and wetland plants has been tested. Humidity-induced convection seems to be the dominant mechanism in *Phragmites* (together with Venturi-induced convention), whereas in floating-leaved aquatic plants like Indian lotus (*Nelumbo nucifera*) and waterlilies, thermal osmosis is apparently also contributing significantly to ventilation (Dacey 1980; Mevi-Schutz and Grosse 1988; Grosse 1996). Since thermal osmosis is less important for wetland plants, I will focus on the other two mechanisms here that were extensively studied on *P. australis*. The interconnected system of gas spaces in this species that supports humidity- and Venturi-induced convection was reviewed by Armstrong et al. (1996a). The resistance of this system to mass flow is low; even diaphragms in the culms and the rhizome do not form a significant barrier. Humidity-induced convection is supposed to originate at the leaf sheaths, since the leaf lamina is separated by a pressure-tight barrier at the pulvinus. Stomatal pores in the sheaths

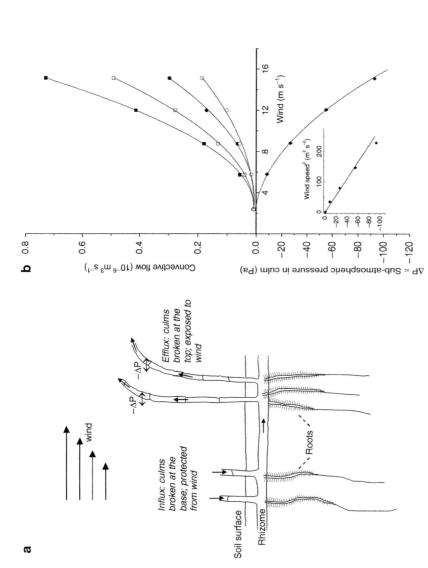

Fig. 1.4 Venturi-induced convection (see sketch in **a**) in *Phragmites* as demonstrated by the effect of wind speed on the gas flow rate at different cross-sectional areas of the pith (■ 0.283 cm^2; □ 0.196 cm^2; ● 0.146 cm^2; ○ 0.096 cm^2) and on the sub-atmospheric pressure generated in the terminal part of culms broken at the top (**b**). Inset, b: Linear relationship between pressure drop in the culm and (wind speed)^2, as predicted from equation 1.8. Data were obtained on rhizomes with cut culms. Adopted from J. Armstrong et al. 1996a, with modifications

were identified as entry points for gases. This may be considered as a weak point, since it is at least questionable if fully open *Phragmites* stomata would form an efficient barrier to Poiseuille flow, as required for humidity-induced convection. Armstrong et al. (1996a) insist that a very small slit between accessory cells forms a bottleneck that is sufficiently small to block Poiseuille flow. It is unknown whether stomatal plugs that have recently been identified in a broad range of mostly woody species (Westhoff et al. 2009) also exist in *Phragmites*; these plugs could form pores of the required size. Substomatal cavities are connected with leaf sheath aerenchyma channels and, in turn, with gas-filled pockets in the nodes of the culm that end at the large, central pith cavities. These cavities extend into the rhizome; the outlet is provided by dead, broken culms that are open to the atmosphere. Roots inserting at the rhizome (but not laterals) also form large aerenchyma, but transport in roots is rather diffusive since no throughflow is possible. Humidity-induced convection has been demonstrated by showing that mass flow is strongly dependent on relative humidity (r.h.) and absent when r.h. is 100% (Fig. 1.3b; Armstrong and Armstrong 1991). Overpressure in excised culms amounted to about 350 Pa; in the intact plant, a value of about 116 Pa was obtained. The flow rate attained maximum values of about $10 \, \mathrm{ml \, min}^{-1}$. Venturi-induced convection resulted in a pressure drop (suction) in decapitated, wind-exposed culms by a few Pa; the pressure drop was dependent on wind speed. Convective flow in these culms was also shown to depend on (wind speed)^2 (Fig. 1.4b) as expected from (1.8). Moreover, ROL from roots was enhanced by either humidity- or Venturi-induced convection compared to pure O_2 diffusion.

While the mechanisms of convection themselves have been studied on *Phragmites* in much detail, surprisingly little work has been performed to demonstrate that convection is indeed beneficial under field conditions. An exception is the work of Vretare and Weisner (2000). When all stems of *Phragmites* plants grown in containers were perforated closely above the water surface to inhibit pressurized ventilation, oxygen concentrations in stem bases and rhizomes decreased. Carbon allocation to below-ground parts and rhizome length were reduced under these conditions, and overall biomass was also affected partially. Evidence was obtained showing that pressurized ventilation improved the performance of *Phragmites*, although results of Vretare and Weisner (2000) were not as clear-cut as one might have anticipated. Obviously, the plant is able to respond by morphological adaptations when convective O_2 supply is inhibited.

Gas ventilation has been studied most extensively on *Phragmites*, but some information is also available on other wetland (but non-aquatic) species that form rhizomes. In the macrophytes *Typha latifolia* and *Typha angustifolia* as well as in the genus *Eleocharis*, similar mechanisms of gas flow seem to operate as in common reed (Bendix et al. 1994; Sorell et al. 1997). There was some debate on O_2 transport in trees adapted to waterlogged soils such as various species of alder. An increase of ethene transport, used as a trace gas, from shoot to root upon exposure to light was interpreted in terms of convective flow driven by thermal osmosis (Grosse and Schröder 1984; Grosse et al. 1993); more recently however, Armstrong and Armstrong (2005) challenged this interpretation and provided evidence that illumination increased photosynthetic O_2 production in the stem, leading to an increase in

O_2 partial pressure and to a moderate increase in downward O_2 transport, that was, according to the authors, purely diffusive. External application of gas pressure to the stem had no such effect on O_2 transport. However, these results do not invalidate those of Grosse, Schröder and co-workers since it remains unclear how enhancement of *ethene* transport reported by these authors should be brought about by O_2 accumulation in the shoot. However, the rather artificial experimental situation (ethene substituting the natural gas composition) may give rise to some convective flow (see above). Obviously, more work is required to clarify this point.

So far, no alternative mechanisms of convective flow to those listed above have been postulated to occur in plants. These established mechanisms all have in common that "macroscopic" pressure gradients are required to drive the mass flow and that separate "inlet" and "outlet" structures are required. However, convective flow could also be initiated at the boundary of the inner wall of the aerenchyma by surface effects, and backflow could occur through the bulk phase, leading to gas circulation. This type of convective flow could be driven e.g. by temperature gradients and could account for the differences between theory and experimental data that are frequently observed when oxygen diffusion was so far identified as the only transport mechanism (e.g. Armstrong et al. 1983).

1.7 O_2 Transport in Plants: Ecological Implications

Efficient mechanisms of O_2 transport in plants do not only help to sustain respiration in the root apex in waterlogged soils. Another important corollary is to keep the rhizosphere at the apex of roots or rhizomes sufficiently oxidized in wetland plants (Armstrong et al. 2006). An anaerobic environment may be harmful to the apex due to toxic substances being produced by microorganisms that prevail under reducing conditions. In order to prevent this, ROL is relatively high at the very apex in wetland species (up to about 0.5 cm away from the root tip). On the other hand, (local) oxidation of the rhizosphere by the root has a strong impact on soil ecology under waterlogging conditions. For a stand of *Carex rostrata*, Mainiero and Kazda (2005) could even show that the bulk O_2 saturation of the soil was higher than in the absence of any vegetation, indicating that O_2 was accumulated in the soil by ROL. Supply of oxygen stimulates the oxidation of ammonium to nitrate and the decomposition of methane to CO_2 (Blom 1999). In plant communities, plant species that are less well adapted to flooding stress will also profit from ROL by wetland plants under waterlogging conditions.

1.8 Open Questions and Directions of Further Research

Progress in our understanding of oxygen transport in plants has been impressive in last decades by combining anatomical and (bio) physical approaches. In that respect, research in this field provides a positive example for other areas of plant science.

Still, however, a number of open questions remain to be answered. It is puzzling that aerenchyma remain gas filled under waterlogging conditions and tend not to be flooded, e.g. upon a local damage of a root or rhizome, even when gas could easily escape via an outlet. Evidence has been obtained that the inner walls of the aerenchyma are hydrophobic (Michael et al. 1999), but more detailed information about surface properties is still lacking. Physical surface properties of the wall may favour convective mass flow at the boundary and lead to gas circulation, even in the absence of "macroscopic" pressure gradients (see above). Embolized xylem vessels may also serve as a pathway for oxygen transport, especially in tree roots. It seems to be counter-intuitive that vessels should cavitate when there is a surplus of water, but radial hydraulic conductance of the root is known to decrease under waterlogging conditions (at least transiently, Gibbs et al. 1998); this would tend to increase xylem tension and, in turn, the probability for vessels to cavitate. Phloem transport could also potentially contribute to O_2 transport from shoot to root, especially when root porosity is low, but to my knowledge this has not been tested yet.

Further research in this area, especially into the genetic basis of O_2 transport and waterlogging tolerance, may also stimulate breeding of cultivars that can cope better with waterlogging conditions. There is a great demand for flooding-tolerant crops worldwide.

Currently, ecological aspects of gas transport in plants are receiving increasing attention. The focus was extended from oxygen transport to the conduction of other gases like methane and CO_2 that are relevant to global climate change; however, this aspect is clearly beyond the scope of this review.

Acknowledgements I received financial support by a grant provided by the "Karlsruhe Institute of Technology" within the framework of the "German excellence initiative". I would also like to thank Prof Tim Colmer for the critical reading of the manuscript.

References

Afreen F, Zobayed SMA, Armstrong J, Armstrong W (2007) Pressure gradients along whole culms and leaf sheaths, and other aspects of humidity-induced gas transport in *Phragmites australis*. J Exp Bot 58:1651–1662

Arber A (1920) Water plants. A study of aquatic angiosperms. Cambridge University Press, Cambridge, UK

Armstrong J, Armstrong W (1991) A convective throughflow of gases in *Phragmites australis* (Cav.) Trin. ex Steud. Aquat Bot 39:75–88

Armstrong J, Armstrong W, Beckett PM (1992) *Phragmites australis*: venturi- and humidity-induced pressure flows enhance rhizome aeration and rhizosphere oxidation. New Phytol 120:197–207

Armstrong J, Armstrong W, Beckett PM, Halder JE, Lythe S, Holt R, Sinclair A (1996a) Pathways for aeration and the mechanisms and beneficial effects of humidity- and Venturi-induce convections in *Phragmites australis* (Cav.) Trin. ex. Steud. Aquat Bot 54:177–197

Armstrong W, Armstrong J, Beckett PM (1996b) Pressurized ventilation in emergent macrophytes: the mechanism and mathematical modelling of humidity-induced convection. Aquat Bot 54:121–135

Armstrong J, Jones RE, Armstrong W (2006) Rhizome phyllosphere oxygenation in *Phragmites* and other species in relation to redox potential, convective gas flow, submergence and aeration pathways. New Phytol 172:719–731

Armstrong W (1964) Oxygen diffusion from the roots of some British bog plants. Nature (Lond) 204:801–802

Armstrong W (1979) Aeration in higher plants. Adv Bot Res 7:225–332

Armstrong W, Armstrong J (2005) Stem photosynthesis not pressurized ventilation is responsible for light-enhanced oxygen supply to submerged roots of alder (*Alnus glutinosa*). Ann Bot 96:591–612

Armstrong W, Beckett PM (1987) Internal aeration and the development of stelar anoxia in submerged roots: a multishelled mathematical model combining axial diffusion of oxygen in the cortex with radial losses to the stele, the wall layers and the rhizosphere. New Phytol 105:221–245

Armstrong W, Wright EJ (1975) Radial oxygem loss from roots: the theoretical basis for the manipulation of flux data obtained by the cylindrical plantinum electrode technique. Physiol Plant 35:21–26

Armstrong W, Healy MT, Webb T (1982) Oxygen diffusion in pea. I Pore space resistance in the primary root. New Phytol 91:647–659

Armstrong W, Healy MT, Lythe S (1983) Oxygen diffusion in pea II. Oxygen concentration in the primary root apex as affected by growth, the production of laterals and radial oxygen loss. New Phytol 94:549–559

Armstrong W, Cousins D, Armstrong J, Turner W, Beckett PM (2000) Oxygen distribution in wetland plant roots and permeability barriers to gas-exchange with the rhizosphere: a micro-electrode and modelling study with *Phragmites australis*. Ann Bot 86:687–703

Armstrong W, Webb T, Darwent M, Beckett PM (2009) Measuring and interpreting critical oxygen pressures in roots. Ann Bot 103:281–293

Bendix M, Tornberg T, Brix H (1994) Internal gas transport in *Typha latifolia* L and *Typha angustifolia* L. I. Humidity-induced pressurization and convective throughflow. Aquat Bot 49:75–90

Blom CWPM (1999) Adaptations to flooding stress: from plant community to single molecule. Plant Biol 1:261–273

Bowling DJF (1973) Measurement of a gradient of oxygen partial pressure across the intact root. Planta 111:323–328

Brix H (1989) Gas exchange through dead culms of reed, *Phragmites australis* (Cav.) Trin. ex Steudel. Aquat Bot 35:81–98

Brix H, Sorrell BK, Schierup H-H (1996) Gas fluxes achieved by in situ convective flow in *Phragmites australis*. Aquat Bot 54:151–163

Colmer TD (2003) Long-distance transport of gases in plants: a perspective on internal aeration and radial oxygen loss from roots. Plant Cell Environ 26:17–36

Colmer TD, Pedersen O (2008) Oxygen dynamics in submerged rice (*Oryza sativa*). New Phytol 178:326–334

Colmer TD, Gibbered MR, Wiengweera A, Tinh TK (1998) The barrier to radial oxygen loss from roots of rice (*Oryza sativa* L) is induced by growth in stagnant solution. J Exp Bot 49:1431–1436

Dacey JWH (1980) Internal winds in water lilies: an adaptation for life in aerobic environments. Science 210:1017–1019

Dedes D, Woermann D (1996) Convective gas flow in plant aeration and thermo-osmosis: a model experiment. Aquat Bot 54:111–120

De Willingen P, van Noordwijk M (1984) Mathematical models on diffusion of oxygen to and within plant roots, with special emphasis on effects of soil-root contact. I. Derivation of the models. Plant Soil 77:215–231

Garthwaite AJ, Armstrong W, Comer TD (2008) Assessment of O_2 diffusivity across the barrier to radial O_2 loss in adventitious roots of *Hordeum marinum*. New Phytol 179:405–416

Gibbs J, Turner DW, Armstrong W, Sivasithamparam K, Greenway H (1998) Response to oxygen deficiency in primary maize roots. II. Development of oxygen deficiency in the stele has limited short-term impact on radial hydraulic conductivity. Aust J Plant Physiol 25:759–763

Greenwood DJ (1967) Studies on oxygen transport through mustard seedlings. New Phytol 66:597–606

Grosse W (1996) The mechanism of thermal transpiration (=thermal osmosis). Aquat Bot 54: 101–110

Grosse W, Schröder P (1984) Oxygen supply of roots by gas transport in alder trees. Zeitschr Naturf 93c:1186–1188

Grosse W, Frye J, Lattermann S (1992) Root aeration in wetland trees by pressurized gas transport. Tree Physiol 10:2285–2295

Grosse W, Schulte A, Fujita H (1993) Pressurized gas transport in two Japanese alder species in relation to their natural habitats. Ecol Res 8:151–158

Jackson MB, Ishizawa K, Ito O (2009) Evolution and mechanism of plant tolerance in flooding stress. Ann Bot 103:137–142

Jackson MB, Armstrong W (1999) Formation of aerenchyma and the process of plant ventilation in relation to soil flooding and submergence. Plant Biol 1:274–287

Justin SHFW, Armstrong W (1987) The anatomical characteristics of roots and plant response to soil flooding. New Phytol 106:465–495

Kaufmann I, Jakob P, Schneider HU, Zimmermann U, Wegner LH (2010) Kinetics of water loss and re-hydration in maize roots exposed to temporal water deprivation. An MR imaging study. New Phytol, submitted

Lambers H, Smakman G (1978) Respiration of the roots of flood-tolerant and flood-intolerant *Senecio* species: affinity for oxygen and resistance to cyanide. Physiol Plant 42:163–166

Lambers H, Steingröver E (1978) Efficiency of root respiration of a flood-tolerant and a flood-intolerant *Senecio* species as affected by low oxygen tension. Physiol Plant 42:179–184

Lambers H, Steingröver E, Smakman G (1978) The significance of oxygen transport and of metabolic adaptation in flood-tolerance of Senecio species. Physiol Plant 43:277–281

Mancuso S, Papeschi G, Marras AM (2000) A polarographic, oxygen-selective vibrating-microelectrode system for the spatial and temporal characterisation of transmembrane oxygen fluxes in plants. Planta 211:384–389

Mainiero R, Kazda M (2005) Effects of *Carex rostrata* on soil oxygen in relation to soil moisture. Plant Soil 270:311–320

Mevi-Schutz J, Grosse W (1988) A two-way gas transport system in *Nelumbo nucifera*. Plant Cell Environ 11:27–34

Michael W, Cholodova VP, Ehwald R (1999) Gas and liquids in intercellular spaces of maize roots. Ann Bot 84:665–673

Pang JY, Newman I, Mendham N, Zhou M, Shabala S (2006) Microelectrode ion and O_2 fluxes measurements reveal differential sensitivity of barley root tissues to hypoxia. Plant Cell Environ 29:1107–1121

Raskin I, Kende H (1983) How does deep water rice solve its aeration problem. Plant Physiol 72:447–454

Seago JL, Marsh LC, Stevens KJ, Soukup A, Votrubova O, Enstone DE (2005) A re-examination of the root cortex in wetland flowering plants with respect to aerenchyma. Ann Bot 96:565–579

Skelton NJ, Allaway WG (1996) Oxygen and pressure changes measured in situ during flooding in roots of the Grey Mangrove *Avicennia marina* (Forssk.) Vierh. Aquat Bot 54:165–175

Sorell BK, Brix H, Orr PT (1997) *Eleocharis sphacelata*: internal gas transport pathways and modelling of aeration by pressure flow and diffusion. New Phytol 136:433–442

Thomson CJ, Greenway H (1991) Metabolic evidence for stelar anoxia in maize roots exposed to low O_2 concentrations. Plant Physiol 96:1294–1301

Van Noordwijk M, De Willingen P (1984) Mathematical models on diffusion of oxygen to and within plant roots, with spezial emphasis on effects of soil-root contact. Plant Soil 77:233–241

Visser EJW, Colmer TD, Blom CWPM, Voesenek LACJ (2000) Changes in growth, porosity, and radial oxygen loss from adventitious roots of selected mono- and dicotyledonous wetland species with contrasting types of aerenchyma. Plant Cell Environ 23:1237–1245

Vretare V, Weisner SEB (2000) Influence of pressurized ventilation on performance of an emergent macrophyte (*Phragmites australis*). J Ecol 88:978–987

Westhoff M, Zimmermann D, Zimmermann G, Gessner P, Wegner LH, Bentrup FW, Zimmermann U (2009) Distribution and function of epistomatal mucilage plugs. Protoplasma 235:101–105

Chapter 2
Waterlogging and Plant Nutrient Uptake

J. Theo M. Elzenga and Hans van Veen

Abstract Waterlogging affects several parameters that determine nutrient uptake from the soil by the roots. We checked systematically, for all the relevant parameters in the nutrient uptake model by Silberbush and Barber (Plant Soil 74:93–100, 1983), how waterlogging changes the magnitude of the parameter, changes that can be both positive and negative for nutrient uptake. If negative effects can be expected we also describe possible specific adaptive responses that could counter the negative effects on these parameters, of plants exposed to waterlogging. We conclude that although most flooding-tolerant species use a hypoxia-avoidance strategy (i.e. increase the supply of oxygen to the root tissue), increasing root length, facilitating infection by mycorrhizal fungi and modification of uptake kinetics could be used by plants to ameliorate the negative effect of flooding on nutrient uptake.

2.1 Introduction

Waterlogging occurs when the rate of infiltration of precipitation exceeds the combined rates of drainage and evaporation of a catchment or when floodwater submerges an area. Waterlogging has a dramatic impact on gas exchange and soil properties (Greenway and Gibbs 2003). Most plants are affected in a negative way by flooded conditions. Shoot and root nutrient content decreases (Ashraf and Rehman 1999; Khabaz-Saberi et al. 2006; Kuiper et al. 1994; Pang et al. 2007a) and carbon assimilation and transpiration are greatly reduced. (Ashraf and Rehman 1999; Kronzucker et al. 1998; Pezeshki 2001). When the ambient oxygen concentration is increased most of the effects of waterlogging are reduced

---

J.T.M. Elzenga (✉) and H. van Veen
Laboratory of Plant Physiology, Center for Ecological and Evolutionary Studies, University of Groningen, P.O. Box 14, Haren NL-9750AA, The Netherlands
e-mail: j.t.m.elzenga@rug.nl

S. Mancuso and S. Shabala (eds.), *Waterlogging Signalling and Tolerance in Plants*,
DOI 10.1007/978-3-642-10305-6_2, © Springer-Verlag Berlin Heidelberg 2010

(Kronzucker et al. 1998). Together with the observation that the effects of water-logging can be reversed by supplying additional nutrients in the form of foliar sprays (Pang et al. 2007a) or ameliorated by application of nitrogen (Arnon 1937; Drew 1988), this suggests that hypoxia leads to severely reduced root functioning and nutrient uptake. The reduced nutrient uptake can be explained by changes that occur at a metabolic and cellular level. These effects are not only important from a fundamental perspective, but also have great relevance to agricultural.

Diffusion of gasses is 10,000 times slower in water than in air and thus in waterlogged soils gas exchange is severely impeded, which leads to a depletion of oxygen and an accumulation of carbon dioxide due to microbial and root respiration. Oxygen levels might drop within hours below the critical oxygen concentration (COC), a level at which the cell shifts from aerobic respiration to glycolysis. The efficiency of the glycolysis in generating ATP is much reduced compared to the aerobic mitochondrial respiration processes: 2 ATP/glucose versus ~38 ATP/glucose, respectively (Jackson and Ricard 2003; Huang et al. 2008). Upon oxygen depletion, some plant species respond by inducing pyruvate carboxylase and alcohol dehydrogenase and increase the rate of glycolysis (Summers et al. 2000; Dolferus et al. 2008). By accelerating glycolysis, plant cells produce at most 37% of the ATP produced under oxygen-sufficient conditions (Greenway and Gibbs 2003). Evidence is also being presented that under anoxic stress part of the metabolic processes shift from ATP as the energy source, to PPi during glycolysis (Huang et al. 2008). The reduction in ATP availability has strong repercussions for root development, root nutrient uptake and root maintenance.

One of the best-studied adaptations of plants to flooding conditions is the formation of aerenchymatic tissue in the root, which provides an alternative path-way for the supply of oxygen to the root tissue (Jackson and Armstrong 1999; Gibberd et al. 2001; Rubinigg et al. 2002). This requires that new, well-adapted, adventitious roots are formed (Visser et al. 1996). Radial oxygen loss is kept to a minimum so that the root tip becomes a well-oxygenated micro-climate (Jackson and Armstrong 1999). Most of the disadvantages for root metabolism imposed by the flooding-induced hypoxic conditions are thereby ameliorated. Plants that are not capable of increasing the oxygen supply through aerenchymous conducts in the root do face a more serious challenge.

In this review, we consider the effect of hypoxia and waterlogging on the nutrient supply of to the plant assuming that the respiration in the root tissue is seriously impaired and that the rhizosphere, the soil immediately surrounding the root is chemically reduced. The delivery of nutrient to the root surface is by either a) interception, the "bumping" of the growing root against the nutrient, b) mass flow, the transport to the root surface by the movement of water in the soil driven by processes such as percolation, transpiration and evaporation), and c) diffusion, the movement of a nutrient along a concentration gradient, which is often created by depletion of the nutrient by the root in the volume of soil in close proximity to the root surface. The rate of nutrient uptake depends on various anatomical, physiological, biochemical, chemical and physical processes in the root and the surrounding soil layer. For this chapter, we have adapted the analysis by Silberbush and Barber (1983, reprinted in Clarkson1985) of the relative effect of changes in the parameters in ion transport and uptake, as a guideline for our analysis. Fig. 2.1 is a graphical representation of this analysis. The parameters that were considered by Silberbush and Barber are:

K	Root elongation rate
r_o	Root diameter
r_i	Inter-root spacing
C_{li}	Initial nutrient concentration
D_e	Diffusion coefficient
B	Buffering strength
k_m	Affinity constant of nutrient uptake system
I_{max}	Maximal nutrient uptake rate
C_{min}	Minimal nutrient concentration that can be taken up from the medium
V_o	Transpiration rate (bulk water flow)

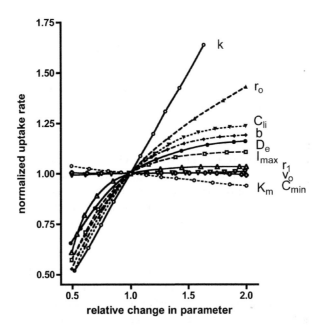

Fig. 2.1 Effects of changing the values of model parameters used for predicting phosphorus uptake by bean plants. It should be noted that the model sensitivity for changes in the parameters are could be strongly affected when the combination of starting values are chosen differently. The standard situation (change ratio = 1) is characterized by a rather high nutrient concentration and a wide spacing of competing roots. k: root elongation rate, r_o: root diameter, C_{li}: initial nutrient concentration, D_e: diffusion coefficient, b: buffering strength, k_m: affinity constant of nutrient uptake system, I_{max}: maximal nutrient uptake rate, C_{min}: minimal nutrient concentration that can be taken up from the medium, V_o: transpiration rate (bulk water flow), r_i: inter-root spacing. (After Silberbush and Barber 1983 as reprinted in Clarkson 1985)

The parameters k, r_o and r_i determine the general architecture of the root system and are essential for the "foraging capacity" of the plant. Foraging by the roots not only involves the utilization of mineral nutrients in the soil, but also applies to the uptake of water. The root system architecture together with the transpiration rate (bulk water flow), V_o, and the initial nutrient concentration, C_{li}, affects the effect of nutrient transport by mass flow. The diffusion coefficient and the buffering strength determine the diffusion of nutrients to the root surface. At the root surface, the uptake system of the plant, consisting of the transporter proteins in the plasma membrane that have specific properties with regard to maximal capacity, I_{max}, affinity, k_m, and minimal nutrient concentration that can be taken up from the medium, C_{min}, finally determines how efficiently the nutrients present at the plasma membrane of the root cortical cells are absorbed. In this chapter, we review our current knowledge how waterlogging affects the different parameters of the Silberbush and Barber analysis, and how plants can respond to these changes and in turn can modify these parameters by physiological, morphological or biochemical changes.

2.2 Effects of Hypoxia on Nutrient Uptake

2.2.1 Physiological Effects of Hypoxia Change Root Elongation Rate, k, and Maximal Nutrient Uptake Rate, I_{max}

One of the immediate effects of flooding and hypoxia on the root physiology is the almost instantaneous arrest of root growth (e.g. Gibbs et al. 1998). The shift from aerobic respiration in the mitochondria to glycolytic generation of ATP does impose a serious reduction of available energy for the maintenance, growth and ion uptake. Although the specific allocation of respiratory cost to these different functions has not received much attention and different studies have yielded somewhat different figures, the consensus opinion is that about 20 to 45% of respiration is involved in growth (Veen 1981; Van der Werf et al. 1988; Poorter et al. 1991; Scheurwater et al. 1998; Scheurwater et al. 1999). Matching energy expenditure to energy generation would therefore demand that the investment of the plant in new root tissue is reduced. Root growth arrest could, however, also be the result of the accumulation of products of anaerobic metabolism. Self-poisoning could occur when protons accumulate in the cytoplasm and the vacuole and lead to a lethal drop in pH in the cytoplasm (Gerendás and Ratcliffe 2002). In *Phragmites australis* low molecular weight, volatile, monocarboxylic acids, like acetic acid, propionic acid, butyric acid and caproic acid, and sulphide, at concentration levels that have been measured in situ, arrested root elongation (Armstrong and Armstrong 2001). Specifically the application of a combination of these compounds was very effective and induced all the symptoms that are associated with root die-back. Sulphide, although it also can be used by the plant as sulphur source for the synthesis of cysteine and methionine, can act as a phytotoxin. In reduced soils the increased availability of ferrous iron and zinc can lead to accumulation in the plant to toxic levels (Pezeshki 2001).

In the analysis by Silberbush and Barber (1983), reduction of new root formation has a strong negative impact on potential nutrient uptake by reducing the interception of nutrients. The rate of root elongation, k, is one of the most important parameters determining the nutrient uptake rate. In more recent modelling excercises aimed at identifying important parameters for nutrient uptake, the foraging capacity, in which the elongation rate is an important factor, proved to be among the most essential properties of the root (e.g. Dunbabin 2006).

Controlled pot experiments have demonstrated that nutrient uptake is not only reduced through limited root growth, but that the uptake is also reduced on a per weight basis (Wiengweera and Greenway 2004; Kuiper et al. 1994). The allocation of respiratory cost to ion uptake is about 50 to 70% (Veen 1981; Van der Werf et al. 1988; Poorter et al. 1991; Scheurwater et al. 1998, 1999). Hypoxia can lead to a strong reduction in the adenylate energy charge (e.g. de Boer 1985; Sieber and Brandle 1991; Nabben 2001; De Simone et al. 2002). Reduction of the AEC can lead to a loss of the nutrient uptake capacity by limiting the supply of ATP to the plasma membrane proton pumping ATPase (de Boer 1985). As a consequence the membrane potential will become less negative and the proton gradient across the membrane will become less steep. Membrane potential and proton gradient form together the proton motive force, which is used to drive the uptake of most nutrients through symporters. It is indeed observed that upon anoxia the cytoplasmic pH drops from 7.4 to 6.9–7.1 and the plasma membrane becomes depolarized (Menegus et al. 1991; Xia and Roberts 1994). However, the possible causes for these effects are still under debate. Acidification could either be a negative by-product of glycolysis, lactic acid and high CO_2 accumulation, or be an adaptive change to optimize ethanol fermentation for energy production (Felle 2005). However, the high costs associated with lactic acid removal and the speed of the response suggests this is an adaptive response to switch to ethanol fermentation as soon as possible. But many adaptive traits to mitigate low pH are also up-regulated suggesting the opposite (Aurisano et al. 1995; Xia and Roberts 1994).

The depolarisation of the plasma membrane can also cause leakage of K^+ out of the cell by activating K^+ channels. In aged beet root disks the inhibition by 55% of Cl^- and K^+ uptake correlated, and was probably caused, by a decrease of the ATP concentration by 80% (Petraglia and Poole 1980). If lack of ATP supply to the plasma membrane ATPase is the cause of the reduction in nutrient acquisition, then nutrient uptake should be (partially) restored by stimulating ethanol fermentation by providing sufficient substrate. This mechanism has indeed been shown to function (Huang et al. 2008; Summers et al. 2000). However, several studies have shown that the anoxia-induced inhibition of nutrient uptake does not correlate with a reduction in the adenylate energy charge (Ishizawa et al. 1999) or with insufficient proton motive force (Zhang and Greenway 1995). These latter results indicate that inhibition of ion uptake is the result of down-regulation of transport systems. Although this down-regulation of the transport activity is not a direct, thermodynamic effect of the reduced synthesis of ATP, it is very likely that the down-regulation of transporter systems is part of the "energy saving" strategy of the plant, enabling the diversion of energy to even more essential processes like the regulation of cytoplasmic pH. A third possibility is provided by the results of Pang et al. (2007b) that show, that under oxygen-sufficient conditions the fermentation products of soil microbes under aerated conditions can mimic the effect of anoxia: membrane depolarisation and reduction of ion uptake.

2.2.2 Waterlogging Leads to Changes in the Availability, C_{li}, and the Effective Diffusion Coefficient, D_e, of Some of the Nutrients in the Soil

As oxygen is depleted soil microbes switch from using O_2 as an electron acceptor to NO_3^-, Fe^{3+} and Mn^{4+} leading to highly reduced conditions. Under severe waterlogging, SO_4^{2-} and CO_2 are also used in respiration (Gambrell and Patrick 1978, Madigan et al. 2003; Tiedje et al. 1984). More reduced conditions can lead to a lower availability of some plant nutrients, specifically nitrogen (Zhang et al. 1990). On the other hand at 100% field capacity the availability of the micronutrients Fe and Mn is strongly improved (Plekhanova 2007). An additional effect of waterlogging is that the phosphorus availability will also increase as result of Fe solubilization (Gambrell and Patrick 1978).

The denitrification as a result of nitrate respiration means that a substantial amount of nitrogen will be lost to the atmosphere as N_2, N_2O and NO (Gambrell and Patrick 1978; Madigan et al. 2003; Tiedje et al. 1984). When the soil chemistry leads to a reduced availability of nitrate, while simultaneously increasing the concentration of ammonia (Ashraf and Rehman 1999), this might have a beneficial effect on the budget of an energy deprived root system. It is assumed that the uptake of nitrate will cost the cell 1 ATP (1:1 stoichiometry of the H^+-NO_3^- symporter and a 1 H^+/ATP transported by the proton pumping ATPase). The subsequent assimilation of nitrate to glutamine consumes 12 ATP. The uptake of nitrogen from ammonia will partly be mediated by the diffusive passage of the membrane by NH_3, and not at the expense of a proton transported (although it should be noted that the acidifying effect of assimilation of ammonia will exacerbate the problem of cytoplasmic acidosis already caused by increased glycolytic activity). The assimilation into glutamine will only cost 2 ATP per ammonia. That in the flooding-sensitive beech (*Fagus sylvatica*) the NH_4^+ uptake is much less affected by flooding than NO_3^- (Kreuzwieser et al. 2002), and might be a consequence of the metabolic difference between these two N-compounds.

The water content of the soil has a pronounced effect on the effective diffusion coefficient D_e of plant nutrients. In soils with a volumetric water content of 0.40 g cm^{-3} only 10 μM P was necessary to achieve the same root uptake as compared with 200 μM P in a soil with a water content of 0.13 g cm^{-3}. For potassium the D_e increases from 2.55 to 4.91 to 6.40×10^{-7} cm^2 s^{-1} in soils with a water content of 0.19, 0.26 and 0.34 g cm^{-3}, respectively (Barber 1984). This effect of soil water content in ion uptake efficiency indicates that, particularly for phosphorus and potassium, a higher water saturation level of the soil has its advantages (Marschner 1995).

Due to the activity of sulphate reducing bacteria, sulphate will be depleted and high amounts of H_2S are produced under anoxic conditions in the pedosphere (Dassonville and Renault 2002). Sulphide can be taken up by the roots cells in analogy to the uptake of atmospheric H_2S through the stomata in the leaf mesophyll (Rennenberg and Polle 1994). Uptake of H_2S by the roots was demonstrated in *P. australis*. The potentially toxic H_2S was used for accumulation of thiols (Fuertig et al. 1996). In flooding tolerant poplar species the decreased allocation of reduced sulphur compounds from the shoot to the root and the accumulation of cysteine in the lateral roots indicated that the uptake of sulphide from reduced soils is a common process (Herschbach et al. 2005). Since the uptake of sulphide is energetically less expensive than the uptake of sulphate waterlogging is not expected to lead to S-deficiency.

2.2.3 In Waterlogged Conditions, Some Plant Species Show More Root Hair Development, Longer and Thinner Roots and Increased Levels of Infection With Mycorhizal Fungi – Effectively Increasing k

Plants have been shown to optimize their foraging strategy for water and nutrients. In a review by Bloom et al. (2003) it was shown that regulation by soil pH and redox potential of root cell division and mechanical properties result in root proliferation in nitrogen rich soil patches. In several studies with different species (tomato, mais) the pattern that emerges is that root growth is more enhanced when the available nitrogen is in the form of NH_4^+ than when it is only available as NO_3^-. Also plants develop thinner and longer roots when ammonium is the main nitrogen source, supposedly an appropriate strategy for scavenging a relatively immobile nutrient. From this example, it becomes clear that plants invest new biomass strategically and economically in structures with a long slender geometry. When roots are waterlogged growth is inhibited. One of the obvious reasons is that the high energy demand of growth cannot be met under anaerobic conditions. To deploy the limited resources under these conditions in an optimal way, would be to allocate them to the thinnest, longest structures possible, in order to maintain a high elongation rate.

Root hairs and mycorrhizal hyphae are thin structures associated with root functioning and presumably with the lowest carbon costs possible per volume of soil explored. To our knowledge no studies on the effects of waterlogging or hypoxia have been published. However, exposing roots to the ethylene precursor cyclopropane-1-carboxylate does, increase root hair formation in *Arabidopsis* roots

(Zhang et al. 2003; Schiefelbein 2000). The plant hormone ethylene is tightly related to flooding-adapted growth. Ethylene, being gaseous, diffuses at a more reduced rate in water than in air, leading to a quick accumulation upon flooding, and is a primary signal that activates water-adapted growth responses in Rumex palustris and deepwater rice (Visser and Voesenek 2004). Low phosphor in the growth medium can also increase higher root hair densities. The low P-induced root hair formation can be inhibited by adding ethylene inhibitors (Zhang et al. 2003). However, in most plants grown hydroponically root hairs are completely absent, caused by suppression of root hair initiation by the same plant hormone ethylene (Goormachtig et al. 2004).

The reported effects of waterlogging on colonization of roots by mycorrhizal fungi are similarly variable. In most publications fungal invection declines with the development of anaerobic conditions (Garcia et al. 2008) or that the increased availability of P under waterlogging depresses arbuscular mycorrhizal (AM) colonization (Stevens et al. 2002). In *Lotus tenuis,* the reduction in fungal infection was due to increased P availability and to a lesser extend to excess water (Garcia et al. 2008). Completely opposite results were found for AM infection in *Pterocarpus officinalis.* Flooded plants had well-developed mycorrhizas and plants that were infected were more flood-tolerant, having a higher growth rate and a higher P acquisition in the leaves (Fougnies et al. 2007). Also in *Casuarina equisetifolia* infected with *Glomus clarum* developed better when flooded than plants that were not infected (Osundina 1998). Ernst (1990) reported that mycorrhiza reduced the accumulation of toxic elements such as Mn and Fe. In *Panicum* and *Leersia* the infection or the roots by AM fungi was negatively affected by waterlogging, however, waterlogged plants that were colonized showed an improvement in phosphorus nutrition over non-colonized plants (Miller and Sharitz 2000). In a field survey that involved the same two grass species, AM colonization was strongly negatively correlated to the extent of waterlogging, but colonization was present in most root samples (Miller 2000). Although the focus has been on the effect of higher levels of phosphorus under waterlogged conditions on the infection rate, also other nutrients, like Zn, Fe and Cu, will be more available under waterlogging and precisely these elements have also been shown to be more readily available for plant roots that have an AM symbiosis (Fageria and Stone 2006). The combination of effects of waterlogging, hypoxia and increased nutrient availability, might partly explain the widely different results.

2.2.4 Waterlogging Decrease Evaporation and Bulk Water Flow, V_o

Many species show a reduction in the transpiration rate upon waterlogging (see Pezeshki 2001 for an overview). Control over the transpiration stream can occur anywhere along the pathway of water through the plant. The reported effects of waterlogging or hypoxia on the hydrolic conductivity of the root systems of different plant species have been summarized by Bramley (2006). Of the eight species listed, one species (*Agave deserti*) did not show a reduction in hydrolic conductivity, another species (*Lycopersicon esculentum*) did not respond in one study and showed a reduced hydrolic conductance in another, and in six species de hydrolic conductance was reduced. In *Arabidopsis* root, the anoxia-induced reduction in hydrolic conductance is the result of a conformational change of water channel proteins (aquaporins) when the cytoplasmic pH drops (Tournaire-Roux et al. 2003). In some plant species stomatal closure has been reported without significant changes in water status (Pezeshki 2001) leading to a reduction of the water flow through the plant, of 40% (Bradford and Hsiao 1982). Reducing the flow of water through the plant does reduce photosynthetic gas exchange and will reduce the nutrient supply to the root as it will decrease mass flow in the soil.

2.2.5 In Response to Waterlogging the Kinetics of Root Transport Systems, k_m and I_{max}, Can Be Modified

Only few studies are available where effects of waterlogging on the kinetic parameters of the root nutrient transport systems have been reported. In *Paspalum dilatatum*, a waterlogging-tolerant grass from South America, both the uptake capacity and the affinity of root transport system for phosphate were affected by waterlogging. The V_{max}, of phosphate uptake increased by more than 100% on root weight basis. While the affinity significantly increases: the k_m was 42 μM under control conditions and 29 under waterlogging conditions (Rubio et al. 1997). This effect of waterlogging on *Paspalum dilatum* is different from rice where no effect of anoxia on P uptake capacity was found (John et al. 1974).

2.3 Summary and Concluding Remarks

Waterlogging has a severe effect on almost all the parameters that affect nutrient uptake from the soil. In Table 2.1 an overview of the effects of waterlogging on these parameters is presented, together with the possible adaptive responses of plants to counteract these effects. We purposely did not include the processes that will take place in rhizosphere of plants that form aerenchyma in the roots, providing a conduct for the transport of O_2 from shoot to root. In these plants most of the effects described above will be completely different: root function will be less affected and the redox potential of the rhizosphere will be much lower. The plants without aerenchyma in the roots generally can no longer use mitochondrial respiration to generate ATP and shift to glycolysis, while reducing energy demanding processes like growth and nutrient uptake. The cytoplasmic pH will become lower

Table 2.1 Summary of the effects of waterlogging on the root parameters used in the model described Silberbush and Barber. In the last column the possible functional adaptive response of roots which counteracts the effect of waterlogging is listed. The question mark following a response indicates that the functional response has only been described occasionally

Parameter		Effect of waterlogging	Functional response
k	Root elongation rate	Reduced	Formation of root hairs, association with AM fungi
r_o	Root diameter		Reduced
C_{li}	Initial nutrient concentration	Higher – lower	
D_e	Diffusion coefficient	Higher	
k_m	Affinity constant uptake system	No effect	Lowered?
I_{max}	Maximal nutrient uptake rate	Reduced	Increased?
C_{min}	Minimal nutrient concentration	No effect	
V_o	Bulk water flow	Reduced	

and hydrolic conductivity of the root will decrease due to acidosis. Stomata will close and the flow of water through the plant will be severely reduced. The production of volatile organic acids in the soil (maybe in combination with low ATP levels) leads to membrane depolarisation and die-back of root tips, resulting in much reduced root functionality. In the soil some nutrients become less available (N), or become available in another form (N, S), but for others (P, Fe, Zn) waterlogging improves their availability. In waterlogged soils also diffusion of mobile nutrients is increased. The literature on most of the possible adaptive responses is not unanimous. Reasons for this are the lack of a general model plant system, the variability in the conditions that are applied (hypoxia, anoxia, waterlogging, long adaptation periods, short exposures, etc.), but also an inherent lack of control over most of the variables involved. When soils are waterlogged a myriad of changes are induced and some of these changes will have opposite effects: waterlogging reduces root uptake capacity, but increase nutrient availability; nutrients become more readily available, but possibly to a toxic level.

But the lack of consensus is most probably another illustration of the fact that waterlogging-tolerance is not due to a single plant attribute, but to an amalgam of processes and features that enhance the survival of the plant.

References

Armstrong J, Armstrong W (2001) An overview of the effects of phytotoxins on *Phragmites australis* in relation to die-back. Aquat Bot 69:251–268

Arnon DI (1937) Ammonium and nitrate nutrition of barley at different seasons in relation to hydrogen ion concentrations, manganese, copper and oxygen supply. Soil Sci 44:91–113

Ashraf M, Rehman H (1999) Interactive effects of nitrate and long term waterlogging on growth, water relation and gaseous exchange properties of maize (*Zea mays* l.). Plant Sci 144:35–43

Aurisano N, Bertani A, Reggiani R (1995) Anaerobic accumulation of 4-aminobutyrate in rice seedlings: causes and significance. Phytochemistry 38:1147–1150

Barber SA (1984) Soil nutrient bioavailability. A mechanistic approach. Wiley, New York, USA

Bloom AJ, Meyerhoff PA, Taylor AR, Rost TL (2003) Root development and absorption of ammonium and nitrate from the rhizosphere. J Plant Growth Regul 21:416–431

Bradford KJ, Hsiao TC (1982) Stomatal behavior and water relations of waterlogged tomato plants. Plant Physiol 70:1508–1513

Bramley H (2006) Water flow in the roots of three crop species: the influence root structure, aquaporin activity and waterloggin. PhD Thesis, University of Western Australia

Clarkson DT (1985) Factors affecting mineral nutrient acquisition by plants. Annu Rev Plant Physiol 36:77–115

Dassonville F, Renault P (2002) Interactions between microbial processes and geochemical transformations under anaerobic conditions: a review. Agronomie 22:51–68

De Boer B (1985) Xylem/symplast ion exchange: mechanism and function in salt-tolerance and growth. Thesis, University of Groningen, Haren, The Netherlands

De Simone O, Haase JK, Mueller E, Junk WJ, Gonsior G, Schmidt W (2002) Impact of root morphology on metabolism and oxygen distribution in roots and rhizosphere of two Central Amazon flooding tree species. Funct Plant Biol 29:1025–1035

Dolferus R, Wolansky M, Carol R, Miyashita Y, Ismond K, Good A (2008) Functional analysis of lactate dehydrogenase during hypoxic stress in *Arabidopsis*. Funct Plant Biol 35:131–140

Drew MC (1988) Effects of flooding and oxygen deficiency on plant mineral nutrition. In: Lauchli A, Tinker PB (eds) Advances in plant nutrition. Praeger, New York, USA, pp 115–159

Dunbabin V (2006) Using the ROOTMAP model of crop root growth to investigate root-soil interactions. In: Turner NC, Acuna T, Johnson RC. Ground-breaking stuff. Proceedings of the 13th Australian Agronomy Conference, 10–14 Sept 2006, Australian Society of Agronomy, Perth, Western Australia

Ernst WHO (1990) Ecophysiology of plants in waterlogged and flooded environments. Aquat Bot 38:73–90

Fageria NK, Stone LF (2006) Physical, chemical and biological changes in the rhizosphere and nutrent availability. J Plant Nutr 29:1327–1356

Felle HH (2005) pH regulation in anoxic plants. Ann Bot 96:519–532

Fougnies L, Renciot S, Muller F, Penchette C, Prin Y, de Faria SM, Bouvet JM, Sylla SN, Dreyfus B, Ba AM (2007) Arbuscular mycorrhizal colonization and nodulation improve flooding tolerance in *Pterocampus officinalis* Jacq. seedlings. Mycorrhiza 17:159–166

Fuertig K, Ruegsegger A, Brunold C, Brandle R (1996) Sulphide utilization and injuries in hypoxic roots and rhizomes of common reed (*Phragmites australis*). Folia Geobotanica Phytotaxon 31:143–151

Gambrell RP, Patrick WH (1978) Chemical and microbiological properties of anaerobic soils and sediments. In: Hook RMM, Crawford C (eds) Plant Life in anaerobic habitats. Ann Arbor Science, USA

Garcia I, Mendoza R, Pomar MC (2008) Deficit and excess of soil water impact on plant growth of *Lotus tenuis* by affectig nutrient uptake and arbuscular mycorrhizal symbiosis. Plant Soil 204:117–131

Gerendás J, Ratcliffe RG (2002) Root pH regulation. In: Waisel Y, Eshel A, Kafkafi U (eds) Plant roots – the hidden half, 3rd edn. Marcel Dekker, New York, pp 553–570

Gibberd MR, Gray JD, Cocks PS, Colmer TD (2001) Waterlogging tolerance among a divers range of *Trifolium* accessions is related to root porosity, lateral root formation and 'aerotropic rooting'. Ann Bot 88:579–589

Gibbs J, Turner DW, Armstrong W, Darwent MJ, Greenway H (1998) Response to oxygen deficiency in primary maize roots. I. Development of oxygen deficiency in the stele reduces radial solute transport to the xylem. Aust J Plant Physiol 25:745–758

Goormachtig S, Capoen W, James EK, Holsters M (2004) Switch from intracellular to intercellular invasion during water stress-tolerant legume nodulation. PNAS (USA) 101:6303–6308

Greenway H, Gibbs J (2003) Mechanisms on anoxia tolerance in plants. II. Energy requirements for maintenance and distribution to essential processes. Funct Plant Biol 30:999–1036

Herschbach C, Mult S, Kreuzwieser J, Kopriva S (2005) Influence of anoxia on whole plant sulphur nutrition of flooding-tolerant poplar (*Populus tremula* x *P. alba*). Plant Cell Environ 28:167–175

Huang S, Colmer TD, Millar AH (2008) Does anoxia involve altering the energy currency to PPi? Trends Plant Sci 13:221–227

Ishizawa K, Murakami S, Kawakami Y, Kuramochi H (1999) Growth and energy status of arrowhead tubers, pondweed turions and rice seedlings under anoxic conditions. Plant Cell Environ 22:505–514

Jackson MB, Armstrong W (1999) Formation of aerenchyma and the processes of plant ventilation in relation to soil flooding and submergence. Plant Biol 1:274–287

Jackson MB, Ricard B (2003) Physiology, biochemistry and molecular biology of plant root systems subjected to flooding of the soil. In: de Kroon H, de Visser EJW (eds) Root ecology. Springer, Berlin, Heidelberg

John CD, Limpinuntana V, Greenway H (1974) Adaptation of rice to anaerobiosis. Aust J Plant Physiol 1:513–520

Khabaz-Saberi H, Setters TL, Waters I (2006) Waterlogging induces high to toxic concentrations of iron, aluminum and manganese wheat varieties on acidic soil. J Plant Nutr 29:899–911

Kreuzwieser J, Fuerniss S, Rennenberg H (2002) Impact of waterlogging on the N-metabolism of flood tolerant and non-tolerant tree species. Plant Cell Environ 25:1039–1049

Kronzucker HJ, Kirk GJD, Siddiqi MY, Glass ADM (1998) Effects of Hypoxia on $^{13}NH_4^+$ fluxes in rice roots, kinetics and compartmental analysis. Plant Physiol 116:581–587

Kuiper PJC, Walton CS, Greenway H (1994) Effects of hypoxia on ion uptake by nodal and seminal wheat roots. Plant Physiol Biochem 32:267–276

Madigan MT, Martinko JM, Parker J (2003) In: Brock (ed) Biology of microorganisms. Prentice Hall, London, UK

Marschner H (1995) Mineral nutrition of higher plants, 2nd edn. Academic, London, p 889

Menegus F, Cattaruzza L, Mattana M, Beffagna N, Ragg E (1991) Response to anoxia in rice and wheat seedlings. Plant Physiol 95:760–767

Miller SP, Sharitz RR (2000) Manipulation of flooding and arbuscular mycorriza formation influences growth and nutrition of two semiaquatic grass species. Funct Ecol 14:738–748

Miller SP (2000) Arbuscular mycorrhizal colonization of semi-aquatic grasses along a wide hydrological gradient. New Phytol 145:145–155

Nabben RHM (2001) Metabolic adaptations to flooding-induced oxygen deficiency and post-anoxia stress in Rumex species. Thesis, Utrecht University, Utrecht, The Netherlands

Osundina MA (1998) Nodulation and growth of mycorrhizal *Casuarina equisetifolia* J.R. and G. First in response to flooding. Biol Fertil Soils 26:95–99

Pang J, Ross J, Zhou M, Mendham N, Shabala S (2007a) Amelioration of detrimental effects of waterlogging by foliar nutrient spray in barley. Funct Plant Biol 34:221–227

Pang JY, Cuin T, Shabala L, Zhou M, Mendham N, Shabala S (2007b) Effect of secondary metabolites associated with anaerobic soil conditions on ion fluxes and electrophysiology in barley roots. Plant Physiol 145:266–267

Petraglia T, Poole RJ (1980) Effect of anoxia on ATP levels and ion transport rates in red beet. Plant Physiol 65:973–974

Pezeshki SR (2001) Wetland plant responses to soil flooding. Environ Exp Bot 46:299–312

Plekhanova IO (2007) Transformation of Fe, Mn, Co and Ni compounds in humic podzols at different moisture. Soil Biol 34:67–75

Poorter H, Van der Werf A, Atkin OK, Lambers H (1991) Respiratory energy requirements of roots vary with the potential growth rate of a plant species. Physiol Plant 83:469–475

Rennenberg H, Polle A (1994) Metabolic consequences of atmospheric sulphur influx into plants. In: Alscher RG, Wellburn AR (eds) Plant responses to the gaseous environment. Chapman and Hall, London, UK, pp 165–180

Rubinigg M, Stulen I, Elzenga JTM, Colmer TD (2002) Spatial patterns of radial oxygen loss and nitrate net flux along adventitious roots of rice raised in aerated or stagnant solution. Funct Plant Biol 29:1475–1481

Rubio G, Oesterheld M, Alvarez CR (1997) Lavado RS Mechanism for the increase in phosphorus uptake of waterlogged plants: soil phosphorus availability, root morphology and uptake kinetics. Oecologia 112:150–155

Scheurwater I, Cornelissen C, Dictus F, Welschen R, Lambers H (1998) Why do fast-and slow-growing grass species differ so little in their rate of root respiration, considering the large differences in rate of growth and ion uptake? Plant Cell Environ 21:995–1005

Scheurwater I, Clarckson DT, Purves JV, Van Rijt G, Saker LR, Welschen R, Lambers H (1999) Relatively large nitrate efflux can account for the high specific respiratory costs for nitrate transport in slow-growing grass species. Plant Soil 215:123–134

Schiefelbein JW (2000) Constructing a plant cell. The genetic control of root hair development. Plant Physiol 124:1525–1531

Sieber M, Brandle R (1991) Energy metabolism in rhizomes of *Acorus calamus* L. and in tubers of *Solanum tuberosum* L. with regard to their anoxia tolerance. Bot Acta 104:79–282

Silberbush M, Barber SA (1983) Sensitivity of simulated phosphorus uptake to parameters used by a mechanistic mathematical model. Plant Soil 74:93–100

Stevens KJ, Spender SW, Peterson RL (2002) Phosphorus, arbuscular mycorrhizal fungi and performance of he the wetland plant *Lythrum salicaria* L. under inundated conditions. Mycorrhiza 12:277–283

Summers JE, Ratcliffe RG, Jackson MB (2000) Anoxia tolerance in the aquatic monocot *Potamogeton pectinatus*: absence of oxygen stimulates elongation in association with unusually large Pasteur effect. J Exp Bot 52:1423–1422

Tiedje TM, Sexstone AJ, Parkin TB, Revsbech NP, Shelton DR (1984) Anaerobic processes in soils. Plant Soil 76:197–212

Tournaire-Roux C, Sutka M, Javot H, Gout E, Gerbeau P, Luu D-T, Bligny R, Maurel C (2003) Cytosolic pH regulates root water transport during anoxic stress through gating of aquaporins. Nature 425:393–397

Veen BW (1981) Relation between root respiration and root activity. Plant Soil 63:73–76

Visser EJW, Bogemann GM, Blom CWPM, Voesenek LACJ (1996) Ethylene accumulation in waterlogged *Rumex* plants promotes formation of adventitious roots. J Exp Bot 47:403–410

Visser EJW, Voesenek LACJ (2004) Acclimation to soil flooding – sensing and signal-transduction. Plant Soil 254:197–214

Wiengweera A, Greenway H (2004) Performance of seminal and nodal roots of wheat in stagnant solution: K+ and P uptake and effects of increasing O_2 partial pressures around the shoot on nodal root elongation. J Exp Bot 55:2121–2129

Van der Werf A, Kooijman A, Welschen R, Lambers H (1988) Respiratory energy costs for the maintenance of biomass, for growth and for ion uptake in roots of *Carex diandra* and *Carex acudformis*. Physiol Plant 72:483–491

Xia JH, Roberts JKM (1994) Improved cytoplasmic pH regulation, Increased lactate efflux, and reduced cytoplamsic lactate levels are biochemical traits expressed in root tips of whole maize seedlings acclimated to a low oxygen environment. Plant Physiol 105:651–657

Zhang B-G, Puard M, Couchat P (1990) Effect of hypoxia, acidity and nitrate on inorganic nutrition in rice plants. Plant Physiol Biochem 28:655–661

Zhang Q, Greenway H (1995) Membrane transport in anoxic rice coleoptiles and storage tissues in beetroot. Aust J Plant Physiol 22:965–975

Zhang Y-J, Lynch JP, Brown KM (2003) Ethylene and phosphorus availability have interacting yet distinct effects on root hair development. J Exp Bot 54:2351–2361

Chapter 3
Strategies for Adaptation to Waterlogging and Hypoxia in Nitrogen Fixing Nodules of Legumes

Daniel M. Roberts, Won Gyu Choi, and Jin Ha Hwang

Abstract Symbiotic nitrogen fixation between legumes and rhizobia bacteria occurs in a microaerobic environment within a specialized organ, the root nodule. The fixation of dinitrogen requires a considerable energy input and a high respiratory rate, but the fundamental nitrogen fixation enzyme, nitrogenase, is inactivated by free oxygen. Because of this apparent conundrum, the diffusion of oxygen into the nodule infection zone is exquisitely regulated in response to multiple environmental cues, and becomes sensitive to alterations in the external rhizosphere oxygen tension. As a result, most legumes are sensitive to waterlogging, showing reductions in nodulation and productivity in flooded soils. Nevertheless, certain legumes have evolved developmental strategies to modulate the pathway of oxygen diffusion to the nodule, the patterns of nodule formation on roots and stems, and altered pathways of bacterial invasion to adapt to flooding conditions. In the present chapter, the regulation of oxygen diffusion and adaptations to waterlogged conditions by nitrogen fixing nodules of flooding-sensitive and flooding-tolerant legumes are discussed.

Abbreviations

ACC   1-Aminocyclopropane-1-carboxylate
GA   Gibberellic acid
IT   Infection thread

D.M. Roberts (✉) and W.G. Choi
Department of Biochemistry and Cellular and Molecular Biology, The University of Tennessee, Knoxville, TN 37996, USA
e-mail: drobert2@utk.edu

D.M. Roberts and J.H. Hwang
Program in Genome Science and Technology, The University of Tennessee, Knoxville, TN 37996, USA

S. Mancuso and S. Shabala (eds.), *Waterlogging Signalling and Tolerance in Plants*,
DOI 10.1007/978-3-642-10305-6_3, © Springer-Verlag Berlin Heidelberg 2010

K_m Michaelis constant
kPa   kilopascal
LRB   Lateral root boundary
O_i Free oxygen concentration within the infected cells
PIP   Plasma membrane intrinsic protein
P_f Osmotic water permeability
RHC   Root hair curling
ROS   Reactive oxygen species
TIP   Tonoplast intrinsic protein

3.1   Introduction: The Oxygen Diffusion Barrier in Nodules

Most terrestrial plants are sensitive to flooding stress, which is principally the result of hypoxia and reduced respiration due to low O_2 diffusion coefficients in water (Ferrell and Himmelblau 1967) and poor radial diffusion of O_2 within the mature root (Ober and Sharp 1996; van Dongen et al. 2003). In many legumes, flooding drastically reduces root growth, suppresses nodulation and nitrogen fixation, and reduces photosynthesis and overall yield (Minchin and Pate 1975; Minchin and Summerfield 1976; Scott et al. 1989; Sung 1993; Linkemer et al. 1998; Bacanamwo and Purcell 1999a). Due to the high energy requirement associated with symbiotic nitrogen fixation, legume root nodules are particularly sensitive to changes in oxygen concentration in the soil, and in general legumes relying on nitrogen fixation are more sensitive to flooding stress than plants grown on other alternative sources of nitrogen, such as nitrate (Minchin and Pate 1975; Buttery 1987; Bacanamwo and Purcell 1999a, b). The problem is exacerbated by the microaerobic environment in which symbiotic nitrogen fixation takes place (Minchin et al. 2008). Nevertheless, all legumes show adaptive strategies that allow tolerance to fluctuations of oxygen concentrations in the rhizosphere by controlling the rate of oxygen diffusion into the core of the nodule, and in extreme cases such as wetland legumes, these strategies have allowed flooding tolerance. Thus, a consideration of the basis of nodule sensitivity to flooding and waterlogging stress must first begin with an analysis of the importance of gas diffusion as a global regulatory mechanism that buffers nitrogen fixation rates in legume nodules in response to metabolic and environmental cues.

3.1.1   Nodule Morphology and the Gas Diffusion Barrier

The basic need for a gas diffusion barrier is apparent from the balancing act that nodules play in: (1) providing adequate oxygen for the substantial respiratory and energy needs of the nitrogen fixation process, while (2) mitigating the damaging effects of free oxygen on the fundamental fixation enzyme, the bacterial nitrogenase. The role of the nodule gas diffusion barrier in the modulation of oxygen entry and the efflux of other gases (e.g., H_2 and CO_2) has been discussed extensively in several reviews (Witty et al. 1986; Hunt and Layzell 1993; Bergersen 1997; Minchin 1997; Layzell 1998; Minchin et al. 2008). For the purpose of this chapter, only the fundamental characteristics of the gas diffusion barrier relevant to adaptation to external oxygen concentrations will be summarized since flooding stress will largely impact these.

The oxygen/gas diffusion barrier is an outcome of the overall cellular and tissue morphology of the legume nodule. In a mature nitrogen-fixing nodule, the infected zone at the core of the nodule consists predominantly of two cell types: (1) the enlarged infected cells, containing the rhizobia bacteroids enclosed within symbiosome organelles, and (2) smaller companion uninfected interstitial cells (Bergersen 1982). Early observations with oxygen electrodes (Tjepkema and Yocum 1974; Witty et al. 1987), followed by accurate estimations of pO_2 by nodule oximetry (based on spectroscopic analysis of fractional leg hemoglobin oxygenation in intact nodules) (Denison and Layzell 1991), revealed that the internal infected zone of legume nodules is microaerobic. For example, the free oxygen concentration within the infected cells (O_i) of soybean nodules is maintained at approximately 20 nM under atmospheric concentrations of oxygen (Layzell et al. 1990). To accommodate the high respiratory need associated with the nitrogen fixation process, several adaptations have evolved to allow high energy metabolism under microaerobic conditions, including the synthesis of leghemoglobin that serves as a free oxygen buffer as well as a facilitated carrier of oxygen for bacteroid respiration (Bergerson, 1982). In addition, the architecture of the infected cell is arranged such that mitochondria are clustered at the periphery of the infected cell near the intercellular spaces which contain higher pO_2 to facilitate host cell respiration (Bergerson, 1982). Also, the terminal oxidase of the bacteroid electron chain has a low K_m (20 nM) for oxygen, allowing respiration under microaerobic conditions (Bergerson and Turner 1993).

The centralized infected zone is surrounded by a cortical tissue layer consisting of several organized cell layers that differ in number and complexity depending upon the nature of the particular legume–rhizobia association (Minchin 1997). The region of the inner cortex surrounding the infected zone consists of small cells with restricted intercellular spaces (Witty et al. 1987; Parsons and Day 1990). Measurements with oxygen microelectrodes reveal that the pO_2 drops precipitously across this inner cortex (Tjepkema and Yocum 1974; Witty et al. 1987). This region has generally been considered the major boundary of the gas diffusion barrier in nitrogen-fixing nodules. However, while the inner cortex is a critical barrier for gas diffusion, O_2 diffusion from the rhizosphere to the bacteroid is likely a more complex process that depends upon multiple additional factors including intercellular gas spaces in the infected zone, the rate of O_2 consumption in mitochondria in the infected cell, and the rates of leghemoglobin-O_2 diffusion in the infected cell cytosol (reviewed in Minchin 1997; Bergersen 1997; Minchin et al. 2008).

3.1.2 Modulation of the Gas Diffusion Barrier

Early support that oxygen diffusion into the nodule is a rate-limiting factor in regulation of nitrogen fixation came from the observation by Pankhurst and Sprent (1975a) that drought stress-induced inhibition of nitrogenase activity can be partially restored by the experimental elevation of external pO_2. The implication was that the reduction of oxygen diffusion into the infected zone due to osmotic stress resulted in an inhibition of the rate of nitrogen fixation based on the reduced availability of oxygen for respiration and energy metabolism. Ultimately, the idea of a variable gaseous diffusion barrier that regulates O_i and serves the dual role of controlling respiratory metabolism/nitrogen fixation while preventing nitrogenase inactivation was proposed (Sheehy et al. 1983).

Subsequent work demonstrated that a wide variety of environmental and metabolic cues modulate the gas diffusion barrier. For example, under conditions of reduced carbon availability, induced by a variety of conditions such as darkness (Drevon et al. 1991), nodule excision (Ralston and Imsande 1982; Sung et al. 1991), stem girdling (Vessey et al. 1988), or detopping (Denison et al. 1992), the lack of carbon induces a rapid reduction in respiration and the rate of nitrogen fixation resulting from a reduction in O_2 diffusion into the infected zone and reduced O_i. Such modulation of the oxygen diffusion barrier is proposed to prevent nitrogenase inactivation by O_2 accumulation in response to reduced oxygen consumption by the bacteroids (Witty et al. 1986). Other environmental factors that regulate the diffusion barrier by restricting oxygen diffusion include nitrate fertilizers, water deficit induced by drought or hyperosmotic stress, and temperature (reviewed in Witty et al. 1986; Hunt and Layzell 1993; Minchin 1997; Layzell 1998). In most cases, experimental elevation of the external pO_2 at least partially restores the inhibitory effects of these treatments, suggesting that a reduction in O_2 flux through the diffusion barrier is in part responsible for the modulation of respiration and nitrogen fixation by these environmental paramenters.

3.1.3 Control of the Gas Diffusion Barrier in Response to Sub-Ambient O_2 and Flooding

Tissues that are naturally hypoxic, even under plant growth in normoxic conditions, are particularly sensitive to oxygen deprivation resulting from flooding stress (Bailey-Serres and Voesenek 2008). Thus, even under mild hypoxic conditions, these tissues can experience severe hypoxia or even anoxia because of the slow diffusion rate of O_2 and the low gas porosity of these naturally hypoxic tissues. As discussed above, the gas diffusion barrier of nitrogen-fixing nodules maintains a microaerobic state (~20 nM O_2) in the infected zone, and analyses of the rate of respiration and nitrogen fixation under normal atmospheric oxygen conditions show that the nitrogenase activity and respiratory metabolism are maintained at a suboptimal level by this concentration of O_2 (Hunt et al. 1989). Exposure of nodules to 10% O_2 results in a rapid reduction of the adenylate energy charge with the bacteroid fraction being particularly sensitive to this hypoxic stress (Kuzma et al. 1999). Nodules show two modes of adaption to changes in external oxygen concentrations: (1) a short term modulation of the gas permeability barrier; and (2) longer term developmental/genetic alterations to enhance gas permeability involving changes in tissue, cellular and subcellular morphology (Minchin 1997).

With respect to the short-term adaptive strategies, reduction of the rhizophere pO_2 results in a rapid reduction in nitrogenase activity that over time eventually recovers to the initial level (Criswell et al. 1976). Comparison of the respiratory rate (CO_2 evolution) and nitrogenase activity (H_2 evolution or acetylene reduction) in intact nodules in response to changes in external oxygen concentrations showed a transient decline in both, but an eventual recovery to rates approximating or even exceeding pretreatment levels (Hunt et al. 1987; 1989; Weisz and Sinclair 1987a). Measurements of internal O_i by nodule oximetry reveal that the infected zone is able to rapidly buffer O_i in response to changes in external oxygen tension. For example, when administered an abrupt change of external oxygen from 20 to 25 kPa, soybean nodules show a transient increase in O_i followed by a rapid (<10 min) return to the original steady state level (King et al. 1988). This change is likely due to reversible modulation of the rate of O_2 flux through the diffusion barrier in response to changes in the external rhizosphere oxygen concentration. In the case of supra-ambient oxygen conditions the nodule resistance to gas diffusion is enhanced compared to normoxic controls (Weisz and Sinclair 1987b; King and Layzell 1991). In contrast, sub-ambient oxygen conditions decrease the resistance of the gas diffusion barrier (Weisz and Sinclair 1987b). Thus, in response to flooding conditions and a drop in oxygen content of the soil, nodules would initially enhance the rate of oxygen uptake by modulation of the gas diffusion barrier, stimulating oxygen uptake as part of a short-term response.

3.1.4 Mechanism of Regulation of the Gas Diffusion Barrier in Response to pO_2

Mechanistic and molecular features of how gas diffusion is regulated in response to fluctuations in external oxygen concentration, and as well as how the oxygen signal is perceived and transduced, remain a subject of debate (Minchin et al. 2008). Nodules exposed to supra-ambient oxygen concentrations show a rapid decrease in intercellular gas spaces within the inner cortex of the nodule that accompanies increased resistance to gas permeability (Witty et al. 1987). This lead to the hypothesis that the osmotically-driven flux of water from inner cortical cells to the apoplastic intercellular space is responsible for restricting the rate of gas diffusion in response to elevated external pO_2 (Hunt and Layzell 1993; Purcell and Sinclair 1994). On the basis of an anatomical investigation, stresses that restrict oxygen flux into the nodule (e.g., supra-ambient pO_2 or salt stress) cause an apparent collapse of the intercellular spaces of the inner cortex that likely further contributes to a restricted pathway for gas diffusion (Serraj et al. 1995). Noting a similarity to other osmocontractile cells, such as pulvini motor cells that undergo turgor-driven changes in cell shape, it was proposed that cells of the nodule inner cortex could respond in an analogous manner to external stimuli that restrict gas entry into the nodule. Electrophysiological observations show that a brief exposure of nodules to elevated pO_2 results in a transient membrane depolarization with normal negative membrane potential re-established upon return to ambient pO_2 conditions (Denison and Kinraide 1995). Based on this observation, an "osmoelectrical" model was proposed through which movement of K^+ into the apoplastic space accompanies the efflux of water from nodule cells which increases the water-filled spaces. This in turn causes a change in cell turgor and shape which increase the resistance to gas flow in response to elevated rhizosphere O_2 (Denison and Kinraide 1995). Hypoxic conditions would be expected to have the reverse effect with low pO_2 causing an increase in gas filled extracellular spaces in the inner cortex. This observation has been confirmed in nodules cultured under sub-ambient concentrations of oxygen (Parsons and Day 1990; Dakora and Atkins 1991).

Additional support for osmotic regulation of gas diffusion through the inner cortex comes from the analysis of the subcellular localization of tonoplast and plasma membrane aquaporins (TIPs and PIPs) in nodules. High densities of these proteins increase the osmotic water permeability (P_f) of biological membranes and are characteristic of membranes associated with the rapid movement of bulk water in response to osmotic and pressure gradients (Maurel et al. 2008). Subcellular localization of γTIP and PIP1 and PIP2 in soybean nodules shows that these aquaporins are expressed at much higher levels on cells of the inner cortex compared to cells in the infected zone and vascular transfer cells (Serraj et al. 1998; Fleurat-Lessard et al. 2005). By analogy to motor cells of the pulvini, Fleurat-Lessard et al. (2005) postulated that this density of aquaporins is necessary to mediate rapid and reversible water movements accompanying ion fluxes to and from the apoplastic space to mediate changes in cell shape, intercellular space water content, and oxygen permeability in response to environmental cues.

Recent measurements of the cortical and infected zone fractions of soybean nodules treated with supra-ambient O_2 (30%) show a change in the localized K^+ concentrations that is consistent with the movement of K^+ from the infected zone to the cortex (Wei and Layzell 2006). It is postulated that water movement into the cortical region accompanying the movement of K^+ ions may flood the intercellular spaces of the inner cortex decreasing the gas permeability of this critical region in response to elevated O_2. This is consistent with observations that artificial elevation of KCl concentrations in the nodule cortex transiently decrease the O_2 permeability of nodules (Purcell and Sinclair 1994). Conversely, Wei and Layzell (2006) showed that hypoxic concentrations of O_2 (10%) result in elevated concentrations of K^+ in the infected zone compared to the nodule cortex. This in turn could result in the movement of water from intercellular spaces of the nodule cortex to cells within the infected zone, increasing the gas permeability of the inner cortex.

3.2 Developmental and Morphological Adaptations of Nitrogen-Fixing Nodules to Low Oxygen Stress

3.2.1 Secondary Aerenchyma Formation

While rapid modulation of the gas diffusion barrier may allow adaptations to reduced soil oxygen accompanying short term flooding, legumes respond to longer term waterlogging and hypoxic oxygen concentrations by triggering a number of developmental adaptations that lead to altered morphology designed to increase the flow of oxygen to the submerged roots and nodules. For example, soybean nodules in waterlogged soils (Pankhurst and Sprent 1975b; Shimamura et al. 2003; Thomas et al. 2005) or grown under conditions of limiting rhizosphere pO_2 (Parsons and Day 1990; Dakora and Atkins 1991) show drastic changes in morphology, with large quantities of lenticels and secondary aerenchyma tissue covering the nodule surface. Secondary aerenchyma differs from the cortical aerenchyma that is commonly formed in root and stem tissues, and is derived from differentiation of secondary meristematic cells (Jackson and Armstrong 1999). Secondary aerenchyma of nodules consists of loosely packed, white parenchymous tissue with large unoccluded intercellular spaces (Parsons and Day 1990) which enhance the gas porosity of the nodule under conditions of low rhizosphere oxygen (Thomas et al. 2005). The formation of secondary aerenchyma in response to waterlogging or reduced concentrations of rhizosphere oxygen is observed in a wide variety of legume nodules (Minchin and Summerfield 1976; Arrese-Igor et al. 1993; Dakora and Atkins 1990a; Pugh et al. 1995; James and Sprent 1999), suggesting that this is a common adaptive response to low pO_2 stress. In contrast, nodules grown under supra-ambient pO_2 conditions show a drastic reduction in the number of aerenchymous lenticels (Parsons and Day 1990; Dakora and Atkins 1991), suggesting that this developmental mechanism is part of the adaptation of the nodule to elevated external pO_2 conditions to restrict oxygen porosity.

In soybean, waterlogging triggers the development of adventitious roots as well as aerenchyma in the hypocotyl, roots and nodule tissues (Pankhurst and Sprent 1975b; Bacanamwo and Purcell 1999b; Shimamura et al. 2003; Thomas et al. 2005). These morphological changes are observed within 1 day after initiation of the flooding, and persist throughout prolonged waterlogging stress (Shimamura et al. 2003; Thomas et al. 2005). Analysis of nitrogen fixation and assimilation based on the xylem sap content of glutamine and ureides (assimilation products) and alanine (a sign of root hypoxia) shows a drastic reduction in the glutamine and ureides, and a concomitant accumulation of alanine within 1 day after waterlogging (Thomas et al. 2005). This trend is reversed by 7 days post flooding, correlating with the development of nodule and root aerenchyma and an increase in the gas porosity of these organs, suggesting a critical role for aerenchyma in adaptation (Thomas et al. 2005).

3.2.2 The Inner Cortex and Infected Zone

In addition to increases in secondary aerenchyma/lenticel formation on the surface of the nodule, additional alterations in the morphology of the inner cortex and the infected zone are also triggered to increase the oxygen porosity of the nodule in response to hypoxic conditions. For example, the analysis of nodules from various legumes cultured under rhizosphere oxygen concentrations below 5% show substantial developmental changes including: (1) reduced size and dry matter of the nodule as well as reduction in the size of the centralized infected zone (Dakora and Atkins 1990a, b; Dakora and Atkins 1991); (2) thickening of the inner cortical region and a more spherical cortical cell morphology resulting in direct connections between the extracellular spaces of adjacent cells (Dakora and Atkins 1990a; Dakora and Atkins 1991; Arrese-Igor et al. 1993); (3) increased extracellular gas spaces in the inner cortex (Parsons and Day 1990; Dakora and Atkins 1990a; Dakora and Atkins 1991; Arrese-Igor et al. 1993); and (4) a reduction in the volume and size of the infected cells, and an increased ratio of uninfected interstitial cells to infected cells (Dakora and Atkins 1990a; Dakora and Atkins 1991; Arrese-Igor et al. 1993). Together with the surface changes and aerechymous lenticel formation discussed above, these alterations are proposed to reduce the resistance of the gas diffusion pathway and account for an enhanced permeability of oxygen into the infected zone observed under conditions of severe hypoxia (Dakora and Atkins 1990a; 1990c). Further, the reduction in the cell size and number within the infected zone would be expected to reduce the respiratory needs of the nodule under conditions of low oxygen.

In addition, changes in the organization of the infected and uninfected cells within the central zone are observed in some legumes under severely hypoxic conditions. For example, when grown under rhizosphere O_2 concentrations of 1%, the morphology of the infected zone of alfalfa nodules is altered with the infected cells clustered in a ring at the periphery of the infected zone, presumably closer to the inner cortex and the oxygen barrier (Arrese-Igor et al. 1993). A comparable strategy is used in white clover (*Trifolium repens* L.) which displays an unusual tolerance to growth under continuously flooded conditions, and actually shows an increase in yield under these conditions compared to aerated control plants (Pugh et al. 1995). Besides the typical formation of lenticels and aerenchyma on the nodule surface, the infected cells of white clover cultured under waterlogged conditions become enlarged and show a more prominent central vacuole compared to aerated controls. The larger central vacuole presses the cytosolic layer to the outer periphery of the infected cell, increasing the cell surface area/volume ratio. This is postulated to increase the accessibility of the infected cell cytosol to oxygen in the intercellular gas spaces (Pugh et al. 1995).

With respect to the inner cortex, additional changes associated with hypoxic rhizosphere oxygen concentrations include alterations in intercellular "occlusions". Ultrastructural analyses showed that nodules accumulate electron dense material that apparently occludes the intercellular gas spaces of the inner cortical region when plants are grown at elevated oxygen concentrations (Parsons and Day 1990; Dakora and Atkins 1990b; 1991). These occlusions become less apparent in nodules from plants cultured at sub-ambient oxygen concentrations. The intercellular spaces of soybean and cowpea nodules of soybean and cowpea contain a glycoprotein that is recognized by a monoclonal antibody (MAC236) that reacts with the carbohydrate moiety (Bradley et al. 1988; VandenBosch et al. 1989), and it has been proposed that this glycoprotein is responsible for the intercellular occlusions. Immunolocalization of the MAC236 glycoprotein antigen in soybean nodules exposed to 40%, ambient, and 10% oxygen shows a correlation between the accumulation of the MAC236-glycoprotein antigen in nodule cortical cells and oxygen concentration (James et al. 1991). At high oxygen concentrations (40%) intercellular spaces become occluded with this glycoprotein and it was proposed that its accumulation and deposition in these spaces, together with the decrease in the actual volume of these spaces, may contribute to the reduction in gas diffusion within the nodule cortex (James et al. 1991). In support of this, a variety of stresses that restrict gas permeability in nodules (e.g., low temperature, darkening, detopping and nitrate fertilizers) also lead to the accumulation of the MAC 236 glycoprotein antigen, as well as an additional lower molecular weight glycoprotein (MAC265 antigen), in the intercellular spaces of soybean (James et al. 2000) and lupin (Ianetta et al. 1993; de Lorenzo et al. 1993) nodules. Extracellular glycoprotein also accumulates at other locations within the nodule, including the intercellular spaces of cells in the infection zone (James et al. 2000), suggesting a possible role in regulation of gas permeability in the central zone of the nodule as well.

3.2.3 Influence of Adaptive Changes on Nitrogen Fixation Under Altered Rhizosphere pO_2 Conditions

Various "flooding sensitive" legumes, show a remarkable ability to adapt to growth at low oxygen within a tolerable range. For example, soybean nodules cultured at subambient pO_2 as low as 2.5% O_2 show similar levels of nitrogen fixation (acetylene reduction), respiration, and the ureide content of phloem sap compared to control nodules grown under ambient oxygen conditions (Dakora and Atkins 1991; Parsons and Day 1990). At extreme hypoxic levels of O_2 (1%) a decrease in these parameters was observed, suggesting that the ability to buffer changes in oxygen content by increasing the gas permeability of the nodule may be exceeded. Studies with cowpea nodules show a similar effect of reduced oxygen, with plants able to adapt to oxygen as low as 10%. At 5% O_2 or lower, a reduction in the rate of nitrogen fixation, as well as a decrease in the nodule and whole plant dry weight was observed (Dakora and Atkins 1990a, b). This may be due to decreased numbers of nodules, and a reduced size of the infected tissue as discussed above. However, the plants partially compensate for this by increasing the efficiency of nitrogen fixation (nitrogenase activity/bacteroid) at extremely low O_2 by an unknown mechanism (Dakora and Atkins 1990a), an observation that was also made in alfalfa plants grown at 1% O_2 (Arrese-Igor et al. 1993). In alfalfa, nodules grown under 8% O_2 show little differences in nitrogenase activity and respiration compared to control nodules (Wycoff et al. 1998). However, plants grown at more severe hypoxic concentrations (1% O_2) show reduced number of infected cells, reduced yield of roots and nodules, and lower nodule respiration based on the rate of CO_2 evolution (Arrese-Igor et al. 1993). Overall these observations suggest that flooding sensitive legumes possess the ability to adapt to mild sub-ambient oxygen conditions but that this capacity is exceeded in conditions of extreme hypoxia.

While nodulated plants show the ability to adapt to long term growth under reduced oxygen conditions, and retain the ability to exert short term adjustments in the gas diffusion barrier to mitigate the effects of modest changes in oxygen concentration (Atkins et al. 1993), plants grown under severe hypoxia ($<2.5\%$ O_2) were much more sensitive to acute increases in rhizosphere O_2 associated with the abrupt return of plants to ambient oxygen (Dakora and Atkins 1990c; Dakora and Atkins 1991). For example, nodulated cowpea plants cultured under 1 or 2.5% O_2 show severe loss of nitrogenase activity upon transfer to ambient O_2 which did not recover for 15 days (Dakora and Atkins 1990c). Thus in addition to severe restriction of oxygen resulting from waterlogging, an additional stress associated with re-oxygenation of the nodule during recovery from hypoxia stress and its effects on nitrogenase also needs to be considered. This re-oxygenation sensitivity phenomena has also been observed for flood-adapted nodules of aquatic legumes such as *Neptunia patens* (James et al. 1992b), suggesting that: (1) the nodules of flood-adapted legumes have a reduced resistance to oxygen diffusion into the nodule; and (2) this higher oxygen permeability may lead to sensitivity to oxidative stress upon rapid transition from a flooded to a well-drained habitat.

3.3 Strategies of Adaptation: Flood-Tolerant Legumes and Oxygen Diffusion

3.3.1 Tropical Wetland Legumes

Although legume nodules in general have short and long term adaptive programs that allow growth over a broad range of rhizosphere oxygen concentrations, the oxygen content of waterlogged soils can lead to severe hypoxia which would exceed this range of adaptation (Pugh et al. 1995), and it is acknowledged that most legumes are sensitive to prolonged flooding (Loureiro et al. 1998). Nevertheless "flooding tolerant" aquatic and semiaquatic legumes have developed additional strategies to allow adaption to extreme conditions associated with growth in wetland environments (Justin and Armstrong 1987; Loureiro et al. 1998; Sprent 1999; James et al. 2001; Koponen et al. 2003; Den Herder et al. 2006). For example, nodulated legume trees and shrubs are widespread in a number of tropical wetlands where they play an important role in providing fixed nitrogen to permanently or seasonally flooded tropical forests (Loureiro et al. 1998; Saur et al. 2000; James et al. 2001; Koponen et al. 2003). Indeed, soils of both seasonally and permanently flooded wetlands are nitrogen deficient because of leaching of nitrogenous compounds, a lack of mineralization of organic matter and increased denitrification (Loureiro et al. 1998). This results in positive selection of nodulated, nitrogen-fixing legumes in these environments.

Tropical wetland legumes adapt to waterlogged environments by using a variety of strategies including: (1) the formation of extensive interconnected aerenchyma or other pathways limiting oxygen diffusion resistance in stem, root and nodule systems; and (2) alterations in the nodulation patterns with the formation of nodules just below the waterline, or on adventitious roots and stems above the waterline, as well as the nodulation of floating root systems. These strategies are discussed extensively for a wide variety of tropical legumes from several wetland ecosystems (Loureiro et al. 1998; Saur et al. 2000; James et al. 2001; Koponen et al. 2003), and for the purpose of this review, examples with representative wetland or aquatic legumes are discussed below.

3.3.1.1 Nodulation of Submerged Stems and Roots: Increased Porosity Mechanisms

Discolobium is a hydrophytic leguminous shrub that is common to the Brazilian Pantanal wetlands where it is found in permanently flooded environments (James et al. 2001). *D. pulchellum* is nodulated on stems and roots under flooding conditions (Loureiro et al. 1994), and unlike avoidance adaptive measures typical of many wetland legumes (see Sect. 3.3.1.2), *D. pulchellum* nodules can be found at depths exceeding 2 m suggesting that they survive and function under severely low oxygen concentrations (Loureiro et al. 1998). To increase air porosity into submerged tissues, *D. pulchellum* stems are hollow, and stem, root and nodule tissues are interconnected by a highly developed network of aerenchyma that penetrates to the mid cortex of the nodule (Loureiro et al. 1994). *D. pulchellum* forms determinant aeschynomenoid type of nodules (Sprent and James 2007) with an infected zone consisting solely of infected cells with large unoccluded intercellular air spaces, and with no interstitial uninfected cells (Loureiro et al. 1994). The internal structures of stem and root nodules of flooded *D. pulchellum* are unique, showing a direct vascular connection between the nodule to stem or root vasculature, and the presence of an internal vascular bundle surrounded by infected cells (Loureiro et al. 1994). An interesting observation is that *D. pulchellum* stem nodules senesce rapidly and aerenchyma tissue collapses upon exposure to air (Loureiro et al. 1994). This suggests that the nodules are terminally adapted to

48 D.M. Roberts et al.

a permanently flooded environment, possibly by maximizing a constitutive low resistance oxygen diffusion pathway from the stem to the nodule. If this is the case, exposure of *D. pulchellum* to atmospheric oxygen concentrations could lead to nitrogenase inactivation or oxidative damage due to the accumulation of reactive oxygen species (ROS).

Development of networks of stem and root aerenchyma, and increases in intercellular air spaces in flooded nodules seem to be a common adaptation mechanism in variety of flood-tolerant, semi-aquatic legumes including *Mimosa pellita* (James et al. 2001), *Neptunia* (Schaede 1940; James et al. 1992a), *Alnus rubra* (Batzli and Dawson 1999), *Pentraclethra macroloba* (Walter and Bien 1989), and *Viminaria juncea* (Walker et al. 1983). These morphological changes are important to maintain nitrogen fixing activity (Walter and Bien 1989; Walker et al. 1983) or for recovery of nitrogen-fixation rates to levels of non-flooded nodules during prolonged waterlogging stress (Batzli and Dawson 1999).

3.3.1.2 Aerial Nodulation of Stems and Adventitious Roots: Avoidance Mechanisms

An additional flooding-tolerance strategy exhibited by semi-aquatic tropical legumes is the alteration in the pattern of root and stem nodulation to avoid the low oxygen environment associated with submergence. For example, under flooding conditions, the nodulation of semi-aquatic species of *Neptunia* occurs at the top of the tap root, as well as on spongy lateral roots that float at the water surface (Schaede 1940; Allen and Allen 1981). In the case of flood-tolerant species of *Aeschynomene* (Alazard 1985) and *Sesbania* (Dreyfus and Dommergues 1981), waterlogging results in the induction of stem nodulation. This commonly occurs on the aerial portion of the stem above the water line (e.g., *S. rostrata* and several species of *Aeschynomene*), and is considered a nodulation strategy to avoid the restricted oxygen environment associated with submersion. The cortical cells of stem nodules on these plants contain chloroplasts and are photosynthetic (James et al. 1998), which may contribute to the high nitrogen fixation activity of stem-nodulated legumes under conditions of flooding (Boivin et al. 1997). In the case of *Aeschynomene fluminensis*, stem nodules form only on submerged roots and show morphological adaptations including large intercellular air spaces in stems that are connected via aerenchymous tissue to the base of the nodules (Loureiro et al. 1995). *A. fluminensis* nodules are also photosynthetic, and it was proposed that photosynthetic evolution of of O_2 could represent an additional mechanism to augment low pO_2 in a submerged environment (Louriero et al. 1995, 1998). In addition, the rhizobia bacteroids in *A. fluminensis* nodules are also photosynthetic and light enhances the nitrogenase activity of the endosymbiont, suggesting an alternative source of biosynthetic energy to drive nitrogen fixation under oxygen limiting conditions (Evans et al. 1990).

3.3.2 Lotus uliginosus: *A Temperate Wetland Legume*

While flooding tolerance is more typically observed in tropical wetland legumes, examples of adaptation to flooding among selected temperate species have also been observed. *Lotus uliginosus* (marsh birdsfoot trefoil) is a wetland forage legume that shows a high tolerance to flooding compared to *L. corniculatus* (birdsfoot trefoil), the more commonly used non-wetland species (Justin and Armstrong 1987). Since *L. uliginosus* is considered to be a parent of the more common *L. corniculatus*, the comparison of the two species has been instructive for examination of adaptive strategies of nodulation and nitrogen fixation in flooding-tolerant legumes. *L. uliginosus* successfully forms active nitrogen fixing nodules upon infection with *Mesorhizobium loti* under waterlogged conditions (James and Crawford 1998; James and Sprent 1999). Comparison of the plant yield and nodulation under conditions of root submergence provided the surprising observation that flooded *L. uliginosus* plants show increased yield (shoot and nodule dry weight, nodule numbers, carbon and nitrogen) when cultured under conditions with reduced dissolved oxygen (roots flooded with N_2-bubbled water for 60 days compared to roots flooded with air-saturated water for 60 days). In comparison, *L. corniculatus* showed a lower yield, higher levels of ethylene gas release, and signs of nodule senescence in response to submergence compared to *L. ulginosus* (James and Crawford 1998).

The morphology of the nodules of the two *Lotus* species grown under low oxygen conditions show typical developmental adaptations including the presence of profuse secondary aerenchyma/lenticel development on the surface of flooded-nodules and development of nodules primarily on the tap root close to the hypocotyl, modifications designed to lower the resistance and distance of gas diffusion (James and Crawford 1998). However, despite the similarity in nodule morphology, the root porosity of *L. ulginosus* nodules cultured under conditions of low oxygen was fourfold higher than *L. corniculatus* nodules. This suggests that a less restricted oxygen diffusion pathway is responsible for increased survival and adaptation to this stress. Consistent with the proposal of a less restricted oxygen diffusion pathway, the cortical cells of *L. uliginosus* nodules showed lower amounts of the MAC265 glycoprotein antigen compared to *L. corniculatus* nodules cultured under low oxygen conditions (James and Crawford 1998).

In addition, *L. ulginosus* plants show adaptations to flooding with respect to the rhizobia infection pathway. Similar to other terrestrial legume species (reviewed in Oldroyd and Downie 2008), *L. ulginosus* is principally infected with *Mesorhizobium loti* through the classical root hair curling (RHC) pathway. However, in flooded plants, the bacteria infect *L. ulginosus* by a "crack-based" infection mechanism (see next section) by accumulating within intercellular spaces in the tap root aerenchyma (James and Sprent 1999). In addition, functional nodules are formed on the adventitious roots which emerge from the stems of flooded plants, as well as on the stem itself (James and Sprent 1999), which likely help obviate the problems of oxygen diffusion to nodules in a flooded environment.

3.4 Strategies of Adaptation: Alternate Nodulation Pathways for Flooding Tolerant Legumes

3.4.1 Intercellular-Based Mechanism of Nodulation: The Lateral Root Boundary Pathway

A hallmark of flood-stressed legumes is the reduction in the numbers of nodules formed on waterlogged roots. Problems that compromise nodulation under conditions of flooding include increased levels of ethylene gas, which inhibit nodulation in several terrestrial legume/rhizobia associations (Penmetsa and Cook 1997; Guinel and Sloetjes 2000), as well as the inhibition of root hair growth. In order to cope with these problems, flooding-tolerant legumes such as *L. ulginosus* (James and Sprent 1999), *Neptunia* sp. (James et al. 1992a; Subba-Rao et al. 1995; Goormachtig et al. 2004a), and *Sesbania rostrata* (Ndoye et al. 1994; Goormachtig et al. 2004a) alter the pathway of rhizobial infection.

Nodulation in most legumes normally takes place by a "root hair curling" (RHC) mechanism (reviewed in Patriarca et al. 2004; Oldroyd and Downie 2008) that involves intracellular entry of the invading rhizobia bacteria through root hairs normally found in a region ("zone 1") immediately above the root tip. Soil rhizobia bind to the tips of the root hairs of the host legume and induce morphological changes including swelling, deformation, and curling, with the later causing entrapment of the bacteria in a microcolony at the root hair tip. Localized cell wall hydrolysis leads to the invasion of the bacterial microcolony and the formation of infection thread (IT) tubes which penetrate the root epidermis and move transcellularly through the root cortex. Interaction of the rhizobia with the root hair simultaneously triggers cortical cell division and the formation of nodule primordia. Upon reaching the nodule primordia, ITs release the rhizobia bacteria which are taken up by endocytosis. Bacterial initiation of these events is stimulated by the release of nodulation factors ("nod factors") which are lipochitooligosaccharide signaling molecules that consist of a tetrameric backbone of β1-4 N-acetyl glucosamine residues, an N-linked fatty acyl group, and a variety of chemical modifications that provide host–bacteria specificity (Geurts and Bisseling 2002; D'Haeze and Holsters 2002; Oldroyd and Downie 2008).

In contrast to the RHC-mechanism, a more rudimentary "lateral root boundary" (LRB) mechanism of infection is commonly used by aquatic legumes. LRB infection is often referred to as "crack invasion" or "intercellular infection", and involves the entry of rhizobia bacteria in breaks in the epidermis, generally at secondary root/primary root boundaries, and accumulation within the extracellular spaces of the root cortex in zones referred to as "infection pockets" (Goormachtig et al. 2004b; Den Herder et al. 2006; Oldroyd and Downie 2008). Similar to the RHC-mechanism, nod factor-induction of cortical cell division and nodule primordia formation take place. Infection of the nodule primorida by bacteria in the infection pocket takes place by the formation of an extracellular IT which, unlike RHC infection, penetrates the root through the intercellular spaces prior to bacterial release into the host cells of the nodule primordium (Goormachtig et al. 2004b). However, depending on the specific legume/rhizobial association, in some cases direct uptake of rhizobia by nodule cells from the cortical infection pocket can occur without IT formation (Goormachtig et al. 2004b).

The primary role of root curling (RHC mechanism) or infection pocket formation (LRB mechanism) is to provide a colony of bacteria entrapped within the plant, presumably to provide a critical "signaling center" to facilitate signal exchange between the plant and symbiont (Goormachtig et al. 2004a). In the case of RHC infection, two legumes have emerged as model systems for the investigation of molecular genetics and functional genomics associated with bacteria/host signal transduction and nodule development: *Lotus japonicus,* a model for legumes that form determinant nodules, and *Medicago truncatula*, a model for legumes that form indeterminant nodules (Udvardi et al. 2005; Stacey et al. 2006; Young and Udvardi 2009). With respect to flood-tolerant species that utilize LRB infection, the tropical legume *Sesbania rostrata* has become a model system for investigation of "crack nodulation" (reviewed in Goormachtig et al. 2004b; Den Herder et al. 2006). Also, since *S. rostrata* exhibits both RHC and LRB nodulation strategies, it also has been an instructive model for how tropical legumes cope with flooding stress by using different infection signaling pathways.

3.4.2 Sesbania rostrata: *A Model Legume for Aquatic Nodulation*

Sesbania rostrata is considered a semi-aquatic tropical legume, and attracted initial interest because of its agronomic potential as a flood tolerant "green manure" legume, especially for nitrogen enrichment of rice fields (Becker et al. 1995). *S. rostrata* can be infected by *Azorhizobium caulinodans* by either the RHC or LRB pathway (Ndoye et al. 1994; Goormachtig et al. 2004a). As part of the adaptation strategy to a semi-aquatic lifestyle, the *S. rostrata* stem contains rows of lateral root primordia located along the stem length. When grown on aerated soils these stem root primordia remain dormant and *S. rostrata* nodulation occurs by the classical RHC pathway, resulting in the formation of indeterminant nodules with a persistent meristerm. However, root submergence triggers the emergence and growth of lateral adventitious roots from these stem root primorida, and the LRB infection pathway is promoted and spherical determinant nodules are formed. Simultaneously, flooding causes a suppression of root hair growth in zone 1 of the root, and the arrest of *A. caulinodans* nodulation by the RHC pathway (Goormachtig et al. 2004a).

LRB nodulation of epidermal breaks at the boundary of the stem and the emerging adventitious roots occurs by a series of steps, including: (1) initiation of a localized cortical cell death response generating the "infection pocket" cavity within the root cortex; (2) entry and colonization of the infection pocket by azorhizobia bacteria; (3) induction of cortical cell division in the root and formation of nodule primordia; (4) formation of intercellular ITs that penetrate the root and release the symbiont into infected cells of the nodule primordia (Den Herder et al. 2006). Infection also takes place at the base of submerged lateral roots by a similar pathway, with *S. rostrata* forming nodules on roots as well as on stems on emerging lateral adventitious roots under waterlogged conditions (Ndoye et al. 1994).

Similar to the RHC-based pathway, LRB infection in *S. rostrata* requires the *A. caulinodans* nod factor (D'Haeze et al. 1998, 2003). However, analysis of *A. caulinodans* mutants that produce nod factors lacking specific modifications of the lipochitinoligosaccharide backbone show that while these mutants are incapable of RHC infection, they can still generate functional nodules through the LRB pathway, although with less efficiency and lower nodule numbers (D'Haeze et al. 2000). This argues that the nod factor receptor and signaling networks for the two infection pathways likely share some properties but that LRB infection has a less stringent nod factor structural requirement than the RHC pathway, and also that the two pathways involve some additional non shared components as well. This is supported by transcriptional profiling analyses. For example, by using cDNA-amplified fragment length polymorphism analysis, Capoen et al. (2007) compared the transcript profiles of roots from the site of *A. caulinodans* infection under hydroponic (LRB) and aeroponic (RHC) conditions. A large number of genes were found that are common to both nodulation pathways. In general these genes are proposed to be associated with common events in both pathways such as nodule primordia formation and later stages of nodule development. However, clusters of genes that were unique to either the RHC or LRB pathways were also identified, several of which encode proteins proposed to be specific to the invasion process of each pathway (Capoen et al. 2007).

The elevation of the levels of ethylene is a common response during flooding stress (Perata and Voesenek 2007; Bailey-Serres and Voesenek 2008), and it is clear that this is a critical signal that reciprocally regulates the path of nodulation in *S. rostrata* (Goormachtig et al. 2004a). Upon flooding of *S. rostrata* roots, the enzymes for ethylene biosynthesis such as *S*-adenosyl methionine synthetase, ACC synthase and ACC oxidase, as well as an ethylene response element transcripion factor, are upregulated suggesting that the production and perception of ethylene is induced (Schroeyers et al. 2004). By using a pharmacological strategy, it was shown that application of ethylene biosynthesis inhibitors results in an increase in zone 1 root hair nodulation whereas the addition of the ethylene precursor ACC shows the opposite effect with zone 1 nodulation inhibited (Goormachtig et al. 2004a). Conversely, inhibitors of ethylene action block nod factor-mediated infection pocket formation and nodule primordia formation, and in general inhibit the nodulation of *S. rostrata* under aquatic growth conditions (D'Haeze et al. 2003). In addition, ethylene apparently controls the type of nodule formed under aquatic and nonaquatic conditions (Fernandez-López et al. 1998). Ethylene biosynthetic inhibitors applied during aquatic *S. rostrata* nodule formation result in the persistence of an indeterminant nodule, typical of RHC infections, suggesting a role of ethylene in regulating the meristematic activity of the nodule and whether an indeterminant or determinant nodule type is formed.

A critical element of LRB nodulation that is lacking in RHC nodulation is the formation of the cortical infection pocket for the accumulation of *A. caulinodans* bacteria. The induction of this pocket involves nod factor and ethylene-induced programmed cell death of cortical cells localized adjacent to the epidermal crack at the lateral root base. In addition, basal root infection was associated with the release of high levels of H_2O_2 which is proposed to mediate the programmed cell death response as part of the infection pocket formation (D'Haeze et al. 2003). Nod factor-induced responses in hydroponic *S. rostrata* are inhibited by the pre-application of H_2O_2 scavengers or inhibitors of ROS production, suggesting that H_2O_2 plays a role as a second messenger of the nod factor signal, selectively in the LRB pathway (D'Haeze et al. 2003).

Besides ethylene gas, additional crosstalk with other hormone-signaling pathways also appears to be important in determining the infection pathway for *S. rostrata* nodulation. For example, a synergistic role in aquatic nodulation has also been proposed for gibberellin in mediating the formation of infection pockets during aquatic nodulation since GA biosynthetic inhibitors block infection pocket and IT formation (Lievens et al. 2005). Also, jasmonic acid has been shown to have the opposite effect, with this hormone triggering the RHC infection pathway while suppressing the LRB pathway (Capoen et al. 2009).

From these observations, a model (Holsters et al. 2005; Den Herder et al. 2006) has been proposed for the regulation of the path of rhizobial infection of semi-aquatic legumes. The RHC pathway is proposed to be the default pathway in flood-tolerant legumes under normal growth in well-aerated media. However, upon flooding, ethylene accumulates due to: (1) the low diffusion co-efficient of ethylene in aqueous media; and (2) increased induction of genes coding for ethylene biosynthetic enzymes and signaling proteins as a general response to flooding stress (Schroeyers et al 2004). This triggers a shift in nod factor signaling from a RHC-based intracellular infection mechanism through epidermal root hairs, to a LRB-invasion mechanism through epidermal cracks in adventitious and lateral root boundaries. Since ethylene is commonly involved in the adaptation of a number of growth responses in flood tolerant nonlegumes, it is proposed that flood-tolerant legumes coupled elements of the ethylene response network to traditional nod factor signaling pathways to enable an alternative mode of nodulation. In addition, the ethylene-induced switch to a crack infection pathway avoids the negative effects of ethylene-sensitive arrest of root hair development and RHC-based infection observed in flooding intolerant legumes (Holsters et al. 2005; Den Herder et al. 2006).

3.5 Summary and Concluding Remarks

The nodule represents a unique and specialized organ that has evolved to maintain a microaerobic symbiotic environment designed to simultaneously support a high respiratory rate while preventing the inactivation of oxygen-sensitive nitrogenase.

The regulation of oxygen entry through a high resistance gas diffusion barrier is critical for the maintenance and regulation of this environment. The diffusion rate is modulated both by short term and long term adaptations in response to environmental and metabolic cues, and this represents a critical control of the energetically expensive nitrogen fixation process in response to stress and metabolic need. From the perspective of waterlogging stress, the rhizosphere oxygen concentration is a key signal that rapidly and reversibly stimulates oxygen diffusion in response to hypoxia or inhibits diffusion in response to elevated pO_2.

With respect to short term regulation of gas diffusion, osmotically-driven water flow between the cells of the inner cortex and the apoplastic spaces has emerged as a promising model to explain the reversible modulation of gas diffusion through the inner nodule cortex. However, mechanistic details of the molecular components of this apparatus and the signaling process through which oxygen signal is sensed and transduced are lacking. Questions remain regarding the relative contribution of the inner cortex and the infected zone to the control of oxygen diffusion to the bacteroid under conditions of limiting rhizosphere oxygen and flooding. In addition, the induction of the synthesis and deposition of extracellular glycoproteins appear to be correlated with the degree of resistance of the gas diffusion pathway, however how these proteins aid in short and long term regulation of gas diffusion through the intercellular cortical and infected cell spaces remains to be addressed.

With respect to long-term adaptations, both flooding-tolerant and flooding-sensitive legumes have general developmental programs to adapt to waterlogged conditions. In response to flooding and/or reduced external oxygen concentrations, an increase in the gas porosity of the nodule by the formation of secondary aerenchyma and surface lenticels is triggered. In the case of aquatic wetland legumes, this network of aerenchyma is extensive and is continguous with stems, roots and nodules. In extreme cases, such as the tropical aquatic legume *Discolobium*, this may have led to an "obligate flooding" nodule, which is terminally adapted to a submerged state and cannot survive return to ambient oxygen conditions. Flood-tolerant legumes have also developed additional developmental strategies including the formation of nodules on stems and adventitious roots that avoid hypoxia by growing on aerial tissues or on floating root or stem systems. In these cases, an alternative nodulation strategy involving "crack infection" of epidermal breaks at the point of lateral root or adventitious root emergence has been employed. Interestingly nod factor signaling is also involved in this process but the pathway is distinct from the classical nod factor/RHC signal transduction pathway, and involves crosstalk with other signaling pathways, particularly with the stress hormone ethylene. *Sesbania rostrata* has emerged as the model legume for this flooding-tolerant nodulation pathway, and remains a promising system for future elucidation of the mechanistic features of crack nodulation, and the role of ethylene signaling in switching the mechanism of nodulation that is critical to adaptation to waterlogging stress.

Acknowledgments Supported by National Science Foundation grant MCB-0618075 to DMR.

References

Alazard D (1985) Stem and root nodulation in *Aeschynomene* spp. Appl Environ Microbiol 50:732–734

Allen ON, Allen EK (1981) The Leguminosae: a source book of characteristics, uses, and nodulation. The University of Wisconsin Press, Madison, WI, USA

Arrese-Igor C, Royuela M, de Lorenzo C, de Felipe MR, Aparicio-Tejo PM (1993) Effect of low rhizosphere oxygen on growth, nitrogen fixation and nodule morphology in lucerne. Physiol Plant 89:55–63

Atkins CA, Hunt S, Layzell DB (1993) Gaseous diffusive properties of soybean nodules cultured with non-ambient pO_2. Physiol Plant 87:89–95

Bacanamwo M, Purcell LC (1999a) Soybean dry matter and N accumulation responses to flooding stress, N sources and hypoxia. J Exp Bot 50:689–696

Bacanamwo M, Purcell LC (1999b) Soybean root morphological and anatomical traits associated with acclimation to flooding. Crop Sci 39:143–149

Bailey-Serres J, Voesenek LA (2008) Flooding stress: acclimations and genetic diversity. Annu Rev Plant Biol 59:313–339

Batzli JM, Dawson JO (1999) Development of flood-induced nodule lenticels on red alder during the restoration of nitrogenase activity. Can J Bot 77:1373–1377

Becker M, Ladha JK, Ali M (1995) Green manure technology: potential, usage, and limitations. A case study for lowland rice. Plant Soil 174:181–194

Bergersen FJ (1982) Root nodules of legumes: structure and functions. Wiley, New York

Bergersen FJ (1997) Regulation of nitrogen fixation in infected cells of leguminous root nodules in relation to O_2 supply. Plant Soil 191:189–203

Bergersen FJ, Turner GL (1993) Effects of concentrations of substrates supplied to N_2-fixing soybean bacteroids in flow chamber reactions. Proc R Soc Lond B 245:59–64

Boivin C, Ndoye I, Molouba F, Lajudie P, Dupuy N, Dreyfus B (1997) Stem nodulation in legumes: diversity, mechanisms, and unusual characteristics. Crit Rev Plant Sci 16:1–30

Bradley DJ, Wood EA, Larkins AP, Galfre G, Butcher GW, Brewin NJ (1988) Isolation of monoclonal antibodies reacting with peribacteroid membranes and other components of pea root nodules containing *Rhizobium leguminosarum*. Planta 173:149–160

Buttery BR (1987) Some effects of waterlogging and supply of combined nitrogen on soybean growth. Can J Plant Sci 67:69–77

Capoen W, Den Herder J, Rombauts S, De Gussem J, De Keyser A, Holsters M, Goormachtig S (2007) Comparative transcriptome analysis reveals common and specific tags for root hair and crack-entry invasion in *Sesbania rostrata*. Plant Physiol 144:1878–1889

Capoen W, Den Herder J, Sun J, Verplancke C, De Keyser A, De Rycke R, Goormachtig S, Oldroyd G, Holsters M (2009) Calcium spiking patterns and the role of the calcium/calmodulin-dependent kinase CCaMK in lateral root base nodulation of *Sesbania rostrata*. Plant Cell 21:1526–1540

Criswell JG, Havelka UD, Quebedeaux B, Hardy RWF (1976) Adaptation of nitrogen fixation by intact soybean nodules to altered rhizosphere pO_2. Plant Physiol 58:622–625

D'Haeze W, Gao M, De Rycke R, Van Montagu M, Engler G, Holsters M (1998) Roles for azorhizobial nod factors and surface polysaccharides in intercellular invasion and nodule penetration, respectively. Mol Plant-Microbe Interact 11:999–1008

D'Haeze W, Mergaert P, Promé JC, Holsters M (2000) Nod factor requirements for efficient stem and root nodulation of the tropical legume *Sesbania rostrata*. J Biol Chem 275:15676–15684

D'Haeze W, Holsters M (2002) Nod factor structures, responses, and perception during initiation of nodule development. Glycobiology 12:79–105

D'Haeze W, De Rycke R, Mathis R, Goormachtig S, Pagnotta S, Verplancke C, Capoen W, Holsters M (2003) Reactive oxygen species and ethylene play a positive role in lateral root base nodulation of a semiaquatic legume. Proc Natl Acad Sci USA 100:11789–11794

Dakora FD, Atkins CA (1990a) Morphological and structural adaptation of nodules of cowpea to functioning under sub- and supra-ambient oxygen pressure. Planta 182:572–582

Dakora FD, Atkins CA (1990b) Effect of pO_2 on growth and nodule functioning of symbiotic cowpea (*Vigna unguiculata* L. Walp.). Plant Physiol 93:948–955

Dakora FD, Atkins CA (1990c) Effect of pO_2 during growth on the gaseous diffusional properties of nodules of cowpea (*Vigna unguiculata* L. Walp.). Plant Physiol 93:956–961

Dakora FD, Atkins CA (1991) Adaptation of nodulated soybean (*Glycine max* L. Merr.) to growth in rhizospheres containing nonambient pO_2. Plant Physiol 96:728–736

de Lorenzo C, Iannetta PPM, Fernandez-Pascual M, James EK, Lucas MM, Sprent JI, Witty JF, Minchin FR, de Felipe MR (1993) Oxygen diffusion in lupin nodules II. Mechanisms of diffusion barrier operation. J Exp Bot 44:1469–1474

Den Herder J, Schroeyers K, Holsters M, Goormachtig S (2006) Signaling and gene expression for water-tolerant legume nodulation. Crit Rev Plant Sci 25:367–380

Denison RF, Kinraide TB (1995) Oxygen-induced membrane depolarizations in legume root nodules (Possible evidence for an osmoelectrical mechanism controlling nodule gas permeability). Plant Physiol 108:235–240

Denison RF, Layzell DB (1991) Measurement of legume nodule respiration and photometry of leghemoglobin. Plant Physiol 96:137–143

Denison RF, Hunt S, Layzell DB (1992) Nitrogenase activity, nodule respiration, and O(2) permeability following detopping of Alfalfa and Birdsfoot trefoil. Plant Physiol 98:894–900

Drevon JJ, Gaudillère JP, Bernoud JP, Jardinet F, Evrard M (1991) Influence of photon flux density and fluctuation on the nitrogen fixing *Glycine max* (L Merr)-*Bradyrhizobium japonicum* symbiosis in a controlled environment. Agronomie 11:193–199

Dreyfus BM, Dommergues YR (1981) Nitrogen-fixing nodules induced by Rhizobium on the stem of the tropical legume *Sesbania rostrata*. FEMS Microbiol Lett 10:313–317

Evans WR, Fleischman DE, Calvert HE, Pyati PV, Alter GM, Subba Rao NS (1990) Bacteriochlorophyll and photosynthetic reaction centers in *Rhizobium* Strain BTAi 1. Appl Environ Microbiol 56:3445–3449

Fernandez-Lopez M, Goormachtig S, Gao M, D'Haeze W, Van Montagu M, Holsters M (1998) Ethylene-mediated phenotypic plasticity in root nodule development on *Sesbania rostrata*. Proc Natl Acad Sci USA 95:12724–12728

Ferrell RT, Himmelblau DM (1967) Diffusion coefficients of nitrogen and oxygen in water. J Chem Eng Data 12:111–115

Fleurat-Lessard P, Michonneau P, Maeshima M, Drevon JJ, Serraj R (2005) The distribution of aquaporin subtypes (PIP1, PIP2 and gamma-TIP) is tissue dependent in soybean (Glycine max) root nodules. Ann Bot 96:457–460

Goormachtig S, Capoen W, James EK, Holsters M (2004a) Switch from intracellular to intercellular invasion during water stress-tolerant legume nodulation. Proc Natl Acad Sci USA 101:6303–6308

Goormachtig S, Capoen W, Holsters M (2004b) Rhizobium infection: lessons from the versatile nodulation behaviour of water-tolerant legumes. Trends Plant Sci 9:518–522

Geurts R, Bisseling T (2002) Rhizobium nod factor perception and signalling. Plant Cell 14:239–249

Guinel FC, Sloetjes LL (2000) Ethylene is involved in the nodulation phenotype of *Pisum sativum* R50 (sym 16), a pleiotropic mutant that nodulates poorly and has pale green leaves. J Exp Bot 51:885–894

Holsters H, Capoen W, Den Herder J, Goormachtig S (2005) Signaling for nodulation in a water-tolerant legume. In: Wang Y-P, Lin M, Tian ZX, Elmerich C, Newton WE (eds) Biological nitrogen fixation, sustainable agriculture, and the environment. Proceedings of the 14th International Nitrogen Fixation Congress, Springer, The Netherlands, pp 161–164

Hunt S, King BJ, Canvin DT, Layzell DB (1987) Steady and nonsteady state gas exchange characteristics of soybean nodules in relation to the oxygen diffusion barrier. Plant Physiol 84:164–172

Hunt S, King BJ, Layzell DB (1989) Effects of gradual increases in O_2 concentration on nodule activity in soybean. Plant Physiol 91:315–321

Hunt S, Layzell DB (1993) Gas exchange of legume nodules and the regulation of nitrogenase activity. Annu Rev Plant Physiol Plant Mol Biol 44:483–511

Ianetta PPM, De Lorenzo C, Iannetta PPM, James EK, Fernandez-Pascual M, Sprent JI, Lucas MM, Witty JF, De Felipe MR, Minchin FR (1993) Oxygen diffusion in lupin nodules I. Visualization of diffusion barrier operation. J Exp Bot 44:1461–1467

Jackson MB, Armstrong W (1999) Formation of aerenchyma and the processes of plant ventilation in relation to soil flooding and submergence. Plant Biol 1:274–287

James EK, Sprent JI, Minchin FR, Brewin NJ (1991) Intercellular location of glycoprotein in soybean nodules: effect of altered rhizosphere oxygen concentration. Plant Cell Environ 14:467–476

James EK, Sprent JI, Sutherland JM, McInroy SG, Minchin FR (1992a) The structure of nitrogen fixing root nodules on the aquatic mimosoid legume *Neptunia plena*. Ann Bot 69:173–180

James EK, Minchin FR, Sprent JI (1992b) The physiology and nitrogen-fixing capability of aquatically and terrestrially grown *Neptunia plena*: the importance of nodule oxygen supply. Ann Bot 69:181–187

James EK, Crawford RMM (1998) Effect of oxygen availability on nitrogen fixation by two *Lotus* species under flooded conditions. J Exp Bot 49:599–609

James EK, Minchin FR, Oxborough K, Cookson A, Baker NR, Witty JF, Crawford RMM, Sprent JI (1998) Photosynthetic oxygen evolution within *Sesbania rostrata* stem nodules. Plant J 13:29–38

James EK, Sprent JI (1999) Development of N_2-fixing nodules on the wetland legume *Lotus uliginosus* exposed to conditions of flooding. New Phytol 142:219–231

James EK, Iannetta PPM, Deeks L, Sprent JI, Minchin FR (2000) Detopping causes production of intercellular space occlusions in both the cortex and infected region of soybean nodules. Plant Cell Environ 23:377–386

James EK, Loureiro MF, Pott A, Pott VIJ, Martins CM, Franco AA, Sprent JI (2001) Flooding-tolerant legume symbioses from the Brazilian Pantanal. New Phytol 150:723–738

Justin SHFW, Armstrong W (1987) The anatomical characteristics of roots and plant response to soil flooding. New Phytol 106:465–495

King BJ, Hunt S, Weagle GE, Walsh KB, Pottier RH, Canvin DT, Layzell DB (1988) Regulation of O_2 concentration in soybean nodules observed by in situ spectroscopic measurement of leghemoglobin oxygenation. Plant Physiol 87:296–299

King BJ, Layzell DB (1991) Effect of increases in oxygen concentration during the argon-induced decline in nitrogenase activity in root nodules of soybean. Plant Physiol 96:376–381

Koponen P, Nygren P, Domenach AM, Le Roux C, Saur E, Roggy JC (2003) Nodulation and dinitrogen fixation of legume trees in a tropical freshwater swamp forest in French Guiana. J Trop Ecol 19:655–666

Kuzma MM, Winter H, Storer P, Oresnik I, Atkins CA, Layzell DB (1999) The site of oxygen limitation in soybean nodules. Plant Physiol 119:399–408

Layzell DB (1998) Oxygen and the control of nodule metabolism and N_2 fixation. In: Emerich C, Kondorosi A, Newton WE (eds) Biological nitrogen fixation for the 21st century. Kluwer Academic, Dordrecht, The Netherlands, pp 435–440

Layzell DB, Hunt S, Palmer GR (1990) Mechanism of nitrogenase inhibition in soybean nodules: Pulse-modulated spectroscopy indicates that nitrogenase activity Is limited by O_2. Plant Physiol 92:1101–1107

Lievens S, Goormachtig S, Den Herder J, Capoen W, Mathis R, Hedden P, Holsters M (2005) Gibberellins are involved in nodulation of *Sesbania rostrata*. Plant Physiol 139:1366–1379

Linkemer G, Board JE, Musgrave ME (1998) Waterlogging effects on growth and yield components in late-planted soybean. Crop Sci 38:1576–1584

Loureiro MF, de Faria SM, James EK, Pott A, Franco AA (1994) Nitrogen-fixing stem nodules of the legume *Discolobium pulchellum* Benth. New Phytologist 128:283–295

Loureiro MF, James EK, Sprent JI, Franco AA (1995) Stem and root nodules on the tropical wetland legume *Aeshynomene fluminensis*. New Phytol 130:531–544

Loureiro MF, James EK, Franco AA (1998) Nitrogen fixation by legumes in flooded regions. In: Scarano FR, Franco AC (eds) Oecologia Brasiliensis, vol IV. PPGE-UFRJ, Rio de Janeiro, Brazil, pp 195–233

Maurel C, Verdoucq L, Luu DT, Santoni V (2008) Plant aquaporins: membrane channels with multiple integrated functions. Annu Rev Plant Biol 59:595–624

Minchin FR, Pate JS (1975) Effects of water, aeration, and salt regime on nitrogen fixation in a nodulated legume. Definition of an optimum root environment. J Exp Bot 26:60–69

Minchin FR, Summerfield RJ (1976) Symbiotic nitrogen fixation and vegetative growth of cowpea (*Vigna unguiculata* (L.) walp.) in waterlogged conditions. Plant Soil 45:113–127

Minchin FR (1997) Regulation of oxygen diffusion in legume nodules. Soil Biol Biochem 29:881–888

Minchin FR, James EK, Becana M (2008) Oxygen diffusion, production of reactive oxygen and nitrogen species, and antioxidants in legume nodules. In: Dilworth MJ, James EK, Sprent JI, Newton WE (eds) Nitrogen-fixing leguminous symbioses. Springer, The Netherlands, pp 321–362

Ndoye I, de Billy F, Vasse J, Dreyfus B, Truchet G (1994) Root nodulation of *Sesbania rostrata*. J Bacteriol 176:1060–1068

Ober ES, Sharp RE (1996) A microsensor for direct measurement of O_2 partial pressure within plant tissues. J Exp Bot 44:447–454

Oldroyd GE, Downie JA (2008) Coordinating nodule morphogenesis with rhizobial infection in legumes. Annu Rev Plant Biol 59:519–546

Pankhurst CE, Sprent JI (1975a) Effects of water stress on the respiratory and nitrogen-fixing activity of soybean root nodules. J Exp Bot 26:287–304

Pankhurst CE, Sprent JI (1975b) Surface features of soybean root nodules. Protoplasma 85:85–98

Parsons R, Day DA (1990) Mechanism of soybean nodule adaptation to different oxygen pressures. Plant Cell Environ 13:501–512

Patriarca E, Tate R, Ferraioli S, Iaccarrino M (2004) Organogenesis of legume root nodules. Int Rev Cytol 234:201–263

Penmetsa RV, Cook DR (1997) A legume ethylene-insensitive mutant hyperinfected by its rhizobial symbiont. Science 275:527–530

Perata P, Voesenek LA (2007) Submergence tolerance in rice requires Sub1A, an ethylene-response-factor-like gene. Trends Plant Sci 12:43–46

Purcell LC, Sinclair TR (1994) An osmotic hypothesis for the regulation of oxygen permeability in soybean nodules. Plant Cell Environ 17:837–843

Pugh R, Witty JF, Mytton LR, Minchin FR (1995) The effect of waterlogging on nitrogen fixation and nodule morphology in soil-grown white clover (*Trifolium repens* L.). J Exp Bot 46:285–290

Ralston EJ, Imsande J (1982) Entry of oxygen and nitrogen into intact soybean nodules. J Exp Bot 33:208–214

Saur E, Carcelle S, Guezennec S, Rousteau A (2000) Nodulation of legume species in wetlands of Guadeloupe (Lesser Antilles). Wetlands 20:730–734

Scott HD, DeAngulo J, Daniels MB, Wood LS (1989) Flood duration effects on soybean growth and yield. Agron J 81:631–636

Schaede R (1940) Die Knöllchen der adventiven Wasserwurzeln Von Neptunia oleracea und ihre Bakteriensymbiose. Planta 31:1–21

Schroeyers K, Chaparro C, Goormachtig S, Holsters M (2004) Nodulation-enhanced sequences from the water stress-tolerant tropical legume *Sesbania rostrata*. Plant Sci 167:207–216

Serraj R, Fleurat-Lessard P, Jaillard B, Drevon JJ (1995) Structural changes in the inner-cortex cells of soybean root-nodules are induced by short-term exposure to high salt or oxygen concentrations. Plant Cell Environ 18:455–462

3 Strategies for Adaptation to Waterlogging and Hypoxia in Nitrogen Fixing Nodules

Serraj R, Frangne N, Maeshima M, Fleurat-Lessard P, Drevon JJ (1998) γ-TIP cross-reacting protein is abundant in the cortex of soybean N_2-fixing nodules. Planta 206:681–684

Sheehy JE, Minchin FR, Witty JF (1983) Biological control of the resistance to oxygen flux in nodules. Ann Bot 60:345–351

Shimamura S, Mochizuki T, Nada Y, Fukuyama M (2003) Formation and function of secondary aerenchyma in hypocotyl, roots and nodules of soybean (*Glycine max*) under flooded conditions. Plant Soil 251:351–359

Sprent JI (1999) Nitrogen fixation and growth of non-crop legume species in diverse environments. Perspect Plant Ecol Evol Syst 2:149–162

Sprent JI, James EK (2007) Legume evolution: where do nodules and mycorrhizas fit in? Plant Physiol 144:575–581

Stacey G, Libault M, Brechenmacher L, Wan J, May GD (2006) Genetics and functional genomics of legume nodulation. Curr Opin Plant Sci 9:110–121

Subba-Rao NS, Mateos PF, Baker D, Pankratz HS, Palma J, Dazzo FB, Sprent JI (1995) The unique root-nodule symbiosis between *Rhizobium* and the aquatic legume, *Neptunia natans* (L. f.) Druce. Planta 196:311–320

Sung L, Moloney AH, Hunt S, Layzell DB (1991) The effect of excision on O_2 diffusion and metabolism in soybean nodules. Physiol Plant 83:67–74

Sung FJM (1993) Waterlogging effect on nodule nitrogenase and leaf nitrate reductase activities in soybean. Field Crops Res 35:183–189

Thomas AL, Guerreiro SMC, Sodek L (2005) Aerenchyma formation and recovery from hypoxia of the flooded root system of nodulated soybean. Ann Bot 96:1191–1198

Tjepkema JD, Yocum CS (1974) Measurement of oxygen partial pressure within soybean nodules by oxygen microelectrodes. Planta 119:351–360

Udvardi MK, Tabata S, Parniske M, Stougaard J (2005) *Lotus japonicus*: legume research in the fast lane. Trends Plant Sci 10:222–228

van Dongen JT, Schurr U, Pfister M, Geigenberger P (2003) Phloem metabolism and function have to cope with low internal oxygen. Plant Physiol 131:1529–1543

VandenBosch KA, Bradley DJ, Knox JP, Perotto S, Butcher GW, Brewin NJ (1989) Common components of the infection thread matrix and the intercellular space identified by immunocytochemical analysis of pea nodules and uninfected roots. EMBO J 8:335–342

Vessey JK, Walsh KB, Layzell DB (1988) Oxygen limitation of N_2 fixation in stem-girdled and nitrate-treated soybean. Physiol Plant 73:113–121

Walker BA, Pate JS, Kuo J (1983) Nitrogen fixation by nodulated roots of *Viminaria juncea* (Schrad. & Wendl.) Hoffmans, (Fabaceae) when submerged in water. Aust J Plant Physiol 10:409–421

Walter CA, Bien A (1989) Aerial root nodules in the tropical legume, *Pentaclethra macroloba*. Oceologia 80:27–31

Wei H, Layzell DB (2006) Adenylate-coupled ion movement. A mechanism for the control of nodule permeability to O_2 diffusion. Plant Physiol 141:280–287

Weisz PR, Sinclair TR (1987a) Regulation of soybean nitrogen fixation in response to rhizosphere oxygen: I. Role of nodule respiration. Plant Physiol 84:900–906

Weisz PR, Sinclair TR (1987b) Regulation of soybean nitrogen fixation in response to rhizosphere oxygen: II. Quantification of nodule gas permeability. Plant Physiol 84:906–910

Witty JF, Minchin FR, Skøt L, Sheehy JE (1986) Nitrogen fixation and oxygen in legume root nodules. Oxford Surv Plant Mol Cell Biol 3:275–314

Witty JF, Skøt L, Revsbech NP (1987) Direct evidence for changes in the resistance of legume root nodules to O_2 resistance. J. Exp Bot 38:1129–1140

Wycoff KL, Hunt S, Gonzales MB, VandenBosch KA, Layzell DB, Hirsch AM (1998) Effects of oxygen on nodule physiology and expression of nodulins in alfalfa. Plant Physiol 117:385–395

Young ND, Udvardi M (2009) Translating *Medicago truncatula* genomics to crop legumes. Curr Opin Plant Biol 12:193–201

Chapter 4
Oxygen Transport in the Sapwood of Trees

Sergio Mugnai and Stefano Mancuso

Abstract Gas composition inside large woody stems differs significantly from that of the ambient atmosphere because of cellular respiration in the xylem, phloem and cambium. Oxygen is required for oxidative respiration, which under most conditions provides the energy for plant cells. The gaseous environment within the woody stems is enriched in CO_2 and depleted in O_2 as a result of the net effects of respiration, exchange with the atmosphere via diffusion through bark, and exchange with the transpiration stream through the dissolution of gases. Oxygen concentration normally declines from the cambium toward the heartwood boundary during stem respiration, but no values below 3–5% gaseous mole fraction (corresponding to approximately 15–25% of air O_2 content) were ever measured. It is clear that the gas diffusivity of wood is necessary to supply live sapwood with oxygen and that, given the strong effect of water content on diffusion coefficient, not enough oxygen would diffuse through fully saturated wood, illustrating that the xylem cell walls present a major barrier to gas diffusion. This supports the hypothesis that heartwood formation can be triggered by low oxygen concentrations, or even anoxia, in the innermost sapwood, as colored heartwood seems to be produced only in the presence of oxygen.

Abbreviation

D	Diffusion coefficient (diffusivity)
K_{m}	Michaelis constant
LVDT	Linear variable differential transformers

S. Mugnai and S. Mancuso (✉)
Dpt. Plant, Soil and Environmental Science, University of Florence, viale delle Idee 30, Sesto Fiorentino 50019, Italy
e-mail: sergio.mugnai@unifi.it; stefano.mancuso@unifi.it

S. Mancuso and S. Shabala (eds.), *Waterlogging Signalling and Tolerance in Plants*,
DOI 10.1007/978-3-642-10305-6_4, © Springer-Verlag Berlin Heidelberg 2010

4.1 Brief Anatomy of a Woody Stem

Stems of woody plants have different functions: they support the crown; store water, carbohydrates and minerals; conduct water and minerals upward from the roots to the leaves; transport hormones from the synthesis region to where they are used in growth. A stem of a mature tree typically consists of a column of wood (secondary xylem) composed of a series of layers (annual increments), placed one above the other, enclosed in a cover made of bark. Between the bark and wood, the vascular cambium which is a thin, sheathing lateral meristem, is located (Fig. 4.1).

Wood is normally classified into sapwood and heartwood. Young xylem or sapwood conducts the sap, strengthens the stem, and to some extent serves as a storage reservoir for food. Sapwood is composed by both living and nonliving parenchyma cells, which consists of horizontally oriented ray cells and, in many species of woody plants, vertically oriented axial parenchyma cells as well. Sapwood contains 5–35% living parenchyma by volume depending on the species (Panshin and de Zeeuw 1980) with conifers ranging from about 5–8% and angiosperms from 10–35%. Parenchyma cells within the secondary xylem can be extremely long-lived. Individual cells may live for 2–200 years with longevities that are specific to a species but are also shaped by the environment (Spicer and Holbrook 2007a). Living parenchyma cells form an important point of exchange between the symplast and the apoplast in woody tissue, operating in different processes such as carbohydrate storage, transport and wound response. Rather than remaining functional throughout the entire life of a tree, sapwood is gradually compartmentalized in the center of the stem to form a physiologically inactive core, the "heartwood,"

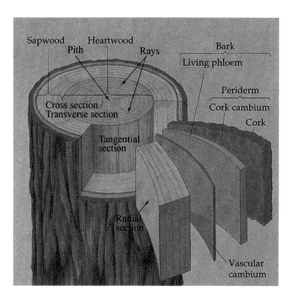

Fig. 4.1 Typologies of wood found in a typical woody stem

a central cylinder of dark-colored dead tissue which continues to provide mechanical support but is no longer involved in physiological processes. This process occurs in nearly every tree species, and is vital to the maintenance of a balance between the volume of autotrophic and heterotrophic tissue.

Heartwood formation is defined by the death of parenchyma cells. Stem cross sections of many species show a distinct "transition zone," a narrow region between sapwood and heartwood, usually less than 1 cm wide (Shain and MacKay 1973; Bowman et al. 2005). In some species, this "transition zone," which typically is lighter in color than the heartwood, is not readily recognized or does not exist.

At the moment, the cause of parenchyma cell death is not exactly known and understood, but there are evidences for a decreased metabolic activity in the innermost sapwood. This situation may explain parenchyma ageing as a gradual, passive decline in metabolism that terminates in cell death. In fact, a large number of studies report a decline in sapwood respiration, the main parameter used to assess parenchyma cell metabolism, towards the sapwood/heartwood boundary.

4.2 Atmosphere Inside a Stem: Gas Composition and its Effects on Respiration

Gas composition inside large woody stems differs significantly from that of the ambient atmosphere because of cellular respiration in the xylem, phloem and cambium. Oxygen is required for oxidative respiration, which under most conditions provides the energy for plant cells. The gaseous environment within the woody stems is enriched in CO_2 and depleted in O_2 as a result of the net effects of respiration, exchange with the atmosphere via diffusion through bark, and exchange with the transpiration stream through the dissolution of gases. A recent review of gas content in wood found an average gas volume of 18% in sapwood and 50% in heartwood in 26 softwood species, 26% in both sapwood and heartwood in 31 temperate hardwood species, and 18% (sapwood and heartwood not distinguished) in 52 tropical hardwoods, with substantial variation between species (Gartner et al. 2004). Sorz and Hietz (2006) believed that an even more important function of gas in stems may be to provide oxygen through storage and faster diffusion, taking into account the strong effect of gas content on its diffusion. Sapwood with 25% by volume of cell wall material, 50% water, and 25% gas in equilibrium with oxygen concentration of ambient air at 20°C is able to store 0.14 mol O_2 m^{-3} in the aqueous phase (solubility at 20°C is 9 g O_2 m^{-3}) and 2.18 mol m^{-3} in the gas phase (air has 8.73 mol O_2 m^{-3} at 20°C). At full saturation (75% water), only a total of 0.21 mol O_2 m^{-3} can be stored. High rates of stem respiration during the growth period are in the range of 100–200 µmol CO_2 m^{-3} sapwood s^{-1} at 15–20°C (Edwards and Hanson 1996; Lavigne et al. 1996). At 100 µmol m^{-3} s^{-1}, the oxygen stored in the sapwood example cited above with 25% gas will be consumed in 6.4 h. If the water content were 65%, the supply would last for 2.9 h, and at full saturation (0% gas) for only 0.6 h, assuming no oxygen is imported. At zero sapflow, the respiring sapwood will draw oxygen either through bark and cambium, if this path is open, which in many trees does not appear to be (Hook et al. 1972), or from the non-respiring heartwood. All the three models tested by Sorz an Hietz (2006) and previously reported show that the diffusion coefficient D for oxygen diffusion to supply 5 cm of sapwood respiring at a rate of 50 μmol m^{-3} s^{-1} has to be about 10^{-8} m^2 s^{-1}, which is in the range of the values measured for radial diffusion. These models are simplifications and sapwood respiration normally depends on many factors such as distance from the cambium, oxygen concentration, season and temperature. But even allowing for inaccuracies in the estimates, it is clear that the gas diffusivity of wood is necessary to supply live sapwood with oxygen and that, given the strong effect of water content on D, not enough oxygen would diffuse through fully saturated wood, illustrating that the xylem cell walls present a major barrier to gas diffusion. This supports the hypothesis that heartwood formation can be triggered by low oxygen concentrations, or even anoxia, in the innermost sapwood (Eklund and Klintborg 2000; Sorz and Hietz 2008), as colored heartwood seems to be produced only in the presence of oxygen.

Oxygen concentration normally declines from the cambium towards the heartwood boundary during stem respiration, but no values below 3–5% gaseous mole fraction (corresponding to approximately 15–25% of air O_2 content) were ever measured (Spicer and Holbrook 2005). According to Eklund (2000) and Pfanz et al. (2002) the stem internal oxygen concentrations also fluctuate strongly in diurnal as well as annual cycles. In contrast to the considerably high CO_2 concentrations measured inside the stem compared to those present in the external atmosphere (Table 4.1), oxygen concentrations are usually much lower than in ambient air (Table 4.2).

Although oxygen concentrations vary with species (because of the mode of sampling and analytic techniques), average endogenous oxygen concentrations within tree stems generally range far below optimal levels (Pfanz et al. 2002). As respiration clearly depends on the presence of O_2 and may be more or less reduced below a certain threshold, there is the danger of stem internal hypoxia (Bouma et al. 1997; Jahnke 2000; Mancuso and Marras 2003; Spicer and Holbrook 2005). Below a certain O_2 concentration, anaerobic fermentation may even occur. Fermentation would produce ethanol and/or lactic acid, which would have deleterious effects on cellular membranes and enzymes and on cellular pH-stat (Pfanz and Heber 1989).

Table 4.1 CO_2 concentrations measured in intact and detached tree stems. Data are given in % w/v

Tree species	Stem internal CO_2 concentration	References
Acer platanoides	2–4%	Eklund (1993)
Betula pendula	3%	Levy et al. (1999)
Picea abies	10%	Eklund (1990)
Pinus sylvestris	2.5%	Kaipiainen et al. (1998)
Platanus occidentalis	1.6–8.9%	McGuire et al. (2007); Teskey and McGuire (2007)
Populus deltoides	3.9–17.6%	Saveyn et al. (2008)
Quercus robur	1–9%	Eklund (1993)

4 Oxygen Transport in the Sapwood of Trees

Table 4.2 O_2 concentrations measured in intact and detached tree stems. Data are given in % w/v or in μmol L^{-1}. Partially redrawn from Pfanz et al. (2002)

Tree species	Stem internal oxygen concentration	References
Acer platanoides	81–217 μmol L^{-1}	Eklund (1993)
Betula pendula	65–171 μmol L^{-1}	Gansert et al. (2001)
Juglans major	8–15%	McDougal and Working (1933)
Laurus nobilis	172–258 μmol L^{-11}	Del Hierro et al. (2002)
Olea europaea	28–100 μmol L^{-1}	Mancuso and Marras (2003)
Parkinsonia microphylla	5.5–22.3%	McDougal and Working (1933)
Picea abies	2–20%	Eklund (1990)
P. abies	41–285 μmol L^{-1}	Eklund (1993)
Pinus strobus	3–11%	Chase (1934)
Populus deltoides	1–5%	Chase (1934)
Populus tremuloides	0–21.2%	McDougal and Working (1933)
P. macdougalii		
Quercus agrifolia	11.3–18.5%	McDougal and Working (1933)
Q. borealis	12–14%	Chase (1934)
Q. robur	68–231 μmol L^{-1}	Eklund (1993)
Q. rubra	0.4–0.9%	Jensen (1967)
Q. rubra		
Q. coccinea	1–4%	Jensen (1969)
Q. velutina		
Salix lasiolepis	8–14%	McDougal and Working (1933)
Sequoia sempervirens	8.5–17.4%	McDougal and Working (1933)
Ulmus americana	3–6%	Chase (1934)

Oxygen levels within the woody stems are at a minimum level during the growing season (2–8% v/v) and closer to atmospheric levels (15–20%) throughout dormancy (Eklund 1990, 1993; Pruyn et al. 2002b). Diurnal patterns of O_2 are more variable than those of CO_2, but the O_2 concentration of xylem sap ranges from a minimum in the absence of transpiration to a maximum during times of peak flow (del Hierro et al. 2002; Mancuso and Marras 2003), suggesting that transpiration is an important source of O_2. Oxygen is progressively depleted toward the sapwood/heartwood boundary, but may be the lowest near the cambial zone during the early part of the growing season (Pruyn et al. 2002b; van Dongen et al. 2003; Spicer and Holbrook 2005) because of the intensive cambial activity. The lowest O_2 concentrations reported for woody tissue are about 3–5% and may occur in particular situations in both outer and inner sapwood (Eklund 1990; Mancuso and Marras 2003; Spicer and Holbrook 2005). The effects of low O_2 levels on respiration are relatively well understood: respiration decreases with decreasing O_2 concentration, so defining a "hypoxic" range (0.5–5% for many tissues), which does not imply that the tissues are "O_2 deficient." However, the O_2 level at which the inhibition of respiration starts depends on the tissue, its history, and the scale of measurement. For example, two enzymes of the electron transport chain, such as cytochrome oxidase and alternative oxidase, have very high affinities for O_2 (K_m for $O_2 = 0.14$ and 1.7 μmol, respectively; Millar et al. 1994); this suggests that the reaction terminating mitochondrial electron transport becomes substrate-limited when O_2

levels fall below 0.1%. Inhibition of respiration has been monitored at 0.01–0.1% O_2 in plant mitochondria (Rawsthorne and LaRue 1986; Millar et al. 1994), 0.5–2.5% in protoplasts (Lammertyn et al. 2001), 10% in roots grown in air (Saglio et al. 1984), and at 10–20% in tuber slices (Geigenberger et al. 2000), but no data are currently available for woody stems. Little is known about the interacting effects of O_2 and CO_2 on sapwood respiration, as contrasting results have been previously reported. In fact, inhibition due to high CO_2 has been shown both to increase (Kidd 1916) and to decrease (Lammertyn et al. 2001) with decreasing O_2 concentration. Spicer and Holbrook (2007) aimed to clarify whether the low O_2 and/or high CO_2 levels common in woody stems could limit parenchyma respiration to such an extent that internal gas composition might lead to the death of parenchyma during the transition from sapwood to heartwood. Also, they measured the rates of sapwood respiration at physiologically relevant gas compositions, determined whether the effects of O_2 and CO_2 on respiration varied with species or tissue age, and considered whether the air volume fraction of sapwood is a determinant of the effects of either O_2 or CO_2. In general, oxygen level has a significant effect on sapwood respiration (Spicer and Holbrook 2007b), but this effect hardly depends on the level of CO_2.

4.3 Gas Transport and Diffusion

Gas diffuses through the stem tissues with a slow rate because of the considerable barriers posed by bark, cambium and wood (Sorz and Hietz 2006). Plants growing in submerged or waterlogged soil often show anatomical adaptations for the transport of oxygen to below-ground parts, and gas flow can be substantially enhanced by Venturi ventilation and thermo-osmosis. Tree stems are normally exposed to light and air but unless bark is transparent, they do not produce oxygen. In some tree species adapted to waterlogged soil, the cambium has small intercellular spaces permitting oxygen supply through the bark. In other species lacking intercellular spaces, the cambium plus bark appear to be quite impermeable to gases and the transpiration stream is supposed to be the main source of oxygen for the xylem. The gas permeability of wood has been investigated for technical applications (Hansmann et al. 2002), as it affects the speed of drying and the ease of chemical modification of wood. However, for technical applications gas permeation is generally measured under a pressure difference in either steady or unsteady state. In principle, liquids and gases can be transported through wood either by bulk flow along a pressure gradient or by diffusion following a concentration gradient. In living trees, water is always transported along a pressure gradient, and nearly all attempts to measure the transport capacity of wood have been made by applying a pressure and quantifying permeability. However, substances dissolved in water or gas will also diffuse through the wood and particularly under conditions of zero flow this diffusion can be an important factor limiting the supply of oxygen to tissues.

Gas flow under a pressure difference is characterized by Darcy's law (1):

$$Q = \frac{A}{O} \frac{(P_b - P_a)}{L} \tag{4.1}$$

where the total discharge (Q, m s^{-1}) is equal to the product of the permeability (κ units of area, m) of the medium, the cross-sectional area (A) to flow, and the pressure drop ($P_b - P_a$), all divided by the dynamic viscosity μ (kg m^{-1} s^{-1} or Pa s), and the length L over which the pressure drop is taking place. It substantially differs from diffusion as a molecular mass flow under the influence of a concentration gradient, which follows Fick's first law (2):

$$J = -D \frac{\partial \phi}{\partial x} \tag{4.2}$$

where J is the diffusion flux (mol m^{-2} s^{-1}), which measures the amount of substance that will flow through a small area during a small time interval; D is the diffusion coefficient or diffusivity (m^2 s^{-1}); ϕ is the concentration (mol m^{-3}); and x is the position (m). As the measurement of permeation under a pressure gradient is not considered as universal for analyzing oxygen supply to wood in biological systems, greatly depending on wood anatomy, Sorz and Hietz (2006) measured gas diffusion through the wood of two coniferous (*Picea abies* L., *Taxus baccata* L.), two diffuse-porous (*Fagus sylvatica* L., *Carpinus betulus* L.) and two ring-porous species (*Quercus robur* L., *Fraxinus excelsior* L.) in axial and radial direction and at different water content in order to evaluate the effect of wood anatomy and water content on gas transport. They tested the diffusion of air in nitrogen through heartwood in a setup that provided constant gas concentrations at both sides of the wood specimen tested. After the calculation of D from the measured change in the concentration and absolute amount of oxygen, the effect of D on the oxygen supply of respiring wood has been monitored using different models for sapwood respiration. In all species, the diffusion coefficient D strongly increased with the volume of air in the xylem in both axial and radial directions. The relationship between D and gas volume was approximately exponential, but differed between species: at 15% gas volume, D for axial diffusion ranged between about 9.5×10^{-10} m^2 s^{-1} for *Picea*, and 1.3×10^{-8} m^2 s^{-1} for *Quercus*; at 40% gas volume, D increased 5- to 13-fold in *Picea*, *Taxus* and *Quercus*, 36-fold in *Fraxinus* and about 1,000-fold in *Carpinus* and *Fagus*. The model calculation showed that in the range of diffusion coefficients measured, D, and therefore wood gas content, had a very strong influence on the ability to supply live sapwood through radial diffusion. At constant respiration and $D = 10^{-8}$, oxygen reached 5.5 cm sapwood depth, and only above ca. 3×10^{-8} was at least some oxygen reaching the entire 10 cm of sapwood. At $D = 10^{-10}$ m^2 s^{-1}, only the outermost mm could be supplied with oxygen to sustain a respiration rate of 50 µmol m^{-3} s^{-1}. When respiration was reduced at low oxygen concentrations, the supply of deeper layers was slightly better because the oxygen-deficient inner sapwood consumed less. When respiration declined from 100 μmol in the outermost sapwood to 25 μmol at a sapwood depth >5 cm, the supply of the sapwood was worse at all but the highest D because the high respiration of the outer layers left little oxygen for the inner layers with less demand.

4.4 Radial and Axial Oxygen Transport to Sapwood

Secondary xylem is typically poor in O_2 and rich in CO_2, with major differences and fluctuations over the season and the daytime. Reported values for O_2 within the xylem during the growing season range from 0.1% to 22% (Table 4.2), with variation due to the differences in time of the year, species and ages studied, and position in the stem. To supply live cells in the sapwood, oxygen can either diffuse radially through periderm, phloem, cambium and wood, or be taken up by the roots and transported towards with the transpiration stream. Radial gaseous diffusion through lenticels and intercellular gas-space continuum has been reported as the major pathway for the oxygen supply of the living cells of sapwood. However, the existence of alternative and more efficient pathways for the supply of oxygen to the sapwood is suggested by: the few intercellular spaces in the sapwood through which gas exchange can be driven by diffusion (Spicer and Holbrook, 2007); the abscence of relationship between bark thickness and anaerobiosis in wood (Kimmerer and Stringer 1988); the fact that cambium consumes most of the oxygen coming from radial diffusion during vegetative activity (Hook et al. 1972). For these rea sons, the transpiration stream was considered as an important aqueous pathway for the supply of oxygen to the sapwood parenchyma since 10 years (Eklu nd 2000; Gansert et al. 2001; del Hierro et al. 2002; Gansert 2003). However, although these experiments provided support for oxygen delivery via the xylem transpiration stream, they were not able to quantify the different pathways of the oxygen supply. Mancuso and Marras (2003) quantified the contribution of different pathways to the oxygen supply of the living cells of the xylem in *Olea europaea* L. plants, by the development of a special Clark-type polarographic microelectrode, composed by a platinum cathode (working electrode), an Ag/AgCl anode (reference anode) and a third electrode (auxiliary electrode) to enhance the sensitivity of the system. The oxygen sensors were inserted into the stem in order to reach the sapwood; at the same time, changes in stem diameter (widening−contraction) were measured using LVDTs (Linear Variable Differential Transformers) to monitor diurnal variation and long-term growth conditions. Results focused on the role of xylem sap flow as the major pathway for the oxygen supply to the wood parenchyma during the daylight hours. This situation reversed during the night when the contribution of the transpiration stream became negligible. As a consequence, other transport mechanisms must be effective in supplying oxygen to the sapwood in the night. Mancuso and Marras (2003) gave consistency to xylary diffusion in the aqueous phase as the main oxygen transport pathway during the night. The sap-filled lumina of tracheids, whose mass flow during daylight represents

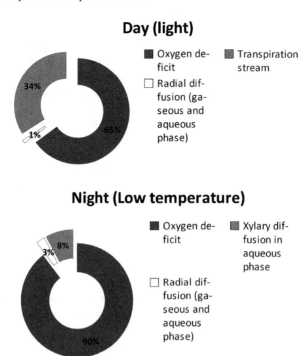

Fig. 4.2 Different pathways of oxygen transport in the sapwood of *Olea europaea* L. The contribution of the different transport pathways during daylight and night hours, together with the oxygen deficit, is shown relative to the oxygen concentration of an aqueous solution saturated with air (set at 100). Redrawn from Mancuso and Marras (2003)

the principal source of oxygen for the living cells of the wood, become during the night the aqueous pathway through which the oxygen taken up by the root diffuses up to the sapwood. This longitudinal diffusion accounts for 23–25 mmol O_2 L^{-1}, corresponding to around 8% of the oxygen present in a saturated aqueous solution and representing the source of 87% of all the oxygen present in the sapwood during the night (Fig. 4.2).

The remaining 13% (corresponding to 3–4 mmol O_2 L^{-1}) of the "nocturnal" oxygen is supplied to the sapwood via radial diffusion. Both pathways are substantially driven by the low temperatures, according to Henry's law (3):

$$p = k_H c \qquad (4.3)$$

where p is the partial pressure of the solute, c is the concentration, k_H is a constant depending on the solute, the solvent and the temperature. For oxygen, k_H in water at 298°K is 769.2 L atm mol^{-1}.

Regardless of the different pathways of oxygen supply to the living cells of the wood, it is clear that in *O. europaea* a condition of oxygen deficiency during the growing season in the range between 65% (with respect to an aqueous solution saturated with air) in the daytime and 89% in the night is a physiological characteristic (Fig. 4.2). During the night, the low oxygen consumption of the wood parenchyma because of the lower temperature of the stem is, in part, an aid in allowing the cells to survive a daily oxygen shortage. The changes in magnitude of the oxygen concentration between day and night were accompanied by an alternation between two different main oxygen transport systems. In the daytime, the transpiration stream transports nearly all the oxygen available for the sapwood tissues, whereas in the night diffusion in the aqueous phase of the xylem and, to a minor extent, radial diffusion assures the necessary oxygen. Pfanz et al. (2002) suggested that the oxygen evolved during wood or pith photosynthesis might play a role in avoiding or reducing stem internal anaerobiosis. However, there has been no strong supporting evidence until now.

4.5 Sapwood Respiration

To allow the normal maintenance respiration of pit tissues and ray parenchyma, oxygen must be delivered to these tissues and also assured in the case of adverse events. In fact, respiration of sapwood naturally depends on many external and internal factors and is therefore quantitatively extremely variable in the different seasons. In addition, higher respiration rates occur in fast-growing tissues and as a consequence of environmental stresses such as frost, wounding or infections. Multiple reports suggest that sapwood respiration declines towards the sapwood/heartwood boundary (Pruyn et al. 2002a, b; Pruyn et al. 2003; Pruyn et al. 2005), although there have been reports of no change (Bowman et al. 2005) or even an increase in respiration in the "transition zone," a narrow region between the sapwood and the heartwood (see Sect. 4.1). The vast majority of these studies have been performed in conifers, and reports on angiosperms mainly refer to whether respiration is reduced in inner sapwood (Pruyn et al. 2003). At a cellular level, the shrinkage of parenchyma nuclei and the decline in the number of mitochondria with tissue age suggest a strong reduction in metabolic activity. In contrast, the ability of xylem parenchyma to de-differentiate and form callus tissue is not affected by age in some *Pinus* species (Allen and Hiatt 1994). Moreover, some Krebs cycle enzymes, such as malic and succinate dehydrogenase, and glucose-6-phosphate dehydrogenase are more active in the innermost sapwood of *Pinus radiata* (Hauch and Magel 1998). This latter enzyme catalyzes the first reaction of the oxidative pentose phosphate pathway, which supplies intermediates for phenolic synthesis, an activity that characterizes heartwood formation in many species. Although most evidence points towards a reduction in parenchyma cell metabolism with age, there are logical reasons to contrast this hypothesis. As the majority of work demonstrating a reduction in respiration with tissue age has been reported in conifers, and that conifers gradually lose parenchyma living cells with tissue age (Gartner et al. 2000; Nakaba et al. 2006), this metabolic decline may be suppressed when respiration is expressed on a *per* live cell volume basis. Secondly, parenchyma cells play an active role in the transition phase from sapwood to heartwood. In fact, these cells are known to synthesize complex polyphenolic compounds (Magel et al. 2000), produce cellulosic material to form tyloses and tylosoids (organic compunds blocking angiosperm vessels and conifer resin canals, respectively; Saitoh et al. 1993) and secrete gums (Saitoh et al. 1993).

Spicer and Holbrook (2007a) tested the hypothesis that parenchyma in secondary xylem respires at the same rate, regardless of age, by expressing respiration on a live cell volume basis. For this reason, they measured sapwood respiration measurements on mature trees of two conifer (*Tsuga canadensis* (L.) Carr. and *Pinus strobus* L.) and three angiosperm (*Acer rubrum* L., *Fraxinus americana* L., *Quercus rubra* L.) by a needle-tipped fiber optic O_2 probe, which operates using the principles of fluorescence quenching in the presence of oxygen. The effect of radial position was highly significant such that the youngest, outermost sapwood respired at a higher rate than the oldest, innermost sapwood in all species except *P. strobus* when expressed on a per tissue volume basis. Large species differences in parenchyma content and vitality were reported (Fig. 4.3), with conifers having less than one-third the parenchyma volume of angiosperms. When respiration was expressed on a per live cell volume basis, there were large species differences such that conifers had almost three times the respiration rate of angiosperms (Fig. 4.4). The effect of radial position on live cell respiration depended on species. Parenchyma in the outermost sapwood respired at almost twice the rate as that of inner sapwood for all the three angiosperms, whereas for conifers there was no significant difference between the two radial positions.

Spicer and Holbrook (2007a) found little evidence for a decline in parenchyma cell respiration that is inherent to the process of cellular ageing. Although species-specific cellular ageing is certainly possible, it seems unlikely given the range of

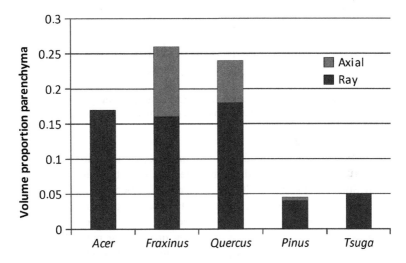

Fig. 4.3 Volume proportion of ray and axial parenchyma. Adapted from Spicer and Holbrook (2007a)

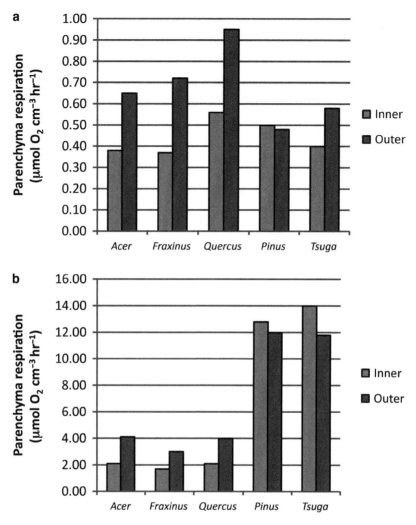

Fig. 4.4 Parenchyma respiration expressed as rate of O_2 consumption (**a**) per unit fresh sapwood volume and (**b**) per unit live parenchyma volume. Adapted from Spicer and Holbrook (2007a)

maximum parenchyma ages found both within a species and within an individual (>15 year differences in sapwood age between the stem base and top have been observed by Spicer and Gartner (2001), Domec et al. (2005), Spicer and Holbrook (2007a)). Species differences in respiration rate appeared to be much larger when expressed on a live cell basis than on a tissue volume basis. These data suggest that bulk tissue respiration cannot be estimated by quantifying parenchyma volume because there may be large species differences in cellular respiration rates. Sapwood, rather than parenchyma, may behave uniformly across species: here, the large parenchyma volume in angiosperms (18–25%) combined with a low cellular respiration rate (about one-third that of conifers) produce tissue respiration rates roughly similar to those of conifers. These smaller differences in bulk tissue respiration may be normalized at the whole-tree level by varying volumes of sapwood. The results reported suggest that bulk sapwood respiration is quite low, roughly an order of magnitude lower than cambial tissue and half that of secondary phloem and cambial tissue combined (Pruyn et al. 2002a, b; Pruyn et al. 2003; Bowman et al. 2005; Pruyn et al. 2005). By expressing respiration on a live cell volume basis, Spicer and Holbrook (2007a) were able to ask whether this low tissue respiration rate is simply a function of the small proportion of living cells, or whether parenchyma cells have a low metabolic rate as well.

References

Allen RM, Hiatt EN (1994) Tissue culture of secondary xylem parenchyma of four species of southern pines. Wood Fiber Sci 26:294–302

Bouma TJ, Nielsen KL, Eissenstat DM, Lynch JP (1997) Soil CO_2 concentration does not affect growth or root respiration in bean or citrus. Plant Cell Environ 18:575–581

Bowman WP, Barbour MM, Turnbull MH, Tissue DT, Whitehead D, Griffin K (2005) Sap flow rates and sapwood density are critical factors in within- and between-tree variation in CO_2 efflux from stems of mature *Dacrydium cupressinum* trees. New Phytol 167:815–828

Chase WW (1934) The composition, quantity, and physiological significance of gases in tree stems. Tech Bull 99, University of Minnesota

del Hierro AM, Kronberger W, Hietz P, Offenthaler I, Richter H (2002) A new method to determine the oxygen content inside the sapwood of trees. J Exp Bot 53:559–563

Domec JC, Pruyn ML, Gartner BL (2005) Axial and radial profiles in conductivities, water storage and native embolism in trunks of young and old-growth ponderosa pine trees. Plant Cell Environ 28:1103–1113

Edwards NT, Hanson PJ (1996) Stem respiration in a closed-canopy upland oak forest. Tree Physiol 16:433–439

Eklund L (1990) Endogenous levels of oxygen, carbon dioxide and ethylene in stems of Norway spruce trees during one growing season. Trees 4:150–154

Eklund L (1993) Seasonal variations of O_2, CO_2, and ethylene in oak and maple stems. Can J For Res 23:2608–2610

Eklund L (2000) Internal oxygen levels decrease during the growing season and with increasing stem height. Trees 14:177–180

Eklund L, Klintborg A (2000) Ethylene, oxygen and carbon dioxide in woody stems during growth and dormancy. In: Savidge R, Barnett J, Napier R (eds) Cell and molecular biology of wood formation. BIOS Scientific, Oxford, pp 43–56

Gansert D (2003) Xylem sap flow as a major pathway for oxygen supply to the sapwood of birch (*Betula pubescens* Ehr.). Plant Cell Environ 26:1803–1814

Gansert D, Burgdorf M, Lösch R (2001) A novel approach to the in situ measurement of oxygen concentrations in the sapwood of woody species. Plant Cell Environ 24:1055–1064

Gartner BL, Baker DC, Spicer R (2000) Distribution and vitality of xylem rays in relation to tree leaf area in Douglas-fir. IAWA Bull 21:389–401

Gartner L, Moore JR, Gardiner BA (2004) Gas in stems: abundance and potential consequences for tree biomechanics. Tree Physiol 24:1239–1250

Geigenberger P, Fernie AR, Gibon Y, Christ M, Stitt M (2000) Metabolic activity decreases as an adaptive response to low internal oxygen in growing potato tubers. Biol Chem 381:723–740

Hansmann C, Gindl W, Wimmer R, Teischinger A (2002) Permeability of wood – a review. Wood Res 47:1–16

Hauch S, Magel E (1998) Extractable activities and protein content of sucrose-phosphate synthase, sucrose synthase and neutral invertase in trunk wood tissues of *Robinia pseudoacaia* L. are related to cambial wood production and heartwood formation. Planta 207:266–274

Hook DD, Brown CL, Wetmore RH (1972) Aeration in trees. Bot Gaz 133:433–454

Jahnke S (2000) Is there any direct effect of elevated CO_2 on dark respiration in leaves? Plant Physiol Biochem 38(Suppl):s134

Jensen KF (1967) Measuring oxygen and carbon dioxide in red oak trees. (US Forest Service research note NE-74) US Forest Service

Jensen KF (1969) Oxygen and carbon dioxide concentrations in sound and decaying red oak trees. For Sci 15:246–251.

Kaipiainen LK, Sofronova GI, Hari P, Yalynskaya EE (1998) The role of xylem in CO_2 exchange in *Pinus sylvestris* woody stems. Russ J Plant Physiol 45:500–505

Kidd F (1916) The controlling influence of carbon dioxide. Part III. The retarding effect of carbon dioxide on respiration. Proc R Soc Lond B 89:136–156

Kimmerer TW, Stringer MA (1988) Alcohol dehydrogenase and ethanol in the stems of trees: evidence for anaerobic metabolism in the vascular cambium. Plant Physiol 87:693–697

Lammertyn J, Franck C, Verlinden BE, Nicolai BM (2001) Comparative study of the O_2, CO_2 and temperature effect on respiration between 'Conference' pear cell protoplasts in suspension and intact pears. J Exp Bot 52:1769–1777

Lavigne MB, Franklin SE, Hunt ER (1996) Estimating stem maintenance respiration rates of dissimilar balsam fir stands. Tree Physiol 16:687–696

Levy PE, Meir P, Allen SJ, Jarvis PG (1999) The effect of aqueous transport of CO_2 in xylem sap on gas exchange in woody plants. Tree Physiol 19:53–58

Magel EA, Jay-Allemand C, Ziegler H (2000) Biochemistry and physiology of heartwood formation. In: Savidge R, Barnett J, Napier R (eds) Cell and molecular biology of wood formation. BIOS Scientific, Oxford, UK, pp 363–376

Mancuso S, Marras AM (2003) Different pathways of the oxygen supply in the sapwood of young *Olea europaea* trees. Planta 216:1028–1033

McDougal DT, Working EB (1933) The pneumatic system of plants, especially trees. Carnegie Institution Washington Publication, 441

McGuire MA, Cerasoli S, Teskey RO (2007) CO_2 fluxes and respiration of branch segments of sycamore (*Platanus occidentalis* L.) examined at different sap velocities, branch diameters, and temperatures. J Exp Bot 58:2159–2168

Millar AH, Bergersen FJ, Day DA (1994) Oxygen affinity of terminal oxidases in soybean mitochondria. Plant Physiol Biochem 32:847–852

Nakaba S, Sano Y, Kubo T, Funada R (2006) The positional distribution of cell death of ray parenchyma in a conifer. Plant Cell Rep 25:1143–1148

Panshin AJ, de Zeeuw C (1980) Textbook of wood technology. Part 1. Formation, anatomy, and properties of wood, 4th edn. McGraw-Hill, New York, USA, pp 11–404

Pfanz H, Heber U (1989) Determination of extra- and intracellular pH values in relation to the action of acidic gases on cells. In: Linskens HF, Jackson JF (eds) Modern methods of plant analysis. Gases in plant and microbial cells, vol 9. Springer, Berlin, pp 322–343

Pfanz H, Aschan G, Langenfeld-Heyser R, Wittmann C, Loose M (2002) Ecology and ecophysiology of tree stems: corticular and wood photosynthesis. Naturwissenschaften 89:147–162

Pruyn ML, Gartner BL, Harmon ME (2002a) Respiratory potential in sapwood of old versus young ponderosa pine trees in the Pacific Northwest. Tree Physiol 22:105–116

Pruyn ML, Gartner BL, Harmon ME (2002b) Within-stem variation of respiration in *Pseudotsuga menziesii* (Douglas-fir) trees. New Phytol 154:359–372

Pruyn ML, Harmon ME, Gartner BL (2003) Stem respiratory potential in six softwood and four hardwood tree species in the central cascades of Oregon. Oecologia 137:10–21

Pruyn ML, Gartner BL, Harmon ME (2005) Storage versus substrate limitation to bole respiratory potential in two coniferous tree species of contrasting sapwood width. J Exp Bot 56:2637–2649

Rawsthorne S, LaRue TA (1986) Metabolism under microaerobic conditions of mitochondria from cowpea nodules. Plant Physiol 81:1097–1102

Saglio PH, Rancillac M, Bruzan F, Pradet A (1984) Critical oxygen pressure for growth and respiration of excised and intact roots. Plant Physiol 76:151–154

Saitoh T, Ohtani J, Fukazawa K (1993) The occurrence and morphology of tyloses and gums in the vessels of Japanese hardwood. IAWA Bull 14:359–372

Saveyn A, Steppe K, McGuire MA, Lemeur R, Teskey RO (2008) Stem respiration and carbon dioxide efflux of young *Populus deltoides* trees in relation to temperature and xylem carbon dioxide concentration. Oecologia 154:637–649

Shain L, MacKay JFG (1973) Seasonal fluctuation in respiration of aging xylem in relation to heartwood formation in *Pinus radiata*. Can J Bot 51:737–741

Sorz J, Hietz P (2006) Gas diffusion through wood: implications for oxygen supply. Trees 20:34–41

Sorz J, Hietz P (2008) Is oxygen involved in beech (*Fagus sylvatica*) red heartwood formation? Trees 22:175–185

Spicer R, Gartner BL (2001) The effects of cambial age and position within the stem on specific conductivity in Douglas-fir sapwood. Trees 15:222–229

Spicer R, Holbrook NM (2005) Within-stem oxygen concentration and sap flow in four temperate tree species: does long-lived xylem parenchyma experience hypoxia? Plant Cell Environ 28:192–201

Spicer R, Holbrook NM (2007a) Parenchyma cell respiration and survival in secondary xylem: does metabolic activity decline with cell age? Plant Cell Environ 30:934–943

Spicer R, Holbrook NM (2007b) Effects of carbon dioxide and oxygen on sapwood respiration in five respiration tree species. J Exp Bot 58:1313–1320

Teskey RO, McGuire MA (2007) Measurement of stem respiration of sycamore (*Platanus occidentalis* L.) trees involves internal and external fluxes of CO_2 and possible transport of CO_2 from roots. Plant Cell Environ 30:570–579

van Dongen JT, Schurr U, Pfister M, Geigenberger P (2003) Phloem metabolism and function have to cope with low internal oxygen. Plant Physiol 131:1529–1543

Part II
Intracellular Signalling

Chapter 5
pH Signaling During Anoxia

Hubert H. Felle

Abstract Plant cells that are exposed to anoxia run into an energy crisis. As a result of this, compartmental transmembrane gradients break down (sooner or later) leading ultimately to cell death. Regulation of intracellular pH and intercompartmental pH signaling may play a critical role in tolerating anoxia for some time. An early consequence of anoxia is a cytoplasmic pH drop which, according to the Davis-Roberts hypothesis, arises from the lactate formation. H^+ leakage from the vacuole or H^+ arising from nucleotide triphosphate hydrolysis have been also suggested. Research has focused on the assumption that the anoxic cytoplasmic pH change is an "error signal" to which the cell should respond to avoid cell-damaging acidosis. This view is challenged here. It is argued that pH under anoxia represents a new set point required for the anaerobic metabolism and for gene activation. It is concluded that acidosis does not occur because of H^+ leaking through membranes or because of the production of acids, but because of energy shortage which prevents the maintenance of transmembrane gradients; the leveling of the pH gradients and subsequent cytoplasmic acidosis are a consequence thereof.

5.1 Introduction

Higher plants as aerobic organisms occasionally experience situations of reduced oxygen or total anoxia because of environmental factors. Waterlogging blocks the transfer of O_2 and other gases between plant organs and the atmosphere. As a result of this, plants or parts thereof become O_2 deficient and run into an energy crisis (Greenway and Gibbs 2003), primarily caused by a reduced or, in case of anoxia, a complete shutdown of the oxidative respiration. Unlike facultative bacteria, most

H.H. Felle
Botanisches Institut I, Justus-Liebig-Universität, Gießen, Senckenbergstr 17, 35390 Gießen, Germany
e-mail: Hubert.Felle@bio.uni-giessen.de

S. Mancuso and S. Shabala (eds.), *Waterlogging Signalling and Tolerance in Plants,*
DOI 10.1007/978-3-642-10305-6_5, © Springer-Verlag Berlin Heidelberg 2010

higher plants cannot withstand anoxia for long, but some species have developed strategies to tolerate anaerobiosis to a high degree. There are quite a number of plant responses to anoxia (reviewed by Perata and Alpi 1993; Ratcliffe 1999; Vartapetian 2005; Greenway et al. 2006; Drew 1997) from which mainly the pH response will be focused on in this chapter.

The fast drop in cytoplasmic pH, which occurs within seconds following oxygen shortage, is one of the earliest measurable features of anoxia, the extent of which depends on how well the organism is adapted to oxygen stress. The development of cytoplasmic pH as a consequence of anoxia can shortly be described as follows (Fig. 5.1): upon anoxia, normoxic well-regulated cytoplasmic pH (Smith and Raven 1979; Davies 1986; Felle 1988) drops to a new stable level, which is required by anaerobic metabolism and is mainly set by H^+-consuming and -producing (enzymatic) processes, i.e. by a biochemical pH-stat (Davies 1986; Felle 2005). This relatively stable pH level can be maintained as long as sufficient energy can be provided to maintain transmembrane gradients, depending on adaptation, acclimation, etc. In case of enduring energy shortage, cytoplasmic pH will decrease further eventually leading to cell-damaging acidosis.

With respect to pH development under anoxia, research has focused on major questions: (1) where does the anoxic acidification come from? (2) To avoid acidosis, must cells under anoxia restore the normoxic pH? (3) Is acidosis the cause for anoxia intolerance? (4) Does an improved pH regulation ameliorate anoxic effects? Taking most of the experimental data and observations into consideration, it is attempted in this chapter to judge some of the observations from a

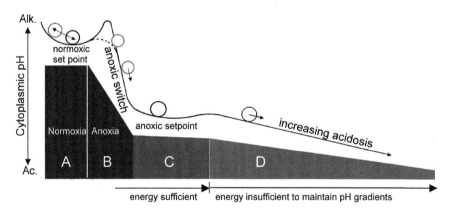

Fig. 5.1 A model to illustrate cytoplasmic pH development with respect to energy availability before and during anoxia. (**a**) Normoxic conditions and stable, well-regulated cytoplasmic pH. (**b**) Switch from normoxia to anoxia which involves a massive drain in energy and a pH drop to another stable value (anoxic set point). (**c**) Phase of relative stability during which energy is harvested from resources or from non-affected photosynthetic organs. The duration of this phase varies strongly, being relatively short for anoxia-intolerant plants. (**d**) Failure to harvest sufficient energy to keep up substrate gradients, including pH. Declining energy availability and subsequent increasing acidosis lead to cell death

slightly different point of view which, however, may result in causalities that are not always totally congruent with the current opinion.

5.2 pH, Signal and Regulator

The pH of an aqueous compartmental solution represents the basic condition to which many processes therein, e.g. enzymatic reactions as well as membrane transport, have to succumb to enzymatic reactions. Since protons can be at the same time substrate, activator and product, it is mostly difficult to pinpoint the primary effect and to distinguish it from the subsequent processes, and often this may not even be possible or reasonable. For instance, the activity of an H^+ATPase that transports H^+ across a membrane depends chemically on the ambient pH (binding, optimum). This H^+ transport generates or changes a transmembrane pH gradient which in turn will influence the activity of this transporter for thermodynamic reasons. Since in all cellular compartments these principles hold for a variety of processes that involve protons in one way or the other, for a given pH change the term "signal" appears problematic (Felle 2001). Undoubtedly, a pH change may signal a process that has already happened or is just about to take place, but it represents much more. In Fig. 5.2, an attempt is made to shortly outline the problem: a pH change is multifunctional and retroactive with respect to its own cause (a self-regulating aspect). This problem is enhanced by the fact that a pH change need not be the result of an H^+ shift across some barrier (membrane) or the result of a biochemical reaction. pH changes may well occur through net shifts of other (strong) ions like K^+, Ca^{2+}, Cl^- (strong ion difference; Stewart 1983; Ullrich and Novacki 1990) or simply through ion exchange, within cell wall constituents like glucoronic acid. If we consider pH changes as signals, we have to ask how and where information is transferred to. With regard to anoxia, this would be intracellular in the first place, but it is also systemic, i.e. from one organ (e.g. root) to another (e.g. shoot; leaf) or vice versa.

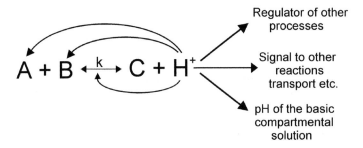

Fig. 5.2 A simplified scheme to illustrate the problem of overlapping and retroactive involvements of H^+ in cellular processes. H^+ produced (or consumed) by cellular reactions have a tendency to shift the basic pH of the respective compartment, influence other reactions and membrane transport, but also have a retroactive effect on the reagents A, B, C and the enzyme (k)

5.2.1 pH as Systemic Signal

In case of flooding, the root is the most likely organ that experiences shortage of oxygen first; a stress event that is signaled to the shoot. There is some evidence that electrical long-distance signals are forerunners which signal "stress" to the shoot or the leaves (Shvetsova et al. 2002; Felle and Zimmermann 2007). There is also evidence that such electrical signals cause a relatively unspecific upregulation of a variety of genes and the production of stress-related proteins (e.g. Wildon et al. 1992; Stanković and Davies 1997). More specific information as to the kind of stress has to take place as mass transfer, both through the xylem and, to some extent (albeit slower), through the phloem. Whereas the xylem pathway appears unproblematic, because it is fuelled by the transpiration (providing it still exists under anoxia), the inverse direction from shoot to root is not so simple, because mass flow within the sieve tubes depends on source-sink principles, which may be disturbed during anoxia. Additionally, since the pH of the sieve tube cytoplasm (as in any cell) is well regulated, it would not permit substantial pH shifts, and therefore is not suited for the transmission of large pH signals (e.g. 1 unit or more). This brings us to the question "how large a pH shift must be to be sensed as a signal?" Physiological normoxic pH changes exceeding 0.3 units are untypical for the cytoplasm. Would this be sufficient to bring across a message? The answer is "yes"! A pH change of 0.3 means a doubling of the H^+ activity which is easily recognized and responded to by enzymes and transporters. Anoxic pH shifts are mostly larger (see Table 5.1). As a rule of thumb, a cytoplasmic acidification of 0.6 units can be expected within the first few minutes of oxygen shortage in a cell that experiences sudden anoxia, albeit

Table 5.1 Selection of measured cytoplasmic pH changes of different plant (organs) responding to anoxia. "Time interval" denotes the period in which the pH change takes place

Plant (organ)	pH change (ΔpH)	Time interval	References
Maize (root tips)	7.4–6.8 (0.6)	First few min	Roberts et al. (1984a, 1984b)
			Fox et al. (1995)
Rice (coleoptiles)	(0.2)		Fan et al. (1992)
Rice (shoots)	7.4–7.1 (0.3)		Menegus et al. (1991)
Wheat (shoots)	7.4–6.5 (0.9)		Menegus et al. (1991)
Medicago (root hairs)	7.32–6.8 (0.52)	2 min	Felle (1996)
Maize (root tips)	7.55–7.0 (0.55)	30 min	Ratcliffe (1992)
Article I. *Potamogeton* (tuber)	7.48–7.32 (0.16)	60 min	Summers et al. (2000)
Pea (internodes)	7.48–6.8 (0.58)	30 min	Summers et al. (2000)
Acer susp. (cultured cells)	7.5–7.1 (0.4)		Gout et al. 2001
Nicotiana			
a. Roots	7.57–6.27 (1.3)	105 min	Stoimenova et al. (2003)
b. Transformant lacking nitrate reductase	7.50–6.48 (1.02)	105 min	
Arabidopsis			
a. WT	8.03 – 7.29 (0.74)	6 h	Mattana et al. (2007)
b. Transgene	7.98 – 7.07 (0.91)	6 h	

smaller as well as larger changes have been reported (see below). The question must be: is the anoxic pH change suffered at the root in any way signaled to the shoot? If yes, what is the nature of the transfer?

5.2.2 The Nature of pH Transmission

As demonstrated recently, flooding of barley roots which caused hypoxia had no effect on the apoplastic pH of shoot tissue within the experimental period of 2 h. On the other hand, anoxia, imposed to the same system (caused by N_2) led to a strong alkalinization within the leaf apoplast right away, the signal traveling at 2–5 cm min^{-1} (Felle 2006). However, whereas the cytoplasmic pH change was an acidification, the pH change measured within the normoxic leaf apoplast was an alkalinization, meaning that the pH shift was transformed into a stress signal to be transmitted within the xylem. This is nothing out of the ordinary as all kinds of stress seem to cause apoplastic pH increase (Wilkinson 1999; Felle et al. 2005).

Since the velocity of pH transmission from root to shoot is related to the aperture of stomata, one would expect that such a pH increase should be transmitted along with mass flow within the xylem, e.g. as protonated/deprotonated weak acids. This does not seem to be the case! pH changes fed directly into the xylem of a dissected leaf petiole, are either not or at best marginally (at about one-tenth of any pH unit altered) transferred to the point of measurement (leaf blade). pH changes of several units, externally applied directly to the roots of an intact plant (barley), were not transferred to the shoot at all! (Felle 2005). This means that other transfer mechanisms must apply. Root-to-shoot studies revealed that apoplastic pH readily changed and was transmitted when the apoplastic activity of inorganic ions, e.g. K$^+$ or Ca^{2+} was altered (Felle 2005). As a matter of fact, apoplastic changes in pH occurred simultaneously with K$^+$ (Ca^{2+}) activity, i.e. an increase in apoplastic pH goes along with an increase in K$^+$ or Ca^{2+} activity. This could explain the conversion of the anoxic cytoplasmic acidification into apoplastic alkalinization. Owing to H$^+$ pump deactivation under anoxia and the subsequent decrease in inwardly directed driving force for cations, K$^+$ ions leave the cell and, according to the strong ion principles (Stewart 1983; Ullrich and Novacki 1990; Gerendas and Schurr 1999), alkalize the apoplast. Subsequently, K$^+$ is transported in the xylem into the shoot and thus transmits the pH change. These principles also hold for anoxic situations. Whereas flooding of roots does not necessarily lead to a rapid change in the apoplastic leaf pH (within the first 2 h), anoxia under N_2 does right away (Felle 2005). This indicates that the stress signal "pH increase" seems only released and transmitted after oxygen availability has fallen short of a critical value.

5.2.3 What is the Information?

What information can there be in a pH shift? Obviously, the direction of the pH shift is different in the apoplast (alkalinization; exceptions see below) and in the

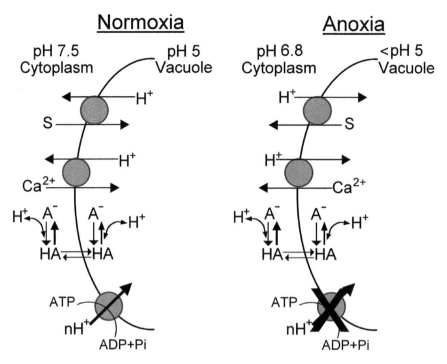

Fig. 5.3 Vacuolar pH and the switch from normoxia to anoxia. (1) Under anoxia, the activity of the Tp-ATPase is negligible (X). As a consequence thereof, merely passive processes prevail. (2) Cotransported (H$^+$ antiport) substrates (S) and ions (Ca^{2+}) that were accumulated in the vacuole flow back into the cytoplasm. Because of the antiport, H$^+$ will be translocated into the vacuole. (3) Weak acids (organic acids) dissociate according to their pKs and the existing pH value. Anoxia causes a cytoplasmic acidification, i.e. an increase in protonated weak acid(s). Since HA is better permeable than A$^-$, HA will flow into the vacuole, dissociate there into A$^-$ and H$^+$. Thus, as long as transmembrane gradients exist, the vacuolar pH will have the tendency to acidify under anoxia (see text)

cytoplasm (acidification). The latter means "activation" but this holds primarily for cells that are not directly affected by anoxia. Activation also means the upregulation of a broad array of genes, the products of which will ameliorate the immediate effects of the energy shortage in the affected organs. Activation also means the upregulation of transport and enzyme activities that are involved in gaining and transporting energy-rich compounds. It is difficult, however, to explain how the apoplastic alkalinization that arrives in the shoot (leaf) can be transformed into a cytoplasmic acidification. One possibility is anion channel activation. As discussed above, pH transmission is not carried by compounds with dissociable groups like weak acids, but indirectly through the mass transport of K$^+$, Na$^+$ or especially Ca^{2+} within the xylem. Apart from the resulting alkalinization, the increase in the apoplastic Ca^{2+} activity could well lead to an activation of Ca^{2+}channels: the subsequent influx of Ca^{2+} elevates cytoplasmic Ca^{2+} and activates anion channels, as frequently shown (e.g. Lewis et al 1997; Felle and Zimmermann 2007). These relatively unspecific anion channels do not only release Cl^- and NO_3^-, but also organic acid anions, which due to their weak acid properties, alkalize the apoplast and at the same time acidify the cytoplasm. Obviously, pH signaling cannot be discussed without taking into account changes in ion channel activity, a realization that holds for aspects of pH regulation under anoxia as well.

5.3 Anoxic Energy Crisis and pH Regulation

Depending on their specialization and acclimation to oxygen shortage, plants are able to survive anoxia without damage for a few hours at least, but in many cases much longer. Since under anoxia or inhibition of the oxidative phosphorylation with cyanide cytoplasmic ATP drops to about 20% or less within seconds or a minute at the most (Felle 1981), survival is obviously not linked to normoxic ATP concentrations, but to much lower levels, which will have to be generated and maintained by anaerobic processes. Energywise, there are three main problems during anoxia. Firstly, owing to the elimination of oxidative phosphorylation only a fraction of the energy stored chemically within glucose can be harvested. Secondly, owing to the energy shortage and partial deactivation of the H^+ ATPases, the H^+ circulation across membranes which drives cotransport is massively impaired. Thirdly, the driving force for H^+- cotransportable substrates is reduced. Therefore, extrusion of H^+ by the H^+ pumps does not serve mitigation of cytoplasmic acidosis, but to keep up H^+ cotransport, which is the only linkage to any of the scarce energy sources left.

5.3.1 The Davis-Roberts-Hypothesis: Aspects of pH Signaling

In short, this hypothesis says: the anoxia-induced cytoplasmic acidification is due to lactic acid formation/accumulation; the stabilization of the cytoplasmic pH is due to a metabolic shift to ethanolic fermentation, which is mediated and pH stimulated by the pyruvate decarboxylase, having its pH optimum in a more acid range than the lactate dehydrogenase; this will result in a redirection of carbon flow to ethanol synthesis with no further H^+ production (Davies et al. 1974; Davies 1986; Roberts et al. 1984a, b). Although this hypothesis has been challenged, more or less successfully, it is not the issue of this article to prove or disprove the validity of this hypothesis nor its critics. It is rather the question "whether pH signaling is a component therein?" If we take the cytoplasmic acidification to be the result of the lactic acid production, altered pH would have a retroactive effect on the lactate dehydrogenase (to reduce activity), but would also affect the pyruvate decarboxylase which responds accordingly with the activity increase to stabilize the anoxic pH on a new level. This is a typical feature of the biochemical pH regulation: the pH shift from one enzyme is a result, but at the same time also an error signal to the other enzyme, which responds and produces/consumes H^+ because of its pH optimum.

In spite of the reports which show that the Davies–Roberts hypothesis may not be universal, it should be stressed that pretreatments with acetic acid which decreased the cytoplasmic pH of maize root tips (1) prevented the transient formation of lactate, and (2) eliminated the lag in the production of ethanol (Roberts et al. 1984a). Fox et al. (1995) extended this work using methylamine to increase cytoplasmic pH and found that ethanol production ceased as cytoplasmic pH increased. Both studies showed that pH was a signaling component in the pathway to activate or deactivate the respective enzymes.

There are a number of examples indicating that there is no single or universal mechanism for the induction of lactic/ethanol fermentation and the pH development do not always follow lactate formation. Sakano et al. (1997) demonstrated in suspension-cultured *Catharanthus* cells that accumulation of lactate was not related to cytoplasmic acidification. Tadege et al. (1998) showed that in transgenic tobacco concentration of lactate remains at the limit of detection under anoxia while ethanol increases dramatically, indicating that no obvious interrelation between lactate formation and cytoplasmic pH need to exist. Kato-Noguchi (2000) found in lettuce that ethanolic fermentation was activated without preceding activation of lactate fermentation. Saint-Ges et al. (1991) argued that hydrolysis of nucleotide triphosphates was correlated to pH development under anoxia rather than lactate formation, which continued in spite of a stabilized pH. These are only some aspects of the observation that pH development under anoxia is the result of several metabolic steps that work differently under oxygen shortage than under normoxia. The question whether pH or the pH change is a signal therein must be investigated for each plant and enzyme individually, taking the state of respective adaptation to anoxic situations into account. Whereas one can assume that the basic enzyme pattern is essentially similar in higher plants, already minor shifts in the pH optima of the involved enzymes may result in set points that differ in a few tenths of a pH unit. It might be an error, however, to assume that the new pH level reached under anoxia is the result of an active process, such as membrane transport. Once the mitochondrial energy production has ceased, energy gets scarce and ions and molecules are transported because of their transmembrane gradients until a new more or less stable dynamic equilibrium is established; protons are a part thereof and cytoplasmic pH will rest (for some time) at a new set point.

5.3.2 Cytoplasmic Acidification, ATP and Membrane Potential

It has been suggested that cytoplasmic acidification was causally linked to the cytoplasmic nucleotide triphosphate potential rather than to lactate formation (Saint-Ges et al. 1991; Gout et al. 2001). The problem with the kinetics given in these studies is that there is just about 1 measuring point per min. As shown here in Fig. 5.4a, the ATP decay can be very fast and levels off already 1 min after

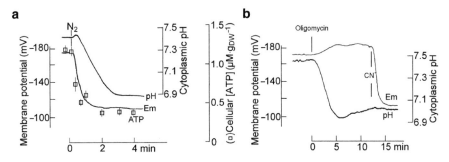

Fig. 5.4 Kinetics of cytoplasmic pH, [ATP] and membrane depolarization following anoxia (**a**; N_2) or inhibition of the oxidative phosphorylation by oligomycin (**b**) (see text)

imposing anoxia (N_2). Whereas kinetics of membrane depolarization seems to follow the ATP-decay (indicating H^+ pumps as major ATP consumers), there is no correlation with the pH kinetics. The deviation from the results given by Saint-Ges et al. (1991) and Gout et al. (2001) probably arises from the resolution of the pH measurement. Anoxia, suddenly imposed to cells, may yield pH and membrane potential kinetics that apparently correspond with each other. The problem with such an approach is that potentially different onsets of the kinetics become indistinct. One way out of this would be to impose a gradual O_2 decrease or use sub-threshold respiratory chain inhibitors, which then will show up whether pH, ATP, and membrane depolarization start to change, either simultaneously or not. Fig. 5.4b in fact reveals that cytoplasmic pH already starts to decrease well before the membrane potential started to depolarize (Fig. 5.4b; Felle 1996), indicating that the level of ATP available or pump activity were in no causal relationship with the cytoplasmic pH or its anoxic shift.

5.3.3 Cytoplasmic pH (Change), An Error Signal?

Under normoxic conditions, any aberration from the pH set point is experienced as "error" to which it is reacted to, whether biophysically (membrane transport) or biochemically. For instance, H^+ ATPases will respond to cytoplasmic acidification with an increased activity, because H^+ is transport substrate or because the enzyme optimum requires it (Felle 1991). Additionally, the biochemical pH-stat gets active and consumes H^+ by converting malate into pyruvate. Although one could debate which of the two mechanisms are the more effective, this, however, is of no consequence to the question raised here. The same holds for anoxia, although under quite different proportions for all those processes that consume metabolic energy like H^+ATPases. Under anoxia, the activity of the H^+ ATPase is rather low, which is indicated by the fact that the membrane potential drops to the level of the so-called diffusion potential. Therefore, the H^+ATPase activity would not contribute much to restore the pH. But even so, the basic question must be: is the pH drop following anoxia (Table 5.1) experienced as error signal at all? Most authors who have dealt with this problem in the past appear to be affirmative of the "error" argument, and try to prove that the affected cells attempt to restore the pH to its original (normoxic) level (to mitigate acid loads). This view ought to be reconsidered! A partial pH recovery following the initial pH decrease is often taken as an indication of a back regulation. Apart from the possibility that the system investigated may have experienced hypoxia only (under which partial recovery is quite the rule) rather than anoxia, this transient behavior is nothing else but an adjustment to a new set point, whereby the system recovers from the rapid pH change (overshoot). Following anoxia, cytoplasmic pH drops because of a shift in chemical equilibria, whereby lactose formation may be one, but not the only factor, as the different investigations show. Upon oxygen shortage, pH drops to that level which is set by all H^+-consuming and H^+-producing processes that contribute to establish a new equilibrium under anaerobic conditions. Thus, the anoxic pH is not an error signal therein but the optimal value for the given metabolic conditions to which the cell has to succumb. Differences in this set point between different plants experiencing anoxia may arise from enzyme patterns that might work at slightly different optima as well as with different efficiencies and largely represent different grades of adaptation to anoxic situations.

5.4 pH Interactions Between the (Major) Compartments During Anoxia

Major cellular compartments with respect to H^+ are cytoplasm, vacuole and apoplast. It should be noted, however, that during normoxia chloroplasts can temporarily accumulate rather high amounts of H^+, a fact that has been demonstrated with pH-sensitive microelectrodes. Following light-off or even a reduction of light intensity, the cytoplasmic pH decreases transiently up to 0.3 units, to which the H^+ ATPase responds with a temporary accelerated H^+ extrusion and hyperpolarization, whereas light-on is responded to with the opposite. Taking the cytoplasmic buffer capacity of 50 µeq H^+/pH into account (Felle 1987; Guern et al. 1991), this means that chloroplasts are able to store a substantial amount of H^+. During anoxia, there is no pH reaction to light/dark changes, which means that pH signaling between cytoplasm and chloroplasts need not be considered under these conditions.

5.4.1 The pH Trans-Tonoplast pH Gradient

The vacuole is a more or less acidic compartment, and it is interconnected with the cytoplasm by a variety of H^+ cotransporters which mediate substrate translocation between the two compartments. Since there is very little voltage across the tonoplast (10–20 mV), any change in the pH gradient across the tonoplast will largely result in a likewise change in H^+ driving force (pmf). Thus, the more acidic the vacuole the steeper the H^+ gradient (which in citrus fruits can reach easily a factor of 10^5, e.g. $pH_{Vac} \approx 2$; Echeverria et al. 1992), the more energy may be needed to hold the trans-tonoplast H^+ gradient. The question is: to what extent does the anoxic drop in cytoplasmic pH affect the tonoplast/vacuole in the short-term and to what extent is the energy shortage responsible for the long-term acidosis? A priori, the anoxic drop in cytoplasmic pH will reduce the H^+ gradient from the vacuole to the cytoplasm, which will reduce the H^+ pressure toward the cytoplasm to some extent. However, despite the remaining steep H^+ gradient, a passive "H^+ leak" from the vacuole to the cytoplasm may be rather small at first. One reason is that biomembranes are not very H^+ permeable (Deamer 1987), i.e. there are no H^+ channels that would permit such a leak as would be for other ions. It is a chemiosmotic principle that protons crossing a membrane do this either via a primary active system (a pump), through cotransport or as weak acids (or bases). Figure 5.3 explains this and one comes to the conclusion that (at least for some time) after anoxic conditions have set in, vacuolar pH should actually become more acidic. Indeed, upon anoxia and shortage of energy substrates and ions that have been cotransported (H^+ antiport) under normoxia and were accumulated within the vacuole will now flow down their gradient (leak back into the cytoplasm) and thus drive H^+ into the vacuole until gradients of substrate and H^+ inversely match. Clearly, the substances involved would be very different from plant to plant and from cell to cell, but since the vacuole is a storage compartment holding energy-rich compounds, for some cells the reactivation of those molecules would be a primary task. Therefore, the assumption that anoxia should lead to an increase in the vacuolar pH right away is – at least from the transport physiological point of view – not necessarily straightforward. In this context, it is also most important to realize that the H^+ circulation through any membrane is heavily impaired under anoxia (Felle 2005).The reason is that the pump works as a current source, i.e., a substantial amount of pump-exported protons are recycled providing a cotransportable substrate/ion is present. As soon as the pump stalls because of shortage of ATP, the H^+ circulation ceases which drastically reduces the amount of H^+ reentry. All in all, pH signaling between these two compartments would probably not play a major role during phase B (Fig. 5.1) of anoxia. There are reports which indeed show vacuolar acidification (Libourel et al. 2006; Kulichikhin et al. 2007), but there are also reports that show the opposite (Roberts et al. 1984a; Menegus et al. 1991; Dixon 2001; Summers et al. 2000). Since for transport physiological reasons vacuolar pH cannot increase in short-term anoxia, the vacuolar pH increases reported may have metabolic reasons. Kulichikhin et al. (2007) compared the pH development in root tips of wheat and rice and demonstrated that in rice (not in wheat) the vacuolar pH had decreased after 3 h of anoxia but had increased after 6 h of anoxia. Obviously, as soon as the substrate and ion gradients across the tonoplast have collapsed, also H^+ will start leaking into the cytoplasm and cause cellular acidosis.

5.4.2 Cytoplasm and Apoplast

Apart from the energy-rich compounds that are stored within the vacuole, energy has to be provided from other stores; the transport of these compounds will, at least in the last instance before uptake into the cell to be broken down, be apoplastic. Whereas the vacuole is an inner compartment that permits mass storage, the storage capacity of the apoplast is limited; but, since it is linked to other organs, its importance as a transfer organelle for energy-rich compounds during anoxia is high. With respect to the pH, the apoplastic fluid has a buffer capacity of roughly one-tenth of the cytoplasm (Hanstein and Felle 1999; Oja et al. 1999), so apoplastic pH changes can become large (1 pH unit or more) and thus will have signaling characteristics. Although the apoplast is an acidic compartment like the vacuole, the signaling situation at the plasma membrane is different: the transmembrane voltage (membrane potential difference) contributes considerably to the proton motive force (pmf), which is affected much stronger by anoxia than the trans-tonoplast pmf which is largely a pH-pmf. Therefore, H^+ cotransport at the plasma membrane with energy-rich compounds is usually mediated as a symport, i.e. substrate and H^+ move into the same direction. So, any import of energy-rich compounds from the apoplast into the cytoplasm will always have the tendency to further acidify the cytoplasm. However, since under anoxia the cytoplasmic pH is lower than under normoxia and owing to the decreased pmf the symported substrate import is relatively low, this should be of no further consequence to the cytoplasmic pH.

5.4.3 The Apoplast Under Anoxia

As a typical stress response, the pH of the apoplast increases under anoxia, but not under hypoxia (Felle 2006). Whereas this may be the "normal" response of plants to anoxia that are not tolerant to oxygen deprivation, some plant species like *Oryza sativa* L., *Rumex palustris* and others (possessing highly anoxia tolerant organs) display fast shoot elongation when flooded. This elongation growth is due to acid-induced cell wall expansion and under the control of expansins (Cosgrove 1998), which requires an apoplastic acidification. A strong increase in *EXPA2* and *EXPA4* mRNA levels in rice coleoptiles has been reported (Huang et al. 2000). Considering that the organ really experiences anoxia, it is difficult to envisage how a largely deactivated H^+ pump could be responsible for the cell wall acidification,that is required for acid (elongation) growth. Along with this goes the observation that the cytoplasmic pH apparently drops only marginally. It almost appears that, from the pH point of view, the tissue involved did not make the anoxic switch but remained in a quasi-normoxic state. How is this possible? As to the elongation growth, it is well known that acid growth can be mimicked in the laboratory by acidification of the cell wall, which means pump activity is not required. Therefore, other processes contributing to cell wall acidification could be involved, arising, for instance, from a net anion (Cl⁻) efflux or cation/H^+ exchange at glucuronic acid chains.

In some anoxia-tolerant plants, the so-called "Pasteur effect" has been demonstrated (Summers et al. 2000) which is characterized by an acceleration of carbohydrate consumption to increase energy production. Although the ATP production is still 4.4–9 times lower than in air (Gibbs and Greenway 2003), it may be just enough to provide the energy necessary to drive the elongation growth thus enabling the plant a chance to quickly reach areas of higher oxygen. The acidification of the apoplast and the almost constant cytoplasmic pH would not only favor accelerated elongation growth but also increase the pmf to take up carbohydrates more efficiently. The duration of such a scenario would of course depend on the amount of carbohydrates available, but because of the energy used, it could diminish the survival time under anoxia in case higher oxygen was not reached.

5.5 Anoxia Tolerance and pH

"Anoxia tolerance" is generally understood as the ability of plant cells to overcome anoxia for an extended but undefined period of time. Basically, all plant cells can survive anoxia for at least a few hours, but metabolically specialized wetland plants are able to withstand anoxia for much longer. So the terms "short-term tolerance" and "long-term tolerance" would appear to describe the situation more accurately. For instance, emerging leaves of *Scheonoplectus lacustris, Scirpus maritimus, Typha anguistifolia* (Armstrong et al. 1994; Barclay and Crawford 1982) or tubers of *Potamogeton pectinatus* (Summers and Jackson 1996), just to name a few, have been demonstrated to withstand anoxia for weeks or even months. Still, no higher plant as a whole will tolerate total anoxia indefinitely, and even among the given examples long-term tolerance is organ-specific: whereas roots of *Oryza sativa* are as oxygen-sensitive as those of maize, shoots of rice display a remarkable anoxia tolerance.

Frequently, anoxia tolerance is brought into connection with pH, whereby cytoplasmic pH is found not to drop as much as in anoxia-intolerant plants (or organs). From these reports, it is difficult to decide whether this pH behavior is one of the key aspects of anoxia tolerance i.e., a requirement or just a consequence of the different conditions, e.g., metabolic adaptation. Other aspects of this problem are: (1) can one be sure that the plants really experienced full anoxia and not just hypoxia? In fact, hypoxia causes cytoplasmic pH first to drop in a manner similar to anoxia but then recovers to some extent (e.g. Kulichikhin et al. 2007). (2) Are the laboratory tests with the respective plants (or parts thereof) really comparable to field conditions? For instance, root tips in a solution experience anoxia differently than in the soil, as in the latter case the root will not be able to extrude substances to the same extent as in the test tube. A problem may become that excised root tips are without connection to the shoot from which energy-rich compounds could be transported into the root. This may not be so important for tests of short-term tolerance, but in the long run it could yield an entirely different result.

5.5.1 pH as a Stress Signal – Avoidance of Cytoplasmic Acidosis

No doubt, stress of all kind acidifies the cytoplasm and alkalizes the apoplast (Wilkinson 1999; Felle et al. 2004; Felle et al. 2005), which characterizes the pH response as a very general signal that causes the upregulation of a broad array of genes. Despite the apparent concurrence in the pH response, anoxia is not just stress that can be responded to appropriately, but has far-reaching consequences with the potential death of the entire organism. Thus, the avoidance of cytoplasmic acidosis is generally assumed to be one of the main goals of an organism dealing with oxygen stress. For the short term (Fig. 5.1), this view may be questioned. Clearly, acidosis is fatal to a cell; however, pH 6.8 (± 0.1), a value frequently reported as anaerobic in the cytoplasm, is not acidosis, but a pH very close to neutrality. Thus, tolerance of organisms toward oxygen stress may primarily be related not to acidosis *per se*, but rather to what extent the organism has sufficient energy to keep up the dynamic equilibria between the enzymatic reactions, as well as across membranes: as long as sufficient energy is available, pH will be kept stable at a value not harmful to cells, not by actively removing H^+ but by keeping up transmembrane gradients (ions, energy-rich substrates). Although acidosis and energy shortage appear closely interrelated, which makes it difficult under most circumstances to distinguish them temporarily, it seems important to accept that acidosis within any given compartment is nothing else but the breakdown of transmembrane (ion) gradients, where H^+ may not even be involved primarily (strong ion difference). Anoxic cytoplasmic acidification is also brought into connection with the deactivation of the H^+ ATPases across membranes because of apparently coinciding ATP decay and pH decrease. This view is logical, but in most instances likely to be incorrect as the inhibition of the H^+ ATPase(s) under normoxia proves. There is no reason to conclude that upon H^+ ATPase deactivation protons should leak through that membrane in amounts that would overcome the strong pH buffer capacity and acidify the cytoplasm, as measured. As pointed out above, there are no H^+ channels in biomembranes reported at molecular levels, and the amount of H^+ cotransported is small because of the strongly reduced H^+ circulation (Felle 2005).

Amongst the suggested means of the anoxic cell to ameliorate acidosis are: (1) H^+ consumption by succinate synthesis; this idea is based on the observation that long-term anoxia-tolerant rice had a greater succinate synthesis than the short-term tolerant maize leaves (Menegus et al. 1989). Since the pH data were collected from cell sap of squeezed tissues, an interpretation is difficult because, in such samples, the contents of all cellular compartments with different volumes and buffering capacities contribute to the final value. (2) Alanine as an end product of anaerobic fermentation in roots (Good and Muench 1993; Reggiani et al. 1985; Thomson et al.

1989). The ameliorating significance is unclear. In maize root tips, Roberts et al. (1992) observed that enhanced synthesis of alanine did not modify cytoplasmic pH. (3) Pyrophosphate substitutes for ATP at the tonoplast as an energy source. In fact, a 75-fold increase in pyrophosphatase (PPase) activity was observed in anoxic rice seedlings (Carystinos et al. 1995). A decreasing cytoplasmic pH would also favor pyrophosphate as energy source (Davies et al. 1993). In spite of these observations, an effect on cytoplasmic pH of PPase activity under anoxia has not been demonstrated yet. (4) NO_3^- (Fan et al. 1988, 1995; Gibbs and Greenway 2003; Stoimenova et al. 2003) and/or NO_2^- (Libourel et al. 2006) has been suggested as a means of acidosis reduction. In fact, μmolar NO_2^- concentrations in the incubation medium have an ameliorating effect on cytoplasmic pH under anoxia, i.e. pH 6.7 without NO_x vs. pH 6.9 in the presence of NO_x. Question remains, however, whether the difference of 0.2 units really helps the cells to better withstand anoxia. (5) Lactate (or lactic acid) extrusion has been postulated to mitigate negative anoxia effects by reduction of cytoplasmic acidosis (Xia and Saglio 1992; Rivoal and Hanson 1993, 1994; Xia and Roberts 1994, 1996). Although it is demonstrated that maize roots survived longer after acclimation to low oxygen and measured lower cytoplasmic lactate levels compared to non-acclimated plants, the causal links and especially the kind of lactic acid extrusion (transporter?) remain obscure. Since lactic acid has a very acidic pK_S of 3.08, essentially all lactic acid within the cytoplasm exists as anion, regardless of how large the anoxic pH switch becomes. Therefore, passive loss of the protonated acid to the two large ambient compartments (vacuole, apoplast) can be neglected. Export of the anion is also no option because it would always lead to cytoplasmic acidification. The only possibility, an active export of lactic acid (which has not been demonstrated) is energy consuming and thus counterproductive. The observation that substantial amounts of lactic acid actually do appear in the test medium is due to the fact that a lactic acid gradient is kept up between roots and medium. This is not the case in vivo, where lactic acid will leak largely into the apoplast, a small space which very quickly gets saturated with lactic acid and prevents its further export. Even though, a correlation of lactate extrusion and cytoplasmic pH and survival of the maize roots has been demonstrated. The question remains whether these are causal correlations, and did cells survive anoxia (within the tested time) because cytoplasmic pH was 0.3 units less acidic in the acclimated system? An alternative possibility would be an improved energy metabolism (which of course would to some extent also include pH). An unambiguous conclusion cannot be made at this point.

5.6 pH as Signal for Gene Activation

It has been known for some time that cytoplasmic acidification is one of the preconditions for gene activation. This has been demonstrated with elicitors like *N*-acetylchitoheptaose (He et al. 1998) and oligogalacturonides as well as with weak acids like propionic acid (Lapous et al. 1998). This compound rapidly (6–10 min after addition) induced gene expression in suspension cultured rice cells. The levels of mRNAs were up-regulated for acid-responsive genes *EL2, EL3, PAL*, but not for acid-nonresponsive genes. The kinetics of the cytoplasmic acidification induced by propionic acid is quite similar to the anoxic pH drop which could indicate a causal relationship, an idea not too far fetched if one accepts the anoxic pH drop as the result of biochemical reactions that involve organic acids (lactic- and malic acids).

Although anoxia quickly inhibits protein synthesis in general, only a small group of proteins – the anaerobic proteins – continue to be made. Of these transcripts, most are enzymes of glycolysis and fermentation. Clearly, these enzymes are required under normoxic conditions as well, but the observation that these are among the few that are induced under hypoxia and expressed under anoxia, underpins their importance under these life endangering conditions. Among these enzymes, pyruvate-decarboxylase is interesting because its activity increases substantially with lower pH (Morrell et al. 1990).

5.7   pH Signaling and Oxygen Sensing

Bailey-Serres and Chang (2005) have suggested that the rapid decrease in cytosolic pH may reflect indirect oxygen sensing in plant cells. In fact, as Fig. 5.4 shows, cytosolic pH drops within 2–3 min after oxygen shortage has started by several tenths of a pH unit. The fast pH initial response is independent of O_2 once the O_2 level falls short of a level of about 10%. The difference in the response is the pH recovery. As a consequence thereof, cytoplasmic pH would not be an oxygen-sensing signal; also because it is the result of anoxic biochemical responses and not their cause. Therefore, O_2 sensing in plants would be at the level of the respiratory chain; it may be a matter of debate whether this would be real sensing or just a response to the absence or shortage of one reactant (adenylates, ATP etc.).

5.8   Conclusions

No doubt, pH plays a central role during anoxia and when cytoplasmic pH drops below a certain value, cell damage and cell death become the inevitable consequence. However, the conclusion that the *active* maintenance (or regulation) of a certain cytoplasmic pH under anoxia was the means to ameliorate or even to overcome anoxia should be reconsidered. The problem is that pH regulation critically depends on the regulation and activity of other ions; for instance, there is no pH shift because of H^+ transport without charge compensation. Therefore, any disturbance in the ionic composition of a compartment will *nota bene* have an impact on pH as well, meaning that pH changes never occur alone or isolated from but always together with other processes. In other words, a pH change always signals other changes at the same time as well. For anoxia, this means that critical cytoplasmic acidosis signals the breakdown of transmembrane gradients that cannot be maintained because of the energy shortage. It is the maintenance of the transmembrane gradients – ions and energy-rich compounds – that is the foremost problem an anoxic cell has to accomplish. Amongst the various functions, pH is involved in a cell as a signal which informs about the energy status of a certain cellular compartment.

References

Armstrong W, Brändle R, Jackson MB (1994) Mechanisms of flood tolerance in plants. Acta Bot Neerl 43:307–358

Bailey-Serres J, Chang R (2005) Sensing and signalling in response to oxygen deprivation in plants and other organisms. Ann Bot 96:507–518

Barclay HM, Crawford RMM (1982) Plant growth and survival under strict anaerobiosis. J Exp Bot 22:541–549

Carystinos GD, MacDonald HR, Monroy AF, Dhindsa RS, Poole RJ (1995) Vacuolar H^+-locating pyrophosphatase is induced by anoxia or chilling in seedlings of rice. Plant Physiol 108: 641–649

Cosgrove DJ (1998) Cell wall loosening by expansins. Plant Physiol 118:333–339

Davies DD, Grego S, Kenworthy P (1974) The control of the production of lactate and ethanol by higher plants. Planta 118:297–310

Davies DD (1986) The fine control of cytosolic pH. Physiol Plant 67:702–706

Davies JM, Poole RJ, Sanders D (1993) The computed free energy change of hydrolysis of inorganic pyrophosphatase and ATP: apparent significance for inorganic-pyrophosphate-driven reactions of intermediary metabolism. Biochim Biophys Acta 1141:29–36

Deamer DW (1987) Proton permeation of lipid bilayers. J Bioenerg Biomembr 19:457–479

Dixon MH (2001) The anaerobic metabolism of *Potamogeton pectinatus* L, an aquatic monocot with marked tolerance to the prolonged absence of oxygen. PhD thesis, University of Oxford, UK

Drew MC (1997) Oxygen deficiency and root metabolism: injury and acclimation under hypoxia and anoxia. Ann Rev Plant Physiol Plant Mol Biol 48:223–250

Echeverria E, Burns J, Felle H (1992) Compartmentation and cellular conditions controlling sucrose breakdown in mature acid lime fruits. Phytochem 31:4091–4095

Fan TW-M, Higashi RM, Lane AN (1988) An in vivo ^1H and ^{31}P NMR investigation of the effect of nitrate on hypoxic metabolism in maize roots. Arch Biochem Biophys 266:592–606

Fan TWM, Lane AN, Higashi RM (1992) Hypoxia does not affect rate of ATP synthesis and energy metabolism in rice shoot tips as measured by 31P-NMR in vivo. Arch Biochem Biophys 294:314–318

Fan TW-M, Higashi RM, Lane AN (1995) Use of ^{15}N and ^{13}C isotope labeling and multinuclear NMR for exploring nitrate metabolism in anaerobic rice coleoptiles. Plant Physiol 108:S12

Felle H (1981) A study of the current-voltage relationships of electrogenic active and passive membrane elements in *Riccia fluitans*. Biochim Biophys Acta 646:151–160

Felle H (1987) Proton transport and pH control in *Sinapis alba* root hairs: a study carried out with double-barrelled pH micro-electrodes. J Exp Bot 38:340–354

Felle H (1988) Short-term pH regulation in plants. Physiol Plant 74:583–591

Felle HH (1991) The role of the plasma membrane proton pump in short-term pH regulation in the aquatic liverwort *Riccia fluitans* L. J Exp Bot 42:645–652

Felle HH (1996) Control of cytoplasmic pH under anoxic conditions and its implication for plasma membrane proton transport in *Medicago sativa* root hairs. J Exp Bot 47:967–973

Felle HH (2001) pH: signal and messenger in plant cells. Plant Biol 3:577–591

Felle HH (2005) pH regulation in anoxic plants. Ann Bot 96:519–532

Felle HH (2006) Apoplastic pH during low-oxygen stress in barley. Ann Bot 98:1085–1093

Felle HH, Herrmann A, Hanstein S, Hueckelhoven R, Kogel K-H (2004) Apoplastic pH signalling in barley leaves attacked by the powdery mildew fungus *Blumeria graminis* f.sp. *hordei*. Mol Plant Microbe Interact 17:118–123

Felle HH, Herrmann A, Hueckelhoven R, Kogel K-H (2005) Root-to-shoot signalling: apoplastic alkalinization, a general stress signal and defence factor in barley. Protoplasma 227:17–24

Felle HH, Zimmermann MR (2007) Systemic signalling in barley through action potentials. Planta 226:203–214

Fox GG, McCallan NR, Ratcliffe RG (1995) Manipulating cytoplasmic pH under anoxia; a critical test of the role of pH in the switch from aerobic to anaerobic metabolism. Planta 195:324–330

Gerendas J, Schurr U (1999) Physicochemical aspects of ion relations and pH regulation in plants – a quantitative approach. J Exp Bot 50:1101–1114

Gibbs J, Greenway H (2003) Mechanisms of anoxia tolerance in plants: I. Growth, survival and anaerobic catabolism. Funct Plant Biol 30:1–47

Good AG, Muench DG (1993) Long-term anaerobic metabolism in root tissue: metabolic products of pyruvate metabolism. Plant Physiol 101:1163–1168

Gout E, Boisson A-M, Aubert S, Douce R, Bligny R (2001) Origin of the cytoplasmic pH changes during anaerobic stress in higher plant cells. Carbon-13 and phosphorous-31 nuclear magnetic resonance studies. Plant Physiol 125:912–925

Greenway H, Gibbs J (2003) Mechanisms of anoxia tolerance in plants. II. Energy requirements for maintenance and energy distribution to essential processes. Funct Plant Biol 30:999–1036

Greenway H, Armstrong W, Colmer TD (2006) Conditions leading to high CO_2 (>5kPa) in waterlogged-flooded soils and possible effects on root growth and metabolism. Ann Bot 98:9–32

Guern J, Felle H, Mathieu Y, Kurkdijan A (1991) Regulation of intracellular pH in plant cells. Ann Rev Cyt 127:111–173

Hanstein S, Felle HH (1999) The influence of atmospheric NH_3 on the apoplastic pH of green leaves: a noninvasive approach with pH-sensitive microelectrodes. New Phytol 143:333–338

He DY, Yazaki Y, Nishizawa Y, Takai R, Yamada K, Sakano K, Shibuya N, Minami E (1998) Gene activation by cytoplasmic acidification in suspension-cultured rice cells in response to the elicitor, N-acetylchitoheptaose. Mol Plant Microbe Interact 11:1167–1174

Huang J, Takano T, Akita S (2000) Expression of alpha-expansin genes in young seedlings of rice (*Oryza sativa* L.). Planta 211:467–473

Kato-Noguchi H (2000) Evaluation of the importance of lactate for the activation of ethanolic fermentation in lettuce roots in anoxia. Physiol Plant 109:28–33

Kulichikhin KY, Aito O, Chirkova TV, Fagerstedt KV (2007) Effect of oxygen concentration on intracellular pH, glucose-6-phosphate and NTP content in rice (*Oryza sativa*) and wheat (*Triticum aestivum*) root tips: in vivo ^{31}P-NMR study. Physiol Plant 129:507–518

Lapous D, Mathieu Y, Guern J, Laurière C (1998) Increase of defense gene transcripts by cytoplasmic acidification in tobacco cell suspensions. Planta 205:452–458

Lewis BD, Karlin-Neumann G, Davis RW, Spalding EP (1997) Ca^{2+}-activated anion channels and membrane depolarizations induced by blue light and cold in *Arabidopsis* seedlings. Plant Physiol 114:1327–1334

Libourel IGL, van Bodegom PM, Fricker MD, Ratcliffe RG (2006) Nitrite reduces cytoplasmic acidosis under anoxia. Plant Physiol 142:1710–1717

Mattana M, Vannini C, Espen L, Bracale M, Genga A, Marsoni M, Iriti M, Bonazza V, Romagnoli F, Baldoni E, Coraggio I, Locatelli F (2007) The rice Mybleu transcription factor increases tolerance to oxygen deprivation in *Arabidopsis* plants. Physiol Plant 131:106–131

Menegus F, Cattaruzza L, Chersi A, Fronza G (1989) Differences in the anaerobic lactate-succinate production and in the changes of cell sap pH for plants with high and low resistance to anoxia. Plant Physiol 90:29–32

Menegus F, Cattaruzza L, Mattana M, Beffagna N, Ragg E (1991) Response to anoxia in rice and wheat seedlings. Plant Physiol 95:760–767

Morrell S, Greenway H, Davies DD (1990) Regulation of pyruvate decarboxylase in vitro and in vivo. J Exp Bot 41:131–139

Oja V, Savchenko G, Jakob B, Heber U (1999) pH and buffer capacities of apoplastic cell compartments in leaves. Planta 209:239–249

Perata P, Alpi A (1993) Plant responses to anaerobiosis. Plant Sci 93:1–17

Ratcliffe RG (1977) In vivo NMR studies of the metabolic response of plant tissues to anoxia. Ann Bot 79:39–48.

Ratcliffe RG (1999) Intracellular pH regulation in plants under anoxia. In: Egginton S, Taylor EW, Raven JA (eds) Regulation of acid-base status in animals and plants. Soc. Exp. Biol. Seminar Series 68, Cambridge University Press, Cambridge, pp 193–213

Reggiani R, Brambilla I, Bertani A (1985) Effect of exogenous nitrate on anaerobic metabolism in excised rice roots II. Fermentative activity and adenylic charge. J Exp Bot 36:1698–1704

Rivoal J, Hanson AD (1993) Evidence for a large and sustained glycolytic flux to lactate in anoxic roots of some membranes of the halophytic genus *Limonium*. Plant Physiol 101:553–560

Rivoal J, Hanson AD (1994) Overexpression of lactate dehydrogenase in transgenic tomato roots supports the Davies-Roberts hypothesis and points to a critical role for lactate secretion. Plant Physiol 106:1179–1185

Roberts JKM, Callis J, Wemmer D, Walbot V, Jardetzky O (1984a) Mechanism of cytoplasmic pH regulation in hypoxic maize root tips and its role in survival under hypoxia. Proc Natl Acad Sci USA 81:3379–3383

Roberts JKM, Callis J, Jardetzky O, Walbot V, Freeling M (1984b) Cytoplasmic acidosis as determinant of flooding intolerance in plants. Proc Natl Acad Sci USA 81:6029–6033

Roberts JKM, Hooks MA, Miaulis AP, Edwards S, Webster C (1992) Contribution of malate and amino acid metabolism to cytoplasmic pH regulation on hypoxic maize root tips studied using nuclear magnetic resonance spectroscopy. Plant Physiol 98:480–487

Saint-Ges V, Roby C, Bligny R, Pradet A, Douce R (1991) Kinetic studies of the variations of the cytoplasmic pH, nucleotide triphosphates (^{31}P NMR) and lactate during normoxic and anoxic transitions in maize root tips. Eur J Biochem 200:477–482

Sakano K, Kiyota S, Yazaki Y (1997) Acidification and alkalinization of culture medium by *Catharantus roseus* cells – is anoxic production of lactate a cause of cytoplasmic acidification? Plant Cell Physiol 38:1053–1059

Shvetsova T, Mwesigwa J, Labady A, Kelly S, Thomas D, Lewis K, Volkov AG (2002) Soybean electrophysiology: effects of acid rain. Plant Sci 162:723–731

Smith FA, Raven JA (1979) Intracellular pH and its regulation. Ann Rev Plant Physiol 30:289–311

Stanković B, Davies E (1997) Intercellular communication in plants: electrical stimulation of proteinase inhibitor gene expression in tomato. Planta 202:402–406

Stewart PA (1983) Modern quantitative acid-base chemistry. Can J Physiol Pharmacol 61: 1444–1461

Stoimenova M, Libourel IGL, Ratcliffe RG, Kaiser WM (2003) The role of nitrate in the anoxic metabolism of roots. II. Anoxic metabolism of tobacco roots with and without nitrate reductase activity. Plant Soil 253:155–167

Summers JE, Jackson MB (1996) Anaerobic promotion of stem extension in *Potamogeton pectinatus*: roles for carbon dioxide, acidification and hormones. Physiol Plant 96:615–622

Summers JE, Ratcliffe RG, Jackson MB (2000) Anoxia tolerance in the aquatic monocot *Potamogeton pectinatus*: absence of oxygen stimulates elongation in association with an unusually large Pasteur effect. J Exp Bot 51:1413–1422

Tadege M, Brändle R, Kuhlemeier C (1998) Anoxia tolerance in tobacco roots: effect of over-expression of pyruvate decarboxylase. Plant J 14:327–335

Thomson CJ, Atwell BJ, Greenway H (1989) Response of wheat seedlings to low O_2 concentration in nutrient solution. I. Growth, O_2 uptake and synthesis of fermentative end-products by root segments. J Exp Bot 40:985–891

Ullrich CI, Novacki AJ (1990) Extra- and intracellular pH and membrane potential changes induced by K^+, Cl^-, $H2PO4^-$ and $NO3^-$ uptake and fusicoccin in root hairs of *Limnobium stoloniferum*. Plant Physiol 94:1561–1567

Vartapetian BB (2005) Plant anaerobic stress as a novel trend in ecological physiology, biochemistry, and molecular biology: 1. establishment of a new scientific discipline. Russ J Plant Physiol 52:826–844

Wildon DC, Thain JF, Minchin PEH, Gubb IR, Reilly AJ, Skipper YD, Doherty HM, O'Donnell PJ, Bowles DJ (1992) Electrical signalling and systemic proteinase inhibitor induction in the wounded plant. Nature 360:62–65

Wilkinson S (1999) pH as a stress signal. Plant Growth Regul 29:87–99

Xia J-H, Roberts JKM (1994) Improved cytoplasmic pH regulation, increased lactate efflux, and reduced cytoplasmic lactate levels are biochemical traits expressed in root tips of whole maize seedlings acclimated to a low-oxygen environment. Plant Physiol 105:651–657

Xia J-H, Roberts JKM (1996) Regulation of H^+ extrusion and cytoplasmic pH in maize root tips acclimated to a low-oxygen environment. Plant Physiol 111:227–233

Xia J-H, Saglio PH (1992) Lactic acid efflux as a mechanism of hypoxic acclimation of maize root tips to anoxia. Plant Physiol 100:40–46

Chapter 6
Programmed Cell Death and Aerenchyma Formation Under Hypoxia

Kurt V. Fagerstedt

Abstract This chapter describes the sequence of events leading to lysigenous aerenchyma formation in the roots of wetland and dryland plants. The events start from various stimuli that can induce cells to go through programmed cell death (PCD) in mid-cortical regions of the root: lack of oxygen (hypoxia), nitrogen, phosphorus or sulphur starvation and mechanical pressure. Oxygen deprivation results directly in the lowering of the cytoplasmic energy status and affects the cellular redox balance. The initial stimuli lead to secondary signals in cells and tissues such as increases in the free calcium ion levels and in the gaseous plant hormone ethylene. This sequence of signals leads eventually into PCD with eventual rupture of the vacuole, when many hydrolytic enzymes are released and break down the remains of the cell walls. How the cell death is directed to only certain mid-cortical cells and how the lytic cavity formation is restricted to certain areas only, is still an open question.

Abbreviations

ACC	1-Aminocyclopropane-1-carboxylate
ACC oxidase	1-Aminocyclopropane-1-carboxylate oxidase
AIF	Apoptosis-inducing factor
CEP	Cellular elimination process
EGTA	Ethylene glycol-bis(β-aminoethyl ether) N,N,N',N'-tetraacetic acid
GDC	Glycine decarboxylase complex
K-525a	Protein kinase inhibitor
MTP	Mitochondrial transition pore

K.V. Fagerstedt
Division of Plant Biology, Department of Biological and Environmental Sciences, University of Helsinki, Viikki Biocenter 3, POB 65, 00014 Helsinki, Finland
e-mail: kurt.fagerstedt@helsinki.fi

S. Mancuso and S. Shabala (eds.), *Waterlogging Signalling and Tolerance in Plants*,
DOI 10.1007/978-3-642-10305-6_6, © Springer-Verlag Berlin Heidelberg 2010

NO	Nitric oxide
NOS	Nitric oxide synthase
NOX	NADPH oxidase
$ONOO^-$	Peroxynitrite
PTP	Permeability transition pore
RNS	Reactive nitrogen species
ROL	Radial oxygen loss
ROP	Rho-related GTPase
SOD	Superoxide dismutase
TUNEL	Terminal deoxynucleotidyl transferase-mediated dUTP nick-end labelling
VPE	Vacuolar processing enzyme
W-7	Calmodulin antagonist
XET	Xyloglucan endotransglycosidase

6.1 Introduction

Aerenchyma formation in plants has been studied for over a century: We can find information published in Germany and Great Britain in the late 1800s and in the early decades of 1900s (Klinge 1879; Plowman 1906; Kükenthal 1909; Wille 1926) on anatomical structures in many wetland plants with various types of aerenchyma. This includes the extreme development in e.g. *Scirpus* species, where the amount of living tissues in the parenchyma with star-shaped cells is minimal (Kaul 1971). Similar development can be seen in Fig. 6.1 for many species of the *Carex* family with large air spaces in their roots and rhizomes (Crawford 1910; Fagerstedt 1992). There are basically two types of aerenchyma in plants: Lysigenous and schizogenous. Lysigenous aerenchyma results from programmed cell death (PCD) of certain cells in the cortical region of the root to form air-filled cavities (Kawai et al. 1998), while schizogenous aerenchyma is formed through the breakdown of pectic substances in the middle lamellae resulting in cell separation without cell death e.g. in *Rumex* (Laan et al. 1989). In many species, both schizogenous and lysigenous development can be seen. Water is kept from filling in the aerenchymal cavities with hydrophobic compounds lining the cell walls (Woolley 1983).

Even though the development of the microscopical structures was described in admirable detail in these early studies, we do not get any idea how the development of these structures is regulated at the physiological and genetic levels. Neither do we get much of an idea how aerenchyma affects diffusion of gases and hence tissue respiration. We do notice, however, that in many wetland plants, but not in all, aerenchyma development is constitutive, i.e. it develops even though the plants were grown in dryland conditions (and may be enhanced in waterlogged environments) while in most dryland species aerenchyma formation is induced under flooding. However, the picture is not so straightforward: In the dryland plant maize (*Zea mays*), aerenchyma is inducible (He et al. 1996) as well as in the

6 Programmed Cell Death and Aerenchyma Formation Under Hypoxia

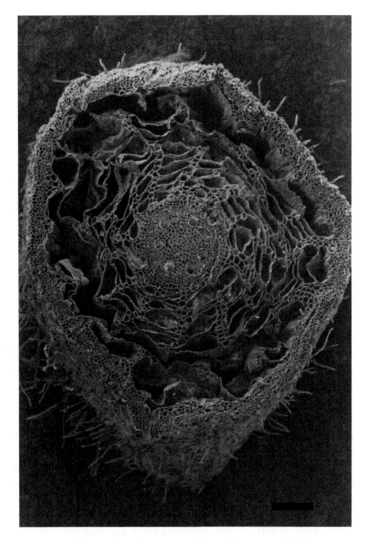

Fig. 6.1 Well-developed aerenchyma in the root of *Carex rostrata*, the beaked sedge, as seen under the scanning electron microscope. Scale bar represents 100 µm. Photo: Kurt Fagerstedt

wetland grass *Spartina patens* (Burdick 1989), while in the dryland teosinthe (*Z. mays* ssp. *mexicana*) and gama grass (*Tripsacum dactyloides*) (Ray et al. 1998) it is constitutive. The same applies for the wetland plants rice (*Oryza sativa*) (Kawai et al. 1998), *Sagittaria lancifolia* (Schussler and Longstreth 2000) and *Juncus effusus* (Visser and Bögemann 2006).

Naturally, the function of aerenchyma has been of great interest as it enables the passage of gases in and out of tissues which are otherwise hindered in their gas exchange e.g. under flooding. In this chapter, we are not concentrating on the function of aerenchyma but on its developmental regulation. Excellent articles on aerenchyma and the diffusion of gases and ventilation systems can be found elsewhere (e.g. Garthwaite et al. 2008; Armstrong and Armstrong 2005).

During the more recent years, information on the regulation of aerenchyma induction has accumulated and now we have detailed knowledge on the effects of low oxygen concentrations, ethylene, and free Ca^{2+}-ions on aerenchyma development.

One of the very interesting points in the development and regulation of aerenchyma formation is the fact that aerenchyma develops only in certain places in the root cortex and not everywhere. Is this due to local differences in oxygen/ethylene concentrations, or are there cells with different sensitivity to the signals in certain places in the root cortex? It seems that reactive oxygen (ROS) and nitrogen species (NO) play a role in the signalling, and hence antioxidants may also affect this development. Recently, we have learned a lot about PCD signalling in plant tissues and hence a whole paragraph has been dedicated on this issue.

6.2 Description of Aerenchyma Formation: Induced and Constitutive

In a study with 91 plant species from wetland, dryland and intermediate species, Justin and Armstrong (1987) described thoroughly the aerenchyma development ability and noted that aerenchyma was induced mainly in those wetland and in intermediate species with an arrangement of cells in cubic and radial manner in the root cortex as seen in cross sections. Aerenchyma rarely developed in plant species with a hexagonal non-radial cell configuration in the cortex, which was mainly found in dryland species. Hence, as a rule of thumb, one can generalize that aerenchyma tends to be inducible in wetland plants with radial rows of cells in the cortex, while in dryland species with hexagonal arrangement in the root cortex, lysigenous of schizogenous aerenchyma rarely develops. There are exceptions in both groups. In the wetland species rice (*O. sativa*) (Jackson et al. 1985; Kawai et al. 1998), aerenchyma formation is not inducible and does not seem to depend on ethylene or the small partial oxygen pressures often encountered during flooding. Similar results have been found with the wetland species *J. effusus* (Visser and Bögemann 2006). Drew and coworkers (1989), He and coworkers (1994) and others have shown that in maize (*Z. mays*) aerenchyma formation is inducible. The same can be said of *S. patens* (Burdick 1989). It remains to be seen how constitutive aerenchyma development is regulated.

In crop species lysigenous aerenchyma formation is widespread as we have seen in maize. In sunflower (*Helianthus annuus*) (Kawase 1979), bean (*Phaseolus vulgaris*) and tomato (*Lycopersicon esculentum*) (Kawase 1981), in wheat (*Triticum aestivum*) (Huang et al. 1994) and barley (*Hordeum vulgare*) (Larsen et al. 1986; Fagerstedt and Crawford 1987) and in *Trifolium subterraneum* (Aschi-Smiti et al. 2003), aerenchyma development seems to be induced by flooding and at least in most cases controlled by low oxygen partial pressures and increased ethylene.

6.3 Evidence for PCD During Lysigenous Aerenchyma Formation

During the past two decades, PCD in plant tissues has been of great interest especially because of the fact that it differs in many details from its counterpart in mammals. An excellent book covering the many aspects of PCD in plants has been published recently (Gray 2004). In plant tissues, PCD is known to occur during many constitutive developmental events such as the development of tracheal elements (TEs) during the hypersensitive response under a pathogen attack (HR) and very similar events take place during aerenchyma formation under environmental stresses such as hypoxia. Other abiotic stresses leading to lysigenous aerenchyma formation include mechanical pressure (He et al. 1996a), high temperature, nitrogen, phosphorus, and sulphur deficiencies (Konings and Verschuren 1980; Bouranis et al. 2003; Fan et al. 2003). These kind of abiotic stresses have some common features: many affect the amount of available oxygen for respiration and as a result of aerenchyma development root respiration decreases thus helping the plant to survive. In the case of nutrient deficiencies, aerenchyma development means less organic material input to root growth while allowing the same surface area for nutrient uptake. It has been noted, however, that during P starvation root respiration decreases much more than what could be expected by just calculating the amount of aerenchyma (Fan et al. 2003). In these experiments, root respiration decreased by 70% during P deficiency while at the same time aerenchyma development reduced the proportion of living cells in the tissues by only 30%. This naturally gives the plant an advantage in exploring the soil for nutrients with less organic material input per root length and lowers the consumption of respiratory substrates during this search.

In mammalian cells, mitochondria have a central role in the PCD cascade. Mitochondria integrate metabolic changes (pH, Ca_{cyt}^{2+}, energy status) with PCD, and respond by increasing membrane permeability and by releasing cytochrome c and an apoptosis-inducing factor (AIF) leading to the activation of caspases or cystein proteases in animal cells. This behaviour has been related to the function of the permeability transition pore (PTP or MTP), which has been studied in detail in mammalian mitochondria but has not been identified as such in plant cells, although we have seen implications that such a pore exists also in plant mitochondria (Fortes et al. 2001; Virolainen et al. 2002). In yeast and plant cells, a differently regulated pore similar to mammalian PTP has been reported. High matrix Ca^{2+} concentration, decrease in mitochondrial membrane potential, and ROS favour PTP open probability (Fortes et al. 2001). Under oxygen deprivation stress, which leads into lowered adenylate energy charge, cytoplasmic acidosis, elevation of Ca_{cyt}^{2+}, lowered antioxidative capacity, and increased probability of ROS formation, plant mitochondria swell and finally disintegrate. Similar swelling and rupture of organelles is typical for plant cell PCD. In our experiments on *Triticum* mitochondrial properties such as membrane potential, Ca^{2+} transport and swelling have been studied under lack of oxygen and discussed in relation to PCD regulation

(Virolainen et al. 2002). In these experiments, the PTP opening and cytochrome c leakage was only evident in cells which suffered from lack of oxygen and were surrounded by high Ca^{2+} concentrations.

A key study to show the hallmark features of PCD in lysigenous aerenchyma formation in maize (*Z. mays* L.) was published by Gunawardena and coworkers (2001a). In this study, several PCD features, viz. terminal deoxynucleotidyl transferase-mediated dUTP nick-end labeling (TUNEL)-positive nuclei and cytoplasmic condensation were detected (Gunawardena et al. 2001a). This piece of work also showed DNA laddering after ethylene treatment of the roots, which has been very difficult to show in other plant tissues going through otherwise clear PCD. Still another feature in maize roots was the presence of seemingly intact organelles within membranous compartments, which resemble the apoptotic bodies found in many cases of animal apoptosis.

Another piece of evidence connecting lysigenous aerenchyma development with a PCD event, viz. tracheary element development, is the *gapped xylem* mutant (*gpx*) in *Arabidopsis* (Turner and Hall 2000). In these mutants, the vessel cell walls are not developing properly and are dissolved after the vacuole ruptures releasing hydrolytic enzymes. This leads to anatomical structures resembling aerenchyma. It remains to be seen whether similar events take place in normal aerenchyma development and whether the same genes that are important for tracheary element development play a role in lysigenous aerenchyma formation. Other similar connections have been discussed together in a review by Kozela and Regan (2003).

Bouranis and coworkers (2007) have described lysigenous aerenchyma formation plausibly as a sequence of events containing the following steps: *The activation process*, during which the cell prepares for PCD and where the mitochondrion plays a central role; *the execution process*, during which many lytic enzymes are released and destroy cellular structures with a major role of the lytic vacuole; *the dissemination process*, during which all the cells destined to die will be executed forming the lytic air cavity; and *the termination process*, during which the boundaries of the air cavities are set restricting the spreading of the PCD process. In my opinion, the last may naturally be determined already at the beginning of the sequence of events as cells where the whole process does not start at all.

In the next chapter, we will concentrate on the events leading to lysigenous aerenchyma formation.

6.4 Description of the Sequence of Events Leading to Induced Lysigenous Aerenchyma Formation

Here, we will first concentrate on the initial signals such as low oxygen, cytosolic free Ca^{2+}, and ethylene, and then continue with the execution process of lysigenous aerenchyma formation, the dissemination and finally the termination process ending up with the finished product: the functional aerenchymatous tissue. The initial signalling events are gathered together in Fig. 6.2.

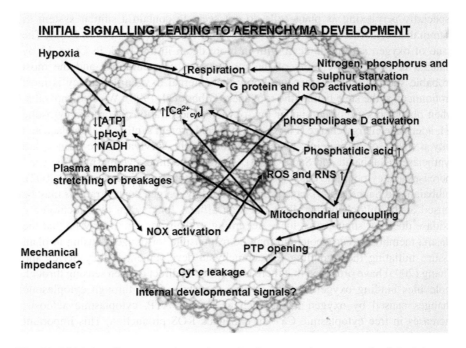

Fig. 6.2 Initial signalling events of aerenchyma development and programmed cell death immediately after induction by hypoxic environment. Background photomicrograph is of a cross section of a young rice root at the stage where cell death has not yet commenced. *NOX* NADPH oxidase, *RNS* reactive nitrogen species, *ROS* reactive oxygen species, *ROP* RHO-like GTPases of plants

6.4.1 Stimuli for Lysigenous Aerenchyma Development (Low Oxygen, Cytosolic Free Calcium, Ethylene, P, N, and S Starvation, and Mechanical Impedance)

The signals or rather environmental stimuli needed for the beginning or activation of the lysigenous aerenchyma development are versatile. Naturally, in the case of the constitutive route for aerenchyma development there must be internal developmental signals that initiate the process of cell elimination. The environmental stimuli include: low oxygen partial pressures (hypoxia), ethylene, P, N, and S starvation, heat stress, and mechanical soil pressure. All of these factors affect cellular respiration, and hence the energy and redox status of the cytoplasm.

6.4.1.1 Oxygen Deprivation

Hypoxic environment has been shown to lead into aerenchyma development in numerous cases (for a recent review, see Evans 2003), but how is the sensing of low oxygen perceived? This question has bothered plant scientists for a long time and is especially perplexing as plant cells do not seem to contain a similar system as mammalian cells (the hypoxia-inducible factor-1 HIF-1, Bergeron et al. 1999). The issue of oxygen sensing is dealt with in more detail in this book in Chap. 11 by Shabala & Pang. If we think of the root oxygen concentrations and the most probable site for sensing, then we know that internal oxygen deficiency is most pronounced at the stele behind the root apex (Gibbs et al. 1998). We also know that when plant roots are placed under low oxygen concentrations, a drop in cytosolic pH, adenylate energy charge (Fox et al. 1995) and protein synthesis takes place and only a small set of proteins, which are vital for the survival of the tissues, are synthesised (Subbaiah and Sachs 2003). At the same time, cytoplasmic free Ca^{2+} increases and cytochrome c is released from mitochondria (Virolainen et al. 2002). Subbaiah and Sachs (2003) have proposed that low oxygen concentrations may be sensed at the mitochondria, albeit the high affinity for oxygen of cytochrome c oxidase does not support this idea. According to their opinion, it may be that the plasma membrane redox potential may be responsible for oxygen sensing in plant tissues initiating the signalling for aerenchyma development. Bailey-Serres and Chang (2005) have proposed that there may either be direct oxygen sensing through molecules binding oxygen, or the sensing may take place through cytoplasmic changes caused by oxygen levels such as loss of ATP, cytoplasmic acidosis, increases in free cytoplasmic Ca^{2+} or enhanced ROS production. This important issue still remains unsolved.

6.4.1.2 Free Cytoplasmic Calcium

It has been shown that the availability of free calcium is paramount for the development of aerenchyma and if it is eliminated by a treatment of the tissues with the calcium binding agent EGTA (ethylene glycol-bis(β-aminoethyl ether) N,N,N',N'-tetraacetic acid) aerenchyma does not develop in maize nodal roots (He et al. 1996b; Drew et al. 2000). Also, if the roots were treated with Ruthenium red, which inhibits calcium release from internal stores (ER and mitochondria), aerenchyma did not develop (He et al. 1996b). But what is the sequence of events? It has been shown that both the rise in intracellular Ca^{2+} and redox alterations follow the rise in H_2O_2 (Rentel et al. 2004). Several other signalling events are dependable on either cytoplasmic or apoplastic free Ca^{2+}, e.g. RHO-like GTPases of plants (ROP) signalling is known to interact with Ca^{2+} (Bailey-Serres and Chang 2005), and cytoplasmic Ca^{2+} activates the NADPH oxidase (NOX) which leads to further production of H_2O_2 (Sagi and Fluhr 2001), and the mitochondrial PTP is partly regulated by free cytoplasmic Ca^{2+} (Petronilli et al. 1994). Here, we should remember that calcium is only one of the signals needed for the drastic event of PTP opening, but once ATP levels are low, ROS levels are high, and mitochondrial membrane potential low, free cytoplasmic Ca^{2+} is the last drop in the ocean of proapoptotic events that opens the PTP (Fortes et al. 2001; Blokhina et al. 2003). This is enhanced by high NO levels which hinder the electron transport chain, and hence the membrane potential is lowered.

The specificity of the calcium signal can be achieved through the calcium signature (for a recent review, see Luan 2009), which will participate in the downstream effects, such as aerenchyma formation, through calmodulins and through many calcium-dependent protein kinases (Harper et al. 2004; Luan 2009). In plants, calmodulin-regulated proteins include catalase, which will protect the cytoplasm against H_2O_2 (Gechev et al. 2004), NAD dikinase (Harding et al. 1997) but also myosin V and kinesin which may take part in membrane movements during PCD, glutamate decarboxylase and Ca^{2+}-ATPases (for a review on calcium signalling, see Lecourieux et al. 2006).

6.4.1.3 Ethylene

It has been known for long that ethylene, the gaseous plant hormone, participates in many cellular events leading to cell death, and hence it was the first hormone to be suspected as being involved in aerenchyma development. Reviews and individual studies have shown that ethylene indeed has a regulatory role in aerenchyma formation (Jackson 1985; Drew et al. 1989; He et al. 1992, 1996a, 1996b; Gunawardena et al. 2001a, 2001b). The sequence of events is envisaged as: 1-aminocyclopropane-1-carboxylate (ACC) synthase is induced during hypoxia within a few hours of its onset (He et al. 1996a, 1996b) and the next enzyme in the production route to ethylene, ACC oxidase, in a reaction where oxygen is needed, should also increase before ethylene is produced. As oxygen is vital for this step, ethylene is not produced under strict anoxia. It has also been suggested that a permeability barrier for ethylene in the roots prevents the loss of ethylene and leads to a high concentration inside the roots (Armstrong et al. 2000). This barrier also prevents radial oxygen loss from the roots into the surrounding medium, and hence facilitates root respiration in rice (Colmer 2003).

Ethylene is perceived in *Arabidopsis* by a family of five ethylene receptors (ETRs), which are related to two-component histidine kinases (for a review, see Bleecher and Kende 2000). They are thought to transmit the signal through interaction with negative regulator Ser/Thr kinase CTR1 (Kieber et al. 1993). Ethylene binding to the receptor/CTR1 complex inhibits signalling through an increase in EIN2 (ethylene insensitive 2) activity and an upregulation of the ethylene-response pathways (Hua and Meyerowitz 1998). EIN2 is thought to work through a transcriptional cascade mediated by a family of EIN3 and ERF1 transcription factors (Solano et al. 1998). The picture of ethylene responses was recently clarified in relation to aerenchyma development in *Arabidopsis* by the finding of the gene LSD1 (Mühlenbock et al. 2007). This gene together with EDS1 and PAD4 operate upstream of ethylene signalling and ROS formation, and they are thought to mediate the metabolic status of the tissues into ethylene signalling (Mühlenbock et al. 2007).

MAPKs have proven to be the mediators of ethylene action in plant tissues. Quaked et al. (2003) have been able to show that, at least in *Medicago* and *Arabidopsis*, certain MAPKs (SIMK and MMK3 in *Medicago* and MPK6 and another MAPK in *Arabidopsis*) are activated by the ethylene precursor ACC, proving that MAPKs are a part of the ethylene signalling pathway in plants. However, it is still not known whether hypoxia may activate any MAPKs in a similar manner as in mammals (Lee et al. 2004). On the other hand, recent evidences gathered together in a review by Hahn and Harter (2009) show that MAPKs take part both in the ethylene signalling pathway as well as in the ethylene biosynthetic pathway, to eventually activate ethylene response genes through dual control of EIN3 (Yoo et al. 2008).

Keeping in mind the recent results above, it is interesting to note that the role of an ethylene receptor (RP-ERS1) has been demonstrated in the response of *Rumex palustris* to submergence (Vriezen et al. 1997). In this wetland species, flooding results in a rapid upregulation of the RP-ERS1 within 20 min, and this may be one of the first events in aerenchyma formation. On the other hand, plant species with constitutive aerenchyma formation do not seem to be sensitive to ethylene, as has been show in *J. effusus* by Visser and Bögemann (2006). In their experiments, treatment with ethylene or with ethylene inhibitors did not lead to any changes in root aerenchyma or in root elongation, pointing also to the fact that in this species inducible aerenchyma does not give any benefit to a wetland plant. As we do not yet know how widespread ethylene insensitivity is in wetland plants, further conclusions are not yet possible.

6.4.1.4 P, N, and S Starvation

Similar to hypoxic stress, phosphorus and nitrogen starvation stresses have been shown to lead to aerenchyma formation independent of hypoxia (Konings and Verschuren 1980; Drew et al. 1989; Fan et al. 2003). Similarly, in well-oxygenated conditions in a nutrient solution sulphur deficiency lead to aerenchyma formation through lysigenous formation of air spaces in the root cortex (Bouranis et al. 2003). It has been noted that the cell death during aerenchyma formation decreases the demand for oxygen as there are less respiring cells per volume, and hence relieves stress under hypoxic conditions (Fan et al. 2003). It seems plausible that the positive effect of aerenchyma formation in the case of phosphorus deficiency is achieved by the smaller amount of living cells needed for soil exploration and nutrient capture (Fan et al. 2003). Naturally, this explanation could apply to any nutrient deficiency, but has so far been shown only with P and S. How this stimulus of low nutrients is translated into the development of lysigenous aerenchyma is still obscure.

6.4.1.5 Energy and Redox Status of the Cell

When the initiation of PCD during aerenchyma development is considered, a careful estimation of the oxygen concentration in the tissues involved is of paramount importance. This is mainly due to the fact that mitochondria act in the production of ATP whenever oxygen, respiratory substrates and ADP are present. The affinity

6 Programmed Cell Death and Aerenchyma Formation Under Hypoxia

for oxygen of cytochrome c oxidase is very high (K_m for oxygen of cytochrome oxidase is approximately 100 nmol L^{-1} (Hill 1998)) which means that the oxygen concentration in the tissues has to be very low before mitochondrial respiration and ATP production cease. Naturally, oxygen concentrations as low as above have been recorded in waterlogged soils, but in aerenchymatous roots this is rare as aerenchyma provides a passage for diffusion of oxygen from the shoots. However, if such a condition occurs, it is possible that one of the signals for lysigenous aerenchyma development goes through membrane disruption (Rawyler et al. 2002).

ROS have been implicated in many stress responses both as signalling molecules (especially H_2O_2) and as the cause for oxidative damage (Blokhina et al. 2003; Foyer and Noctor 2005). The picture is made more complicated by the many antioxidants present in the cells and in the apoplast, which scavenge ROS and protect the tissues from oxidative damage and at the same time modulate the ROS signals as reviewed by Blokhina and coworkers (2003). It has been shown that at least under sulphate stress alterations in ROS and free calcium levels and in pH took place in maize root sectors where aerenchyma was developing and ROS appeared in mid-cortex cells going through cell death (Bouranis et al. 2003, 2006).

It seems that one of the most important enzymes producing ROS under stress and leading to aerenchyma formation, is the plasma membrane-bound NADPH oxidase (NOX) (Sang et al. 2001). NOX produces superoxide radicals which are quickly dismutated to H_2O_2 by superoxide dismutases (SODs). NOX is activated by its cytosolic subunits which in turn are regulated by free calcium ions (Keller et al. 1998). It has been noticed that phospholipase D and a G protein (Munnik et al. 1995; Sang et al. 2001) and most likely small Rho-related GTPases called ROPs (Agrawal et al. 2003) are also involved. It has been shown that the ROP family proteins control many events in plants cells including ethanolic fermentation and they activate and repress signalling cascades leading to changes in H_2O_2 levels (Bailey-Serres and Chang 2005).

In addition to ROS, reactive nitrogen species and especially nitric oxide, NO, also take part in PCD. NO is known to be produced much in the same locations as ROS and a novel NO synthase gene has been identified in plants (Guo and Okamoto 2003). Information on the nitric oxide synthase (NOS) activity of the variant P protein of the mitochondrial glycine decarboxylase complex (GDC) has emerged (Chandok et al. 2003) but that article was retracted in 2004 (Klessig et al. 2004). NO may also act with the thiol groups of proteins causing the formation of nitrosothiols and, what is more, the same proteins may be modified by H_2O_2 (Lindermayr et al. 2005). NO may be enhancing PCD through several reactions as reviewed by Bouranis and coworkers (2007): NO together with superoxide can lead into the formation of peroxynitrite $ONOO^-$, a very reactive nitrogen species causing damage; NO action on the mitochondrial Ca^{2+} channel can enhance the opening of PTP; NO reduces ATP levels by inhibiting the mitochondrial electron transport, which also enhances the opening of the PTP.

Plant cells are not devoid of protecting measures against NO. The oxygenated form of class I haemoglobin reacts with NO converting it into nitrate (Igamberdiev et al. 2005) and then is itself converted back to haemoglobin in a reaction with NAD(P)H.

Interestingly, class I haemoglobin overexpressing alfalfa plant roots do not develop aerenchyma while the wild-type control roots showed cortical cell death (Dordas et al. 2003). There is also a connection to ethylene formation as in an experiment with maize suspension cultured cells with suppressed haemoglobin expression leading to increased ethylene levels (Manac'h et al. 2005). It remains to be seen whether class I haemoglobin acts directly by scavenging NO and preventing PCD, or through a more complicated picture involving ethylene.

Hence, the sequence of events leading to activation of the PCD cascade under severe hypoxic stress is as follows: As mitochondrial respiration declines, ATP levels decrease and due to the takeover of anaerobic metabolism (lactic acid and ethanolic fermentation) NADH level increases and cytosolic pH declines. This leads to possible membrane perturbations and NOX activation, where also G protein, phopholipase D, and free Ca^{2+} play a role. It has been suggested recently that a change in the K^+/Na^+ ratio may be crucial for triggering PCD (Shabala 2009). All this affects intracellular free Ca^{2+} and ROS and reactive nitrogen species (RNS) levels, which in turn lead to thiol residue modifications in many other proteins (Cooper et al. 2002; Dröge 2002). It has been suggested that oxidation of the thiol groups acts in the perception of H_2O_2 levels (Hancock et al. 2006). Finally, as integrators of many cellular signals, mitochondria act and the PTP open probability increases and cytochrome c leakage occurs initiating PCD (Jones 2000). The function of plant mitochondria and the PTP in PCD has been under intense study during the recent years, but we still do not know all the details with the accuracy as is known with mammalian PCD regulation. On the other hand, the NO increase combined with low class 1 haemoglobin levels leads to protein nitrosylations and increase in ethylene concentrations and combined with the radial oxygen loss (ROL) barrier acting also as an ethylene barrier results in a MAPK cascade affecting transcription factors controlling many genes including those that lead to destruction of the remains of the cells in the lysigenous aerenchyma.

6.4.2 PCD and the Clearing of the Cell Debris

The execution of the cellular elimination process (CEP), as reviewed by Bouranis et al. (2007), includes the functioning of an array of enzymes responsible for the removal of all the remains of the cells in the areas of cortex annihilated. This means that the vacuolar processing enzyme (VPE), proteases, lipases and many other esterases, cell wall degrading enzymes, DNases and RNases are needed for the breakdown of the cell constituents. Some of these events in the tissues and inside the cells are clearly visible under the transmission electron microscope or even under the light microscope and have been described in detail. The description is important as it gives us hints on the exact sequence of events during PCD in lytic aerenchyma formation. Downstream signalling events leading to changes in cellular membranes and to the production of hydrolytic enzymes during lysigenous aerenchyma formation are shown in Fig. 6.3.

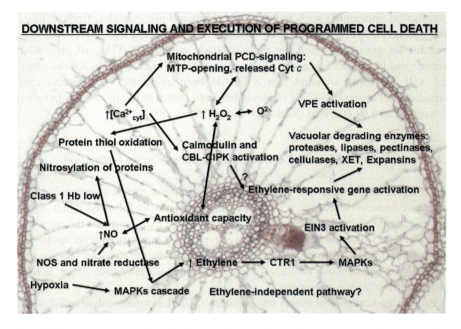

Fig. 6.3 Downstream signalling leading to the execution of programmed cell death during lysigenous aerenchyma formation. Several signals are in action simultaneously and if certain thresholds are exceeded, the signal is passed on and eventually the vacuolar degrading enzymes are activated and released once the tonoplast ruptures. The ethylene-independent pathway, which seems to be in action in some wetland plants with constitutive aerenchyma formation, is not known yet. How certain cells are chosen to die and certain cells to survive is not known either. It may be that some cells are predestined to die and hence are more sensitive to the signals for programmed cell death. CBL *calcineurin B-like proteins*, CIPK CBL-interacting protein kinases, CTR1 a negative regulator of the ethylene-response pathway, EIN3 *ethylene insensitive 3*, MAPKs mitogen-activated protein kinase, NO nitric oxide, NOS nitric oxide synthase, PCD programmed cell death, XET xyloglucan endotransglycosidase

6.4.2.1 Cellular Degrading Enzyme Activities

The vacuolar processing enzyme (VPE) is a typical plant caspase-like enzyme or a cysteine protease similar to the mammalian caspase-1 (Hatsugai et al. 2004), but it is evident that many other proteases are at work in the various compartments of the plant cell during aerenchyma development. This is an area where more research is needed. However, in a study on *T. subterraneum* during hypoxic treatment, a change in the protease composition could be detected pointing to probable lysis-related protease genes having been activated (Aschi-Smiti et al. 2003).

Cell wall degrading enzymes that have been noticed to be upregulated during lysigenous aerenchyma formation are expansins, cellulases, xyloglucan endotransglycosidases (XET) and pectinases (as reviewed by Jackson and Armstrong 1999). Of these, cellulase is induced by ethylene and its activity increases e.g. in maize roots well before aerenchyma is evident (He et al. 1994), while several signal transduction antagonists such as inhibitors of inositol phosphatides, free calcium-binding agent EGTA, and the Ca^{2+} channel blocking agent Ruthenium red, calmodulin antagonist W-7 and the protein kinase inhibitor K-525a, all prevented aerenchyma formation in maize roots in a study by He et al. (1996b). Of the other cell wall degrading enzymes, the XET has been associated with aerenchyma formation and especially XET1 of maize has been shown to be specifically induced by hypoxia and by ethylene (Saab and Sachs 1996).

6.4.2.2 Sequence of Events Leading to Lysigenous Cavity Formation

In an elegant article by Gunawardena et al. (2001a), the sequence of events leading to root lysigenous aerenchyma development in maize was followed under light, fluorescent and electron microscopes. They concluded that the maize tissues went through PCD, which, however, had features typical for both mammalian apoptosis and cytoplasmic cell death. The latter being similar in the fact that during the initiation of aerenchyma formation membrane changes may take place before nuclear ultrastructural alterations.

Briefly, the sequence of events was as follows: Unlike in mammalian cells one of the first symptoms before internucleosomal cleavage, as seen with the TUNEL staining and chromatin condensation, were cytoplasmic changes including plasma membrane invagination and vesicle formation. Later, cellular and chromatin condensation as well as organelles such as mitochondria and Golgi bodies surrounded by membranes resembling apoptotic bodies were seen. It is very interesting to note that the authors state that all these processes concerning the protoplasts were complete before the degradation of the cell walls was apparent (Gunawardena et al. 2001a). In a more recent article, Gunawardena (2008) describes PCD in a larger context as a means for remodeling plant tissues with details on the cell wall dismantling enzymes as well as many references to articles concerning caspase-like enzymes in plants and their role in PCD execution (e.g. Sung et al. 1999; de Jong et al. 2000; Mlejnek and Prochazka 2002; Bozhkov et al. 2004; Hatsugai et al. 2004).

Once the vacuole ruptures, the acidic pH and the many hydrolytic enzymes released will attack the remaining cellular organelles and nuclear DNA finalizing the death of the protoplast (Hara-Nishimura et al. 2005; van Doorn and Woltering 2005). The total cell elimination needed for large air spaces to be formed is achieved by the many cell wall degrading enzymes released (Jones 2000).

6.4.3 What Determines the Architecture of Aerenchyma? – Targeting and Restricting PCD

It has been established a long time ago that aerenchyma develops always in mid-cortical cells in the roots, never nearer the rhizodermis nor deeper near the endodermis, and definitely not in the vascular cylinder (Justin and Armstrong

6 Programmed Cell Death and Aerenchyma Formation Under Hypoxia

1987). Hence, there must be a mechanism targeting the cells which will go through cell death and which will lead to the fine architectural features typical of native wetland species.

It is also well known that aerenchyma develops very often in plants, where the root cortex consists of cells neatly packed cubically in radial rows (Justin and Armstrong 1987), rather than in plant species with hexagonal and non-radial packing of cortex (Drew et al. 2000). The particular feature of aerenchymatous tissue is that certain cells die, while their adjacent cells continue their lives and form the tissues needed e.g. for nutrient and water passage to the stele. We know that cell death does not initiate at the point of lowest oxygen concentration (Evans 2003). The reasons for the differences between the cells are not known; however, it seems probable that there are differences either in the sensitivity of cells to the initial stimuli or in the response pathways leading to PCD (Corpas et al. 2001; Evans 2003).

It has been suggested also that there may be cell death signals, which are diffusing through plasmodesmata from the cells initiating PCD to the surrounding cortex (Drew et al. 2000). The latter, however, still involves differential sensitivity of cells to the initial stimuli. An interesting note is that in the places where lateral roots will emerge through the endodermis, aerenchyma development is hindered (Campbell and Drew 1983; Bouranis et al. 2006), giving a hint that plant hormones other than ethylene may take part in the cell fate decision.

6.5 Future Prospects

As we have seen in the text above, aerenchyma development has been extensively researched, starting from descriptions of the various types of aerenchyma found constitutively formed in hundreds of wetland plants to the induced air spaces in maize and cereals and ending up with the signalling and regulation of PCD in cortical tissues. At the moment, we are facing the very interesting problems of tissue sensitivity for the hypoxic stimulus, which will eventually give us tools in the understanding of the powers that are guiding the architecture of aerenchyma.

References

Agrawal GK, Iwahashi H, Rakwal R (2003) Small GTPase 'Rop': molecular switch for plant defense responses. FEBS Lett 546:173–180

Armstrong W, Armstrong J (2005) Stem photosynthesis not pressurized ventilation is responsible for light-enhanced oxygen supply to submerged roots of alder (*Alnus glutinosa*). Ann Bot 96(4):591–612. (Erratum in: Ann Bot 2005 96(6):1141)

Armstrong W, Cousins D, Armstrong J, Turner DW, Beckett PM (2000) Oxygen distribution in wetland plant roots and permeability barriers to gas exchange with the rhizosphere: a micro-electrode and modelling study with *Phragmites australis*. Ann Bot 86:687–703

Aschi-Smiti S, Chaibi W, Brouquisse R, Ricard B, Saglio P (2003) Assessment of enzyme induction and aerenchyma formation as mechanisms for flooding tolerance in *Trifolium subterraneum* 'Park'. Ann Bot 91:195–204

Bailey-Serres J, Chang R (2005) Sensing and signalling in response to oxygen deprivation in plants and other organisms. Ann Bot 96:507–518

Bergeron M, Yu AY, Solway KE, Semenza GL, Sharp FR (1999) Induction of hypoxia-inducible factor-1 (HIF-1) and its target genes following focal ischaemia in rat brain. Eur J Neurosci 11:4159–4170

Bleecher AB, Kende H (2000) Ethylene: a gaseous signalling molecule in plants. Annu Rev Cell Dev Biol 16:1–18

Blokhina O, Virolainen E, Fagerstedt KV (2003) Antioxidants, oxidative damage and oxygen deprivation stress: a review. Ann Bot 91:179–194

Bouranis DL, Chorianopoulou SN, Kollias C, Maniou P, Protonotarios VE, Siyannis VF, Hawkesford MJ (2006) Dynamics of aerenchyma distribution in the cortex of sulphate-deprived adventitious roots of maize. Ann Bot 97:695–704

Bouranis DL, Chorianopoulou SN, Siyiannis VF, Protonotarios VE, Hawkesford MJ (2003) Aerenchyma formation in roots of maize during sulphate starvation. Planta 217:382–391

Bouranis DL, Chorianopoulou SN, Siyiannis VF, Protonotarios VE, Hawkesford MJ (2007) Lysigenous aerenchyma development in roots – triggers and cross-talks for a cell elimination program. Int J Plant Dev Biol 1:127–140 .

Bozhkov PV, Filinova LH, Suarez MF, Helmerson A, Smertenko AP, Zhivotovsky B, von Arnold S (2004) VEIDase is a principal caspase-like activity involved in programmed cell death and essential for embryonic pattern formation. Cell Death Differ 11:175–182

Burdick DM (1989) Root aerenchyma development in *Spartina patens* in relation to flooding. Am J Bot 76:777–780

Campbell R, Drew MC (1983) Electron microscopy of gas space (aerenchyma formation) in adventitious roots of *Zea mays* L. subjected to oxygen shortage. Planta 157:350–357

Chandok MR, Ytterberg AJ, van Wijk KJ, Klessig DF (2003) The pathogen-inducible nitric oxide synthase (iNOS) in plants is a variant P protein of the glycine decarboxylase complex. Cell 113:469–482 (later retracted, see Klessig 2004)

Colmer TD (2003) Aerenchyma and an inducible barrier to radial oxygen loss facilitate root aeration in upland, paddy and deep-water rice (*Oryza sativa* L.) Ann Bot 91:301–309

Cooper C, Patel RP, Brookes PS, Darley-Usmar VM (2002) Nanotransducers in cellular redox signalling: modification of thiols by reactive oxygen and nitrogen species. Trends Biochem Sci 27:489–492

Corpas FJ, Barroso JB, del Rio LA (2001) Peroxisomes as a source of reactive oxygen species and nitric oxide signal molecules in plant cells. Trends Plant Sci 6:145–150

Crawford FC (1910) Anatomy of the British Carices. Oliver and Boyd, Edinburg

De Jong AJ, Hoeberichts FA, Yakimova ET, Maximova E, Woltering EJ (2000) Chemical-induced apoptotic cell death in tomato cells: involvement of caspase-like proteases. Planta 211:656–662

Dordas C, Hasinoff B, Igamberdiev AU, Manac'h N, Rivoal J, Hill RD (2003) Expression of a stress-induced haemoglobin affects NO levels produced by alfalfa under hypoxic stress. Plant J 35:763–770

Drew MC, He CJ, Morgan PW (1989) Decreased ethylene biosynthesis, and induction of aerenchyma, by nitrogen- or phosphate-starvation in adventitious roots of Zea mays L. Plant Physiol 91:266–271

Drew MC, He CJ, Morgan PW (2000) Programmed cell death and aerenchyma formation in roots. Trends Plant Sci 5:123–127

Dröge W (2002) Free radicals in the physiological control of cell function. Physiol Rev 82:47–95

Evans DE (2003) Aerenchyma development. New Phytol 161:35–49

Fagerstedt KV (1992) Development of aerenchyma in roots and rhizomes of *Carex rostrata* (*Cyperaceae*). Nordic J Bot 12:115–120

6 Programmed Cell Death and Aerenchyma Formation Under Hypoxia

Fagerstedt KV, Crawford RMM (1987) Is anoxia tolerance related to flooding tolerance? Funct Ecol 1:49–55

Fan M, Zhu J, Richards C, Brown KM, Lynch JP (2003) Physiological roles for aerenchyma in phosphorus-stressed roots. Funct Plant Biol 30:493–506

Fortes F, Castilho RF, Catisti R, Carnieri EGS, Vercesi A (2001) Ca^{2+} induces a cyclosporin A-insensitive permeability transition pore in isolated potato tuber mitochondria mediated by reactive oxygen species. J Bioenerg Biomembr 33:43–51

Foyer CH, Noctor G (2005) Redox homeostasis and antioxidant signalling: a metabolic interface between stress perception and physiological responses. Plant Cell 17:1866–1875

Fox GG, McCallan NR, Ratcliffe RG (1995) Manipulating cytoplasmic pH under anoxia – a critical test of the role of pH in the switch from aerobic to anaerobic metabolism. Planta 195:324–330

Garthwaite AJ, Armstrong W, Colmer TD (2008) Assessment of O_2 diffusivity across the barrier to radial O2 loss in adventitious roots of *Hordeum marinum*. New Phytol 179:405–416

Gechev TS, Gadjev I, Hille J (2004) An extensive microarray analysis of AAL-toxin-induced cell death in *Arabidopsis thaliana* brings new insights into the complexity of programmed cell death in plants. Cell Mol Life Sci 61:1185–1197

Gibbs J, Turner DW, Armstrong W, Darwent MJ, Greenway H (1998) Response to oxygen deficiency in primary maize roots. I. Development of oxygen deficiency in the stele reduces radial solute transport to the xylem. Aust J Plant Physiol 25:745–758

Gray J (2004) Programmed cell death in plants. Blackwell, Oxford

Gunawardena AHLAN (2008) Programmed cell death and tissue remodelling in plants. J Exp Bot 59:445–451

Gunawardena AHLAN, Pearce DM, Jackson MB, Hawes CR, Evans DE (2001a) Characterisation of programmed cell death during aerenchyma formation induced by ethylene or hypoxia in roots of maize. Planta 212:205–214

Gunawardena AHLAN, Pearce DME, Jackson MB, Hawes CR, Evans DE (2001b) Rapid changes in cell wall pectic polysaccharides are closely associated with early stages of aerenchyma formation, a spatially localized form of programmed cell death in roots of maize (Zea mays L.) promoted by ethylene. Plant Cell Environ 24:1369–1375

Guo FQ, OkamotoM CNM (2003) Identification of a plant nitric oxide synthase gene involved in hormonal signalling. Science 302:100–103

Hahn A, Harter K (2009) Mitogen-activated protein kinase cascades and ethylene: Signaling, biosynthesis or both? Plant Physiol 149:1207–1210

Hancock J, Desikan R, Harrison J, Bright J, Hooley R, Neil S (2006) Doing the unexpected: proteins involved in hydrogen peroxide perception. J Exp Bot 57:1711–1718

Hara-Nishimura I, Hatsugai N, Nakaune S, Kuroyanagi M, Nishimura M (2005) Vacuolar processing enzyme, an executor of plant cell death. Curr Opin Plant Biol 8:404–408

Harding SA, Oh SH, Roberts DM (1997) Transgenic tobacco expressing a foreign calmodulin gene shows an enhanced production of active oxygen species. EMBO J 16:1137–1144

Harper JF, Breton G, Harmon A (2004) Decoding Ca^{2+} signals through plant protein kinases. Annu Rev Plant Biol 55:263–288

Hatsugai N, Kuroyanagi M, Yamada K, Meshi T, Tsuda S, Kondo M, Nishimura M, Hara-Nishimura I (2004) A plant vacuolar protease, VPE, mediates virus-induced hypersensitive cell death. Science 305:855–858

He C-J, Morgan PW, Drew MC (1992) Enhanced sensitivity to ethylene in nitrogen-starved or phosphate-starved roots of Zea mays L. during aerenchyma formation. Plant Physiol 98:137–142

He C-J, Drew MC, Morgan PW (1994) Induction of enzymes associated with lysigenous aerenchyma formation in roots of *Zea mays* during hypoxia or nitrogen starvation. Plant Physiol 105:861–865

He C-J, Finlayson SA, Drew MC, Jordan WR, Morgan PW (1996a) Ethylene biosynthesis during aerenchyma formation in roots of maize subjected to mechanical impedance and hypoxia. Plant Physiol 112:1679–1685

He C-J, Morgan PW, Drew MC (1996b) Transduction of an ethylene signal is required for cell death and lysis in the root cortex of maize during aerenchyma formation induced by hypoxia. Plant Physiol 112:463–472

Hill RD (1998) What are hemoglobins doing in plants? Can J Bot 76:707–712

Hua J, Meyerowitz EM (1998) Ethylene responses are negatively regulated by a receptor gene family in *Arabidopsis thaliana*. Cell 94:261–271

Huang B, Johnson JW, Nesmith S, Bridges DC (1994) Growth, physiological and anatomical responses of two wheat genotypes to waterlogging and nutrient supply. J Exp Bot 45:193–202

Igamberdiev AU, Baron K, Manac'h-Little N, Stoimenova M, Hill RD (2005) The haemoglobin/nitric oxide cycle: involvement in flooding stress and effects on hormone signalling. Ann Bot 96:557–564

Jackson MB (1985) Ethylene and responses of plants to soil waterlogging and submergence. Annu Rev Plant Physiol 36:145–174

Jackson MB, Armstrong W (1999) Formation of aerenchyma and the processes of plant ventilation in relation to soil flooding and submergence. Plant Biol 1:274–287

Jackson MB, Fenning TM, Jenkins W (1985) Aerenchyma (gas-space) formation in adventitious roots of rice (Oryza sativa L.) is not controlled by ethylene or small partial pressures of oxygen. J Exp Bot 36:1566–1572

Jones A (2000) Does the plant mitochondrion integrate cellular stress and regulate programmed cell death? Trends Plant Sci 5:225–230

Justin SHFW, Armstrong W (1987) The anatomical characteristics of roots and plant response to soil flooding. New Phytol 106:465–495

Kaul RB (1971) Diaphragms and aerenchyma in *Scirpus validus*. Am J Bot 58:808–816

Kawai M, Samarajeewa PK, Barrero RA, Nishiguchi M, Uchimiya H (1998) Cellular dissection of the degradation pattern of cortical cell death during aerenchyma formation of rice roots. Planta 204:277–287

Kawase M (1979) Role of cellulose in aerenchyma development in sunflower. Am J Bot 66:183–190

Kawase M (1981) Effect of ethylene on aerenchyma development. Am J Bot 68:651–658

Keller T, Damude HG, Werner D, Doerner P, Dixon RA, Lamb C (1998) A plant homolog of the neutrophil NADPH oxidase gp91ph_x subunit gene encodes a plasma membrane protein with Ca^{2+} binding motifs. Plant Cell 10:255–266

Kieber JJ, Rothenberg M, Roman G, Feldman KA, Ecker JR (1993) CTR1, a negative regulator of the ethylene-response pathway in *Arabidopsis*, encodes a member of the raf family of protein kinases. Cell 72:427–441

Klessig DF, Ytterberg AJ, van Wijk KJ (2004) Retraction. Cell 119:445

Klinge J (1879) Vergleichend histologische Untersuchung der Gramineen- und Cyperaceen-wurzeln. Mémoires de L'Academie de Sciences de St.Pétersbourg, VIIE Serie, tome XXVI, No 12. St.Pétersbourg

Konings H, Verschuren G (1980) Formation of aerenchyma in roots of *Zea mays* in aerated solutions, and its relation to nutrient supply. Physiol Plant 49:265–270

Kozela C, Regan S (2003) How plants make tubes. Trends Plant Sci 8:159–164

Kükenthal G (1909) Cyperaceae-Caricoideae. In: Engler A (ed) Das Planzenreich, vol Heft 38, IV. 20th edn. Verlag von Wilhelm Engelmann, Lepzig, pp 1–814

Laan P, Berrevoets MJ, Lythe S, Armstrong W, Blom CWPM (1989) Root morphology and aerenchyma formation as indicators of the flood-tolerance of *Rumex* species. J Ecol 77:693–703

Larsen O, Nilsen HG, Aarnes H (1986) Response of young barley plants to waterlogging, as related to concentration of ethylene and ethane. J Plant Physiol 122:365–372

Lee KH, Choi EY, Hyun MS, Kim J-R (2004) Involvement of MAPK pathway in hypoxia-induced up-regulation of urokinase plaminogen activator receptor in a human prostatic cancer cell line, PC3MLN4. Exp Mol Med 36:57–64

Lecourieux D, Rajeva R, Pugin A (2006) Calcium in plant defence-signalling pathways. New Phytol 171:249–269

6 Programmed Cell Death and Aerenchyma Formation Under Hypoxia

Lindermayr C, Saalbach G, Durner J (2005) Proteomic identification of S-nitrosylated proteins in *Arabidopsis*. Plant Physiol 137:921–930

Luan S (2009) The CBL-CIPK network in plant calcium signalling. Trends Plant Sci 14:37–42

Manac'h N, Igamberdiev AU, Hill RD (2005) Hemoglobin expression affects ethylene production in maize cell cultures. Plant Physiol Biochem 43:485–489

Mlejnek P, Prochazka S (2002) Activation of caspase-like proteases and induction of apoptosis by isopentenyladenosine in tobacco BY-2 cells. Planta 215:158–166

Munnik T, Arisz SA, de Vrije T, Musgrave A (1995) G protein activation stimulates phospholipase D signaling in plants. Plant Cell 7:2197–2210

Mühlenbock P, Plaszczyca M, Plaszczyca M, Mellerowicz E, Karpinski S (2007) Lysigenous aerenchyma formation in *Arabidopsis* is controlled by LESION SIMULATING DISEASE1. Plant Cell 19:3819–3830

Petronilli V, Costantini P, Scorrano L, Colonna R, Passamonti S, Bernardi P (1994) The voltage sensor of the mitochondrial permeability transition pore is tuned by the oxidation-reduction state of vicinal thiols. Increase of the gating potential by oxidants and its reversal by reducing agents. J Biol Chem 269:16638–16642

Plowman AB (1906) The comparative anatomy and phylogeny of the *Cyperaceae*. Ann Bot 20:1–30

Quaked F, Rozhon W, Lecourieux D, Hirt H (2003) A MAPK pathway mediates ethylene signaling in plants. EMBO J 22:1282–1288

Rawyler A, Arpagaus S, Braendle R (2002) Impact of oxygen stress and energy availability on membrane stability of plant cells. Ann Bot 90:499–507

Ray JD, Kindiger B, Dewald CL, Sinclair TR (1998) Preliminary survey of root aerenchyma in *Tripsacum*. Maydica 43:49–53

Rentel MC, Lecourieux D, Quaked F, Usher SL, Petersen L, Okamoto H, Knight H, Peck SC, Grierson CS, Hirt H, Knight MR (2004) OXI1 kinase is necessary for oxidative burst-mediated signalling in *Arabidopsis*. Nature 427:858–861

Saab IN, Sachs MM (1996) A flooding-induced xyloglucan endotransglycosylase homolog in maize is responsive to ethylene and associated with aerenchyma. Plant Physiol 112: 385–391

Sagi M, Fluhr R (2001) Superoxide production by plant homologues of the gp91phox NADPH oxidase. Modulation of activity by calcium and by tobacco mosaic virus infection. Plant Physiol 126:1281–1290

Sang Y, Cui D, Wang X (2001) Phospholipase D and phosphatidic acid-mediated generation of superoxide in *Arabidopsis*. Plant Physiol 126:1449–1458

Shabala S (2009) Salinity and programmed cell death: unravelling mechanisms for ion specific signalling. J Exp Bot 60:709–712

Schussler EE, Longstreth DJ (2000) Changes in cell structure during the formation of root aerenchyma in *Sagittaria lancifolia* (Alismataceae). Am J Bot 87:12–19

Solano R, Stepanova A, Chao Q, Ecker JR (1998) Nuclear events in ethylene signalling, a transcriptional cascade mediated by ETHYLENE-INSENSITIVE3 and ETHYLENE-RESPONSE-FACTOR1. Genes Dev 12:3703–3714

Subbaiah CC, Sachs MM (2003) Molecular and cellular adaptations of maize to flooding stress. Ann Bot 91:119–127

Sung YL, Zhao Y, Hong X, Zhai ZH (1999) Cytochrome c release and caspase activation during menadione-induced apoptosis in plants. FEBS Lett 462:317–321

Turner S, Hall M (2000) The *gapped* mutant identifies a common regulatory step in secondary cell wall deposition. Plant J 24:477–488

Van Doorn WG, Woltering EJ (2005) Many ways to exit? Cell death categories in plants. Trends Plant Sci 10:117–122

Virolainen E, Blokhina O, Fagerstedt KV (2002) Ca^{2+}-induced high amplitude swelling and cytochrome c release from wheat (*Triticum aestivum* L.) mitochondria under anoxic stress. Ann Bot 90:509–516

Visser EJW, Bögemann GM (2006) Aerenchyma formation in the wetland plant *Juncus effusus* is independent of ethylene. New Phytol 171:305–314

Vriezen WHVRCPE, Voesenek LAC, Mariani C (1997) A homologue of the *Arabidopsis* ERS gene is actively regulated in *Rumex palustris* upon flooding. Plant J 11:1265–1271

Wille F (1926) Beiträge zur Anatomie des Cyperaceenrhizoms. Beih bot Zbl 43:267–309

Woolley JT (1983) Maintenance of air in intercellular spaces of plants. Plant Physiol 72:989–991

Yoo SD, Cho YH, Tena G, Xiong Y, Sheen J (2008) Dual control of nuclear EIN3 by bifurcate MAPK cascades in C_2H_4 signaling. Nature 451:789–795

Chapter 7
Oxygen Deprivation, Metabolic Adaptations and Oxidative Stress

Olga Blokhina and Kurt V. Fagerstedt

Abstract In this chapter, we discuss the metabolic changes relevant for the production of reactive oxygen and nitrogen species (ROS and RNS) in plant tissues during oxygen deprivation. It is notable too, that at times the oxidative damage does not take place during the oxygen deficiency period but only after the restoration of normal oxygen supply to the tissues. This is mainly due to the fact that ROS may be formed immediately after oxygen re-enters the tissues. The level of oxygen in the tissues naturally depends on the outside concentration and diffusion rate but is also under metabolic control. Two hypotheses on the regulation of internal O_2 concentration in the cells through the control of respiration rely on a regulation of glycolytic pathway and pyruvate availability and mitochondrial electron transport chain by NO and ROS balance. Both adaptive strategies aim to decrease the respiratory capacity and to postpone complete anoxia. This chapter also describes the interaction of the many antioxidants found in plant tissues and oxidative stress and oxidative damage caused by waterlogging and oxygen deprivation. If the antioxidative protection is still capable of detoxifying the ROS formed, damage is minimal, but if not, then considerable damage can take place very suddenly. Many of the ROS and RNS species act as signalling agents acting in the regulation of metabolic events leading to tolerance or, e.g. in the case of aerenchyma development, into programmed cell death. The balance between O_2 transport to hypoxic tissues, regulatory adaptations, anoxic metabolites, ROS–RNS chemistry and signalling, determines the survival of plants under waterlogging and oxygen deprivation.

O. Blokhina and K.V. Fagerstedt (✉)
Division of Plant Biology, Department of Biological and Environmental Sciences, University of Helsinki, Viikki Biocenter 3, POB 65, 00014 Helsinki, Finland
e-mail: olga.blokhina@helsinki.fi; kurt.fagerstedt@helsinki.fi

S. Mancuso and S. Shabala (eds.), *Waterlogging Signalling and Tolerance in Plants*,
DOI 10.1007/978-3-642-10305-6_7, © Springer-Verlag Berlin Heidelberg 2010

Abbreviations

ADH	Alcohol dehydrogenase
AOX	Alternative oxidase
COX	Cytochrome oxidase
ETC	Electron transport chain
FFA	Free fatty acids
LP	Lipid peroxidation
NO	Nitric oxide
$^1O_2$	Singlet oxygen
$O_2^{-\bullet}$	Superoxide anion
$ONOO^-$	Peroxynitrite
PFK-PPi	PPi-dependent phosphofructokinase
PLD	Phospholipase D
PEPC	Phosphoenolpyruvate carboxylase
PPDK	Pyruvate phosphate dikinase
RNS	Reactive nitrogen species
ROS	Reactive oxygen species
SOD	Superoxide dismutase
UCP	Mitochondrial uncoupling protein

7.1 Introduction

Almost two decades of oxidative stress studies in plants under diverse biotic and abiotic stress factors, has yielded a concept of ROS being ubiquitous stress markers and signalling species (Bailey-Serres and Chang 2005; Miller et al. 2008; Van Breusegem et al. 2008). Oxygen deprivation stress stands somewhat apart in the universal and quite concise picture of stress responses. Intrinsic contradiction between low oxygen concentration on the one hand, and the well documented ROS production under these conditions on the other hand, has promoted studies on the biochemistry underlying ROS formation under oxygen deprivation and during reoxygenation (Yan et al. 1996; Blokhina et al. 1999; Biemelt et al. 1998; Biemelt et al. 2000; Blokhina et al. 2001, 2003; Fukao and Bailey-Serres 2004). Such biochemical adjustments are closely related to plant survival under hypoxia and inherently connected with the activation of remaining oxygen. Among the main processes relevant for occurrence of oxidative stress under hypoxia are preservation of energy (Greenway and Gibbs 2003) and regulation of the internal O_2 concentration (Borisjuk et al. 2007; Zabalza et al. 2009). Moreover, during the last years a vast amount of evidence has accumulated on the regulatory role of reactive nitrogen species (RNS) in hypoxic metabolism (Qiao and Fan 2008). These metabolic adjustments have a common crosspoint – mitochondrial metabolism affected by the lack of oxygen and by hypoxic metabolites (Bailey-Serres and Voesenek 2008; Benamar et al. 2008; Igamberdiev and Hill 2009).

Multiple routes for hypoxic ATP production and consumption have been elucidated recently: pyrophosphate-dependent glycolysis (Huang et al. 2008), anaerobic nitrite-dependent ATP synthesis (Stoimenova et al. 2007), and a reverse reaction for ATP synthesis – anaerobic ATP hydrolysis (St-Pierre et al. 2000). The above-mentioned NO can exert a dual effect on cell energetics, depending on the target and localization: inhibition of key mitochondrial enzymes cytochrome oxidase (COX), aconitase and upregulation of alternative oxidase, sustaining of NADH turnover via hemoglobin-dependent reaction cascade, and termination of the lipid peroxidation (LP) cascade (discussed in more detail in the text).

Two hypotheses on the regulation of internal O_2 concentration in the cells through the control of respiration rely on a regulation of glycolytic pathway and pyruvate availability (Geigenberger 2003; Zabalza et al. 2009) and mitochondrial ETC by NO and ROS balance (Borisjuk et al. 2007). Both adaptive strategies aim to decrease the respiratory capacity and to postpone complete anoxia. They also emphasize the key role of adenylate energy charge of the tissue (Bailey-Serres and Voesenek 2008). These routes are inherently connected with the mitochondria and hypoxic oxidative stress. However, several non-mitochondrial enzyme reactions are capable of producing ROS (and RNS) in a stress-dependent manner (Blokhina et al. 2003). The phenomenon of oxidative stress under hypoxia has found its affirmation in microarray studies on the whole genome response to hypoxia. Indeed, expression of a wide range of ROS- and redox-related transcripts is induced by the lack of oxygen (Klok et al. 2002; Branco-Price et al. 2005; Loreti et al. 2005; Branco-Price et al. 2008; van Dongen et al. 2009).

Elegant new methods for the determination of oxygen levels inside plant organs and tissues have given new possibilities for accurate description of the regulation of respiration very precisely. This is discussed further in Sect. 7.4.

7.2 Anoxia: Metabolic Events Relevant for ROS Formation

7.2.1 "Classic" Metabolic Changes Under Oxygen Deprivation Related to ROS Formation

Studies on morphological and metabolic adaptations to oxygen deprivation stress in plants have a long history, and are very well documented, with a number of excellent reviews available on the subject (Gibbs and Greenway 2003; Greenway and Gibbs 2003; Visser et al. 2003; Fukao and Bailey-Serres 2004; Bailey-Serres and Voesenek 2008; Jackson 2008; Sairam et al. 2008). Here we will only discuss changes in metabolic events brought about by oxygen deprivation which can potentially support the progression of oxidative stress in oxygen-limited environments and on reoxygenation.

An induction of alcoholic fermentation can be observed under normoxic conditions as part of regulatory mechanism controlling O_2 consumption and, hence, O_2 concentration in the tissue (Zabalza et al. 2009). Controlling pyruvate availability for respiration underlies the organism's ability to avoid complete depletion of O_2 i.e. anoxia. Energy status of the tissue has been shown to be the key regulatory switch in the regulation of the pyruvate level and, hence, the inhibition of respiration rate (Zabalza et al. 2009), Fig. 7.1. Threshold ATP concentration has been also discussed as a regulatory trigger for membrane integrity, i.e. when ATP level falls below 10 μM membrane lipids are hydrolysed to FFA (Rawyler et al. 2002), the process which can have detrimental consequences for mitochondrial functioning (see Sect. 7.2.2).

It seems that multiple mechanisms for glycolytic flux regulation exist in plants and control adenylate energy charge, O_2 availability, the rate of mitochondrial respiration, and ROS formation in ETC. A switch to pyrophosphate-dependent glycolysis due to upregulation of cytosolic reversible PPi-dependent phosphofructokinase (PFK-PPi) and pyruvate phosphate dikinase (PPDK) in anoxia tolerant plants has been suggested recently as an alternative to conventional glycolysis. Theoretically five ATP vs two ATP molecules are produced in PPi-dependent and conventional glycolysis, respectively (Huang et al. 2008). The dynamic association of glycolytic enzymes with mitochondria, glycolytic support of mitochondrial respiration and regulatory substrate channelling has been shown recently (Graham et al. 2007). Moreover, since both PPi-dependent enzymes are reversible, the increased level of PPi can affect the mitochondrial energy production directly through an engagement of proton pumping inner mitochondrial membrane pyrophosphatase (Vianello and Macrì 1999). Diminished PPi level in transgenic potato expressing *E. coli* pyrophosphatase has resulted in decreased glycolytic activity, lower ATP levels and negatively affected overall survival of plants (Mustroph et al. 2005). In addition to its role in NO formation, nitrite accumulated under oxygen deprivation can act as an electron acceptor in ETC, providing the sink mechanism for reduced NAD(P)H and sustaining ATP synthesis comparable with that of glycolysis (Stoimenova et al. 2007). The additional ATP formed would slow down the demand for mitochondrial ETC, and the concurrent production of mitochondrial ROS.

Regulation of cytoplasmic pH is another control point for anoxic metabolism and a probable upregulation of ROS production. In mammalian models cytoplasmic acidosis accelerated the peroxidation of membrane lipids in Fe^{2+}-dependent manner. Reduced pH favours mobilisation of ferrous ions (e.g. from ferritin) and, therefore, promotes the formation $O_2^{-\bullet}$ in Haber–Weiss reaction or H_2O_2 in Fenton reaction (Pryor et al. 2006; Hassan et al. 2009). It has been shown that more alkaline pH (in the range of pH 7.0–8.8) stimulates ROS production via mitochondrial complex 1 (Turrens and Boveris 1980) and, hence, cytoplasmic acidification (down to pH 6.8) will render ROS formation less probable. However, acidification promotes the formation of another reactive species − NO: non-enzymatically from exogenous nitrite in plants (Bethke et al. 2004) and enzymatically via cytosolic nitrate reductase and plasma membrane-bound nitrite-NO reductase (see Sect. 7.3.2).

7 Oxygen Deprivation, Metabolic Adaptations and Oxidative Stress

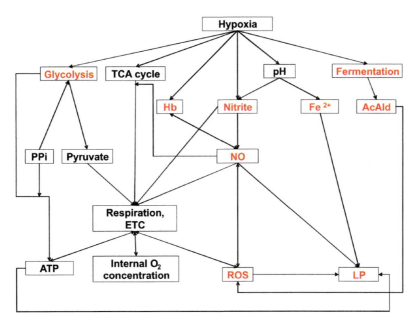

Fig. 7.1 Anoxia-induced metabolic changes regulating ATP level, internal O_2 concentration and favourable for ROS and RNS formation. Upregulated metabolic pathways and accumulating metabolites are highlighted in *red*. AcAld, acetaldehyde; ETC, electron transport chain; LP, lipid peroxidation; Hb, hemoglobin; PPi, pyrophosphate. Enzymes are omitted for clarity of the picture. Decreasing availability of O_2 leads to acidification of cytoplasm (pH down to 6.8) and repression of TCA cycle. Non-functional TCA cycle is unable to sustain respiration when O_2 concentration is not yet limiting for COX. However, accumulating succinate can support ROS-producing reverse e^- transport from mitochondrial complex II to complex I. Acidification optimizes enzymatic production of nitrite and NO and promotes the release of ferrous ions from proteins. NO facilitates this process by acting on FeS clusters of enzymes. Nitrite can serve as an e^- acceptor in mitochondrial ETC, sustaining anoxic ATP synthesis. Accumulating acetaldehyde, one of the products of induced fermentation pathway, serves as a substrate for xanthine oxidoreductase, an enzyme capable for both NO and $O_2^{-\bullet}$ generation. NO formed affects both negatively and positively a number of reactions: inhibitis aconitase (a TCA cycle enzyme), catalase and COX – the latter two enzymes are important for the control of ROS level. Positive effects of NO under anoxia are represented by its interaction with hypoxically-induced hemoglobin in a NADH regenerative cycle and by transcriptional upregulation of AOX. Pyrophosphate-dependent glycolysis is efficient (in terms of theoretical ATP production) in supporting energy status and relies on upregulation of cytosolic reversible PPi-dependent phosphofructokinase and pyruvate phosphate dikinase. Induced conventional (as opposite to PPi-induced) glycolytic pathway is responsible not only for ATP synthesis but also for the regulation of pyruvate available for respiration. High pyruvate stimulates respiration and facilitates the onset of anoxia. Control of pyruvate level can be achieved through pyruvate kinase and phosphoenolpuryvate carboxylase, entry of pyruvate into TCA cycle is regulated by pyruvatedehydrogenase complex. The respiration rate is adjusted according to ATP level and internal O_2 concentration. ETC is the source of ROS when not fully operational. ROS and NO produced in different hypoxia-induced reactions can exert feedback regulation of respiration rate and other signalling functions, directly oxidize biological molecules. The oxidative damage by ROS is exemplified on the graph by lipid peroxidation which is induced by Fe^{2+} ions and also embraces NO-induced lipid modification. ATP is essential for the integrity of lipids and for prevention of FFA release

7.2.2 Changes in Lipid Composition and Role of Free Fatty Acids Under Stress

Functional integrity of membrane lipids, the degree of fatty acid unsaturation, and causes of free fatty acid (FFA) liberation and regulation of LP, are issues closely connected to ROS and RNS metabolism. The maintenance of membrane lipid integrity is an important factor in plant survival under oxygen deprivation. Anoxia-tolerant plants (i.e. rice and *Echinochloa phyllopogon*) are able to synthesize lipids under strict anoxia. Incorporation of radioactive label has been detected in phospholipids, glycolypids and neutral lipids but mainly in saturated fatty acids (Generosova and Vartapetian 2005). The saturation of double bonds of esterified FA has been discussed as an adaptive property (Generosova and Vartapetian 2005). There are two trends in lipid metabolism under anoxia which are important when considering ROS chemistry: increased degree of saturation and dependence of membrane integrity on ATP levels (Chirkova et al. 1989; Pavelic et al. 2000).

Polyunsaturated fatty acids (PUFA) are the targets of lipophilic RNS and ROS (discussed below) accumulating in the membranes. Increased FA unsaturation will affect the progression of both LP and nitration. These oxidative reactions can result in a number of molecular species that restructure the membrane (conjugated dienes and lipid hydroperoxides) that make the membrane less ordered and more hydrophilic). Both FA unsaturation and FA–NO interaction (discussed below) negatively affect LP (O'Donnell and Freeman 2001; Blokhina et al. 2003; Freeman et al. 2008). Falling ATP concentration under anoxia has been shown to act as a threshold regulatory switch for membrane integrity: when ATP concentration decreases below 10 μM the integrity of membrane lipids is no longer preserved and they are hydrolysed to FFA (Pavelic et al. 2000).

Liberation of FFA is indicative of severe membrane damage and manifests cell death. This process also affects mitochondria: FFAs are known to uncouple oxidative phosphorylation via a protonophore mechanism: FFA^- anion forms a complex with $\cdot H^+$ and as a neutral species moves across the membrane via the "flip-flop" mechanism (Skulachev 1998; Arcisio-Miranda et al. 2009). In case of "mild" uncoupling, when the concentration of uncoupler is extremely low, the process can exert an antioxidative function by decreasing the probability of $O_2^{-\bullet}$ formation in the ETC (Skulachev 1998).

In animal systems cytochromes P450 are responsible for epoxidation of non-esterified linolenic acid, and the resulting epoxide has been shown to support mitochondrial respiration under anoxia/reperfusion (Nowak et al. 2004). In plants H_2O_2-dependent fatty acid epoxygenase activity has been detected in broad beans. Interestingly, the products of this reaction – epoxy fatty acids – have been shown to accumulate under the pathogen attack and to exhibit antifungal activity (Hamberg and Fahlstadius 1992).

Close association between ROS, RNS and phospholipase D (PLD) under abiotic stress conditions has been shown (Testerink and Munnik 2005). Indeed, upregulated PLD and its product phosphatidic acid induced plasma membrane NADPH

7 Oxygen Deprivation, Metabolic Adaptations and Oxidative Stress

oxidase under excess copper stress and, hence, were considered responsible for H_2O_2 accumulation and LP (Yu et al. 2008). The formation of adventitious roots in cucumber – one of the main anatomic features in the adaptation to oxygen deprivation – appeared to be mediated by auxin and NO induction of PLD and a concurrent accumulation of phosphatidic acid (Lanteri et al. 2008). Among other activation targets of phosphatidic acid in plants are protein kinases and phosphoenolpuryvate carboxylase (PEPC) (Testerink and Munnik 2005). The latter enzyme catalyzes the phosphorylation of oxaloacetate to form phosphoenolpuryvate and is upregulated during flooding. PEPC is involved in the control of pyruvate level, a metabolite shown to regulate internal O_2 concentration under oxygen deprivation (see above) (Zabalza et al. 2009). Increased PEPC activity has been discussed also as an adaptive mechanism for improved CO_2 assimilation under flooding (Yordanova and Popova 2007).

7.2.3 Modification of Lipids: LP

Oxidative modification of lipids, LP, has long been recognized as the main cause of membrane destabilisation and injury under a range of stimuli in both animals and plants. LP is a result of action of ROS on PUFA in membrane phospholipids. The hydroxyl radical, OH•, is considered as the initiation species for LP due to its high reactivity. Both OH• radical and singlet oxygen $^1O_2$ can react with methylene groups of PUFA producing several products differing in their lifetime: conjugated dienes, lipid peroxy radicals and lipid hydroperoxides (Blokhina et al. 2003). A reaction of ROS (hydroxyl radical and singlet oxygen) with methylene groups of PUFA results in the rearrangement of the double bond and the formation of conjugated dienes (two double bonds sepatrated with a single bond), lipid peroxy (L–OO•) and alkoxy (L–O•) radicals and lipid hydroperoxides (LOOH). In turn, lipid radical species propagate or initiate (branch) a new chain of peroxidative reactions in membrane lipids (Buettner 1993).

ROS and products of LP can build up under hypoxic conditions (Biemelt et al. 2000; Blokhina et al. 2001; Santosa et al. 2007). E.g. the detection of ethane emission from submerged rice seedlings suggests oxidative damage and the formation of ROS already under hypoxia (Santosa et al. 2007) as well as apoplastic H_2O_2 formation (Blokhina et al. 2001). Interaction between small monomeric GTP-binding protein Rop and its negative regulator RopGAP4 (GTPase activating protein) has been shown to control H_2O_2 level and affect gene expression in hypoxic *Arabidopsis* seedlings in Ca^{2+}-dependent manner. Rop-induced increase in H_2O_2 led to the induction of ADH (ethanolic fermentation) and RopGAP4 (a negative regulator of Rop) expression, thus sustaining the negative feedback regulation of oxidative stress, ethanolic fermentation and survivial under hypoxia (Baxter-Burrell et al. 2002; Van Breusegem et al. 2008). However, more prominent oxidative damage occurs upon re-admission of O_2 (Yan et al. 1996; Biemelt et al. 1998; Blokhina et al. 1999; Yordanova and Popova 2007).

The main chain breaking antioxidant in biological membranes undergoing LP is tocopherol. NO can react with alkoxy and peroxy radicals thus terminating the chain reaction of LP (Wink and Mitchell 1998).

$$LOO^\bullet + NO \rightarrow LOONO$$

In soybean, NO donor treatment decreased the level of lipid radicals in chloroplast membranes; however, ONOO treatment resulted in an increase in LP (Jasid et al. 2006). Lipid nitration and the physiological and signalling role of nitro fatty acids have been intensively studied in animal models mostly in connection with vasorelaxation, inflammatory signalling, oxidative injury and apoptosis (Freeman et al. 2008; Rubbo and Radi 2008). The process is sustained by the lipophilicity of molecular O_2 and NO: they can both accumulate three to fourfold in the lipid phase and react to produce oxidants and nitrating species (Moller et al. 2005; Freeman et al. 2008). To our knowledge there are no reports on the existence and role of nitrated lipids in plants. However, some hypoxia-related metabolic changes allow the speculation on the relevance of this process in plants. Accumulation of nitrite and its evolution to NO associated with oxygen deprivation can lead to the formation of secondary NO-derived species ($^\bullet NO_2$, $ONOO^-$, N_2O_2). The onset of cytoplasmic acidification not only favours non-enzymatic formation of NO from nitrite (Bethke et al. 2004) but also promotes protonation of NO_2^- to nitrous acid HNO_2 (a membrane permeable compound) with consequent nitration of proteins and unsaturated fatty acids (Freeman et al. 2008). Increased probability of Fenton reaction under hypoxia relies on the liberation of ferrous ions due to acidification (Navarre et al. 2000) and Fenton chemistry sustains nitrite oxidation to radical $^\bullet NO_2$ (Freeman et al. 2008). Depending on O_2 concentration, $^\bullet NO_2$ can oxidize (LP) or nitrate unsaturated fatty acids in vitro. The nitration products are represented by diverse species and can include alkyl nitrites LONO and nitrolipids LNO_2 (O'Donnell and Freeman 2001). Signalling properties of nitrolipids and nitro-fatty acids rely on their chemistry and encompass NO release and nitroalkylation of protein thiol and histidine residues (Freeman et al. 2008).

7.3 ROS and RNS Chemistry Overview and Sources of Formation Under Lack of Oxygen

7.3.1 Reactive Oxygen Species

During the last two decades, chemistry of oxygen activation and its consequences for biological systems have been extensively reviewed (Mittler 2002; Sorg 2004; Blokhina and Fagerstedt 2006; Halliwell 2006; Pryor et al. 2006; Halliwell 2009,). The first step in O_2 activation (one electron reduction which yields superoxide $O_2^{-\bullet}$) requires energy; the subsequent steps can proceed spontaneously in the presence of proton and/or electron donors. The superoxide $O_2^{-\bullet}$ formed is membrane impermeable and is dismutated to H_2O_2 by compartment specific superoxide dismutase (SOD) isoforms (Bowler et al. 1992). Due to restricted intermembrane mobility, ROS formed in the membrane-enclosed cellular compartments react "on-spot" with their oxidation targets. Therefore, distinct sites of ROS formation under different environmental stimuli might bring about the specificity of ROS signal. However, H_2O_2 is not a radical, is uncharged, has relatively long lifetime and is able to penetrate the membrane. This property makes H_2O_2 a good candidate for long-distance or intercompartment signalling species. An extremely potent oxidant hydroxyl radical $OH^\bullet$ is considered as the most reactive ROS. It is formed in a transition metal-catalysed Fenton reaction, during decomposition of ozone in the presence of protons in the apoplast, and is considered as the species initiating the chain reaction of LP (Ohyashiki and Nunomura 2000; Pryor et al. 2006). Singlet oxygen ($^1O_2$), where one of the electrons on the outer electron sheath has changed its spin, is formed in tissues under UV-exposure and during photoinhibition in chloroplasts (Triantaphylidès and Havaux 2009). Of the ROS, hydrogen peroxide and superoxide are both produced in a number of cellular reactions including the Mehler reaction in the chloroplasts, the iron catalyzed Haber–Weiss and Fenton reactions, photorespiration and by various enzymes such as lipoxygenases, peroxidases, NADPH oxidase, xanthine oxidase and amine oxidase (Bolwell and Wojtaszek 1997; Bolwell et al. 2002; Blokhina et al. 2003; Blokhina and Fagerstedt 2006; Pryor et al. 2006).

Several enzyme reactions have been shown to produce ROS (via the main route or as a side reaction) under the lack of oxygen (Blokhina et al. 2003). Xanthine oxidoreductase can use acetaldehyde accumulating under hypoxia (fermentation product) and hypoxanthine (ATP catabolism product) as an e^- donor to perform primary activation of O_2 and, hence, produce $O_2^{-\bullet}$ and H_2O_2 (Godber et al. 2000; Harrison 2002).

The formation of ROS has been detected virtually in all cellular compartments: mitochondria, chloroplast, peroxisomes, cytoplasm (P450), plasma membrane (NADPH oxidase) and apoplast. For updates on ROS and RNS formation in peroxisomes, see the work by del Rio et al. (2006), and for mitochondria and chloroplasts, see Moller et al. (2007) and Rinalducci et al. (2008).

7.3.2 Reactive Nitrogen Species

The chemical properties of nitric oxide, NO, make this gas a good candidate for a signalling molecule. NO can freely penetrate the lipid bilayer and, hence, can be transported within the cell. NO is quickly produced on demand via inducible enzymatic or non-enzymatic routes. Due to its free radical nature (one unpaired electron) NO has a short half-life (in order of seconds), and can be removed easily when no longer needed (Lamattina et al. 2003; Neill et al. 2003; Besson-Bard et al. 2008; Qiao and Fan 2008). NO can also react either with O_2 or $O_2^{-\bullet}$. The end products, NO_2, N_2O_2, and peroxynitrite $ONOO^-$ all have deleterious consequences in biological systems (Wink and Mitchell 1998).

Metabolic alterations brought about by oxygen deprivation provide a possibility for increased NO levels in hypoxic tissues. At low oxygen tensions NO-generating activity of xanthine oxidoreductase is enhanced. Interestingly, under normoxic conditions xanthine oxidoreductase is capable of both NO and $O_2^{-\bullet}$ formation with consequent production of $ONOO^-$ (Godber et al. 2000). It has been shown that under oxygen deprivation the accumulated nitrite can be converted to NO by several hypoxically induced enzymes, namely cytosolic nitrate reductase (can use nitrite as a substrate and convert it to NO, induced several fold under hypoxia) (Drew 1997; Yamasaki and Sakihama 2000; Rockel et al. 2002), plasma membrane-bound nitrite-NO reductase (acidic conditions under anoxia are favourable for this enzyme with pH 6.1 optimum, hemoglobin (Hb) can act as a physiological e^- donor for this reaction) (Stohr and Stremlau 2006). Indeed, accumulation of NO has been shown in hypoxic alfalfa root cultures and maize cell cultures (Dordas et al. 2004). Anoxic mitochondria are capable of nitrite reduction to NO, in a process specific for root mitochondria and elevated under the lack of O_2 (Gupta et al. 2005; Planchet et al. 2005).

The direct effects of NO on biological targets include reduction of free metal ions or oxidation of metals in protein complexes such as hemoglobin, and Fe-nitrosyl formation resulting in activation of guanylate cyclase and hemoxygenase, inhibition of P450, cytochrome c oxidase, aconitase and catalase, and down-regulation of ferritin (Wink and Mitchell 1998). NO exerts an inhibitory effect on aconitase, a TCA cycle enzyme which catalyses the reversible conversion of citrate to isocitrate. A cytoplasmic isoform of aconitase is also affected: NO promotes the release of iron-sulphur cluster from the active center of the enzyme (Navarre et al. 2000). Possible release of ferrous ions from the cluster, e.g. upon cytoplasmic acidification, can enhance ROS production. NO intervention with TCA cycle enzyme would also have an impact on the energy status of the cell, already affected by COX inhibition. The NO inhibitory effect on COX in potato tuber mitochondria has been shown recently (de Oliveira et al. 2008). Therefore, not only the TCA cycle enzymes but also ETC are targets for NO regulation also under oxygen deprivation stress.

On the other hand, reactions of NO with Hb allow the maintenance of NAD^+ levels for the needs of glycolysis under hypoxic conditions (Igamberdiev and Hill 2004; Igamberdiev et al. 2005). Overexpression of hypoxically induced class 1 hemoglobin in maize cell cultures has resulted in the maintenance of the energy status, and these cells showed less induction of ADH and were able to maintain the redox status of the cells. Transformed alfalfa roots overexpressing Hb have lower levels of NO (Dordas et al. 2004). Hence, the regulation of NO level under oxygen deprivation can be achieved in plants via interaction with class 1 non-symbiotic hemoglobins through several routes. In a reaction with oxyhemoglobin to form nitrate and methemoglobin (Fe^{3+}) with the latter being reduced to hemoglobin (Fe^{2+}) in an NADPH-depending reaction (Igamberdiev and Hill 2004). Another route is interaction of NO with deoxyhemoglobin to form nitrosylhemoglobin (Dordas et al. 2003). Under low oxygen tension, nitrosylhemoglobin will represent a significant part of the Hb pool. The beneficial role of nitrate and nitrite supplementation under anoxia has been attributed to cytoplasmic alkalinisation and to NADH recycling (Stoimenova et al. 2003; Libourel et al. 2006), moreover, anoxic mitochondria have been shown to synthesize ATP in the presence of NAD(P)H and nitrite with concurrent production of NO (Stoimenova et al. 2007).

7.3.3 Plant Mitochondria as ROS Producers: Relevance for Oxygen Deprivation Stress

Aerobic mitochondrial metabolism relies on electron transport along the chain of inner mitochondrial membrane-associated carriers and proton extrusion to create the electrochemical gradient which is the driving force for ATP synthesis with O_2 as terminal e^- acceptor. This tightly coupled redox system is extremely sensitive to inhibition and/or modification of its components by pharmacological agents and different stress factors (Noctor et al. 2007). The redox misbalance leads to over reduction of e-carriers, to e-leakage and formation of ROS. Mitochondria are able to produce superoxide anion $O_2^{-\bullet}$ and the succeeding H_2O_2 (Moller et al. 2007; Murphy 2009) due to electron leakage at the ubiquinone site – ubiquinone: cytochrome b region (Gille and Nohl 2001; Andreyev et al. 2005) – and at the matrix side of complex I (NADH dehydrogenase) (Chakraborti et al. 1999; Moller 2001). Hydrogen peroxide generation by higher plant mitochondria and its regulation by uncoupling of electron transport chain and oxidative phosphorylation have been demonstrated (Braidot et al. 1999; Grabel'nykh et al. 2006).

At the same time, mitochondria can be considered as a tool for ROS level regulation/elimination through AOX, plant uncoupling protein (UCP), and ATP-sensitive plant mitochondrial potassium channel (PmitoKATP) (Grabel'nykh et al. 2006; McDonald and Vanlerberghe 2007; Pastore et al. 2007). An antioxidant role has recently been suggested for mitochondrial uncoupling protein (UCP) which transports fatty acid anions from the inner to the outer leaflet of the membrane (Goglia and Skulachev 2003; Grabel'nykh et al. 2006). First, uncoupling itself lowers mitochondrial ROS production, and second, it is hypothesized that UCP is able to electrophoretically transport fatty acid hydroperoxides from mitochondrial matrix to the intermembrane space (Goglia and Skulachev 2003). Such extrusion preserves mtDNA and matrix proteins from contact with intermediates of LP. Indeed, overexpression of *Arabidopsis* uncoupling protein encoded by AtUCP1 leads to increased oxidative stress tolerance in tobacco (Brandalise et al. 2003).

Mitochondria are also responsible for retrograde ROS signalling. Importance of mitochondrial ETC complexes in the regulation of ROS-mediated abiotic stress responses have been confirmed recently by inactivation and concurrent complementation of mitochondrial pentatricopeptide repeat (PPR) domain protein, PPR40 (Zsigmond et al. 2008). The protein is associated with complex III and its malfunction resulted in ROS accumulation, SOD activation, LP and altered induction of stress-responsive genes, e.g. AOX (Zsigmond et al. 2008).

The alternative oxidase (AOX) present in plant mitochondria catalyzes four-electron reduction of O_2 by ubiquinone and, hence, competes for the electrons with the main respiratory chain. Control of H_2O_2 formation in mitochondrial ETC is one of the functions suggested for AOX. Antisense suppression of AOX in tobacco has resulted in ROS accumulation, while overexpression lead to decreased ROS levels (Maxwell et al. 1999). Under hypoxia the expression of AOX protein has been shown to be translationally regulated, dependent on O_2 concentration (Szal et al. 2003) and activated by pyruvate (Millenaar and Lambers 2003; McDonald and Vanlerberghe 2007).

Mitochondria are capable of e^- transport from complex II to complex 1 (reverse electron flow) supported by succinate, a substrate for complex II (Andreyev et al. 2005). This process has been shown to increase ROS ($O_2^{-\bullet}$) production at complex I and is regulated by ATP hydrolysis via mitochondrial ATPase confirmed by sensitivity to inhibitors of complex 1 (rotenone) and ATPase (oligomycin) (Turrens and Boveris 1980).

Under anoxic conditions mitochondrial ATPase can switch to ATP hydrolysis to maintain the decreasing proton gradient, as is the case in the skeletal muscles of anoxia tolerant frogs (St-Pierre et al. 2000). Under oxygen-limited conditions ATP use in this model can account for 9% from total ATP consumption (St-Pierre et al. 2000). To avoid the complete depletion of already limited ATP supply the enzyme is regulated by inhibitory subunit IF1 of ATPase. The inhibition occurs upon binding of IF1 to F1 subunit, is reversible and takes place under the conditions which limit oxidative phosphorylation (Rouslin 1991). In plants IF1 protein has been isolated from potato (Polgreen et al. 1995) and nucleotide sequences of cDNA for two isoforms of IF1 in rice has been described (Nakazono et al. 2000).

Besides ATP hydrolysis, plant mitochondria exhibit a number of distinct metabolic features under anoxia (reviewed by Igamberdiev and Hill 2009). Anoxic mitochondria have been shown to produce ATP utilizing NADH and nitrite (as an e^- acceptor) with concurrent production of NO (Stoimenova et al. 2007). The organelles are involved in regulation of NO and NADH turnover via hypoxically induced hemoglobins to maintain redox and energy status (Igamberdiev et al. 2005; Hebelstrup et al. 2007). Apparently, the level of NO formed in anoxic mitochondria has to be tightly regulated, since NO affects the activity of ETC and matrix enzymes (as discussed above) and, besides inhibition can increase the Km for oxygen of COX (Cooper et al. 2003) and transcriptionally upregulate AOX (Huang et al. 2002). Nitrite-dependent NO production under hypoxia and the fine regulation of NO and ROS levels in mitochondria have recently been suggested as a main regulatory mechanism slowing down the respiration and preventing tissues from entering complete anoxia (Borisjuk et al. 2007; Benamar et al. 2008).

7.4 O₂ Fluxes in Tissues and Factors Affecting O₂ Concentration In Vivo

At first we have to pay attention to the definition of critical oxygen pressure for respiration (COPR). This is the lowest oxygen concentration to support maximum respiratory rate and below which the rate of O_2 consumption rapidly declines. It has been estimated that the determination of COPR is very difficult through measurements of O_2 consumption rates in intact or excised roots in oxygen free media, and hence a mathematical model and a polarographic method for determining COPR has been published (Armstrong et al. 2009).

In a recent study on oxygen distribution in developing seeds Borisjuk and Rolletschek (2009) argue that NO may be a key player in the oxygen balancing process in seeds and especially in the avoidance of anoxia and fermentation (Borisjuk et al. 2007; Borisjuk and Rolletschek 2009). Increasing NO concentration (synthesised from nitrite) in pea (*Pisum sativum*) and soybean (*Glycine max*) lead to reduced O_2 consumption and decreased ATP availability, and hence biosynthetic activity (Borisjuk et al. 2007; Borisjuk and Rolletschek 2009). This kind of balancing of oxygen diffusion and consumption via NO can be a mechanism for avoidance of anoxia in seeds.

In a very interesting and fundamental study on oxygen consumption rates and the induction of ethanolic fermentation in pea (*P. sativum*) and thale grass (*Arabidopsis thaliana*) alcohol dehydrogenase knockout lines, it has been shown that fermentation is primarily induced by a drop in the energy status of the cells and not by the oxygen concentration *per se* (Zabalza et al. 2009). The fact that O_2 consumption of isolated mitochondria is linear to practically zero O_2 levels due to the high O_2 affinity of the terminal electron acceptor COX, in comparison with the situation in intact tissues, indicates that mitochondrial respiration is regulated (possibly by substrate availability in the TCA cycle). This regulation results in slowing down of mitochondrial respiration under O_2 concentration far from that limiting COX activity. Similarly, in an earlier study by van Dongen and coworkers (2003), oxygen concentrations inside *Ricinus communis* stems were low (7%) even while grown under normal oxygen partial pressures (21%) and resulted in adaptive changes in phloem metabolism and function (van Dongen et al. 2003).

An interesting feature in plant roots grown in stagnant solutions has been noted: A barrier for radial oxygen loss developing in the basal parts of the roots preventing O_2 diffusion into the surrounding O_2 deficient solution (Visser et al. 2000; Colmer 2003; Armstrong et al. 2009). The barrier seems to be constitutive and "tighter" in monocotyledonous species than in dicotyledons but this result has to be confirmed with a wider range of species (Visser et al. 2000). The barrier is formed of suberised cell walls in the exodermis of the roots in rice (Fagerstedt, unpublished data). The ability of anoxia tolerant plant species to maintain an adequate oxygen supply or to restrict oxygen loss from the tissues due to the formation of oxygen-impermeable barrier would certainly imply the formation of ROS or RNS. The reduced cellular environment, together with O_2 availability (and Fe^{2+} liberation) create of favourable conditions for oxidative reactions producing ROS.

In a recent study on root and shoot relation, it has been noticed that root hypoxia leads to increased glycolytic flux and ethanolic fermentation in the roots but not in the leaves of grey poplar (*Populus* x *canescens*) (Kreuzwieser et al. 2009). Similarly, various biosynthetic processes were downregulated in the roots but not in the leaves and shoot growth was not hindered, which suggests that increased glycolytic flux is enough to maintain shoot growth (Kreuzwieser et al. 2009).

7.5 Microarray Experiments in the Study of Hypoxia-Associated Oxidative Stress

The availability of microarray technology advanced the investigation of the whole-genome response to oxygen deprivation stress (Klok et al. 2002; Branco-Price et al. 2005; Liu et al. 2005; Loreti et al. 2005; Lasanthi-Kudahettige et al. 2007; Branco-Price et al. 2008; van Dongen et al. 2009). In a number of studies both metabolic and transcriptomic changes have been assessed. Microarray data have confirmed the upregulation of transcripts coding for anaerobic proteins: about 20 proteins which function mainly in sugar metabolism, glycolysis and fermentation (sucrose synthase, fructokinase, glyceraldehyde-3-phosphate dehydrogenase, puryvate decarboxylase 1 and 2, alcohol dehydrogenase 1, lactate dehydrogenase 1, alanine aminotransferase). Microarray studies have also revealed a number of novel transcripts implicated in signalling (ethylene receptor ETR2, calmodulin-like Ca-binding protein, heat shock proteins, MYB family transcription factors, kinases, zinc finger protein ZAT12, WRKY transcription factors)(Branco-Price et al. 2005). Earlier observations on ROS elevation resulting from oxygen deprivation and upon reoxygenation have been validated by transcriptomic analysis: indeed, expression of a wide range of ROS- and redox-related transcripts are induced by the lack of oxygen. Almost threefold transient upregulation of ascorbate peroxidase has been detected already after 0.5 h of low O_2 treatment, followed by monodehydroascorbate reductase, peroxidase ATP4a and a respiratory burst oxidase protein (Klok et al. 2002; Branco-Price et al. 2005; Liu et al. 2005). Interestingly, catalase expression is downregulated or unaffected under hypoxia and low abundance of the catalase transcript in anoxic rice coleoptiles has been discussed as signalling-related and necessary for H_2O_2 build-up (Lasanthi-Kudahettige et al. 2007).

Regulation of ROS originating from the mitochondrial ETC is achieved by the repression of TCA cycle enzymes and by induction of alternative oxidase (see above) (Klok et al. 2002; Branco-Price et al. 2008). In corroboration with the energy saving strategy the transcripts of genes related to energy-consuming secondary processes has been shown to be downregulated, but genes responsible for ATP-producing reactions shown to be − upregulated (Klok et al. 2002; Loreti et al. 2005; van Dongen et al. 2009).

7 Oxygen Deprivation, Metabolic Adaptations and Oxidative Stress

Transcripts linked to RNS metabolism constitute another group of upregulated redox-related genes: nitrate reductase, a known source of NO under oxygen deprivation (see above) and non-symbiotic hemoglobin 1 implicated in NO-dependent NADH recycling under hypoxia (Klok et al. 2002; Igamberdiev et al. 2005; van Dongen et al. 2009). Induction of class 1 hemoglobin has been detected in all experimental trials under different O_2 concentrations and in a range of durations (van Dongen et al. 2009), emphasizing the importance of hypoxically induced hemoglobins in stress survival.

The response to hypoxia and reoxygenation is regulated not only by induced transcription but also by differential translation of mRNAs: the general trend being translation suppression of the majority of mRNAs, while sustaining the translation of the small group of mRNAs encoding the enzymes relevant for energy maintenance and counteraction of ROS (Ahsan et al. 2007; Bailey-Serres and Voesenek 2008; Branco-Price et al. 2008). Quick metabolic reoxygenation response has found its explanation in a reversal of the sequestered untranslated mRNA and immediate protein synthesis (Branco-Price et al. 2008).

7.6 Update on Antioxidant Protection

The importance of antioxidants during oxidative stress has long been recognised. The positive effect of antioxidant supplementation/upregulation has been documented in numerous studies on metabolic factors underlying anoxia tolerance (Monk et al. 1989; Noctor and Foyer 1998; Scebba et al. 1998; Blokhina et al. 1999; Boo and Jung 1999; Blokhina et al. 2000; Smirnoff 2000; Blokhina et al. 2001, 2003). The complicated antioxidant network is regulated at the site of synthesis and during transport and through interaction with reactive oxygen species. The plant antioxidant defence system is represented by a range of molecules with different chemical nature, and it is, therefore, capable of exerting the antioxidant function in both aqueous and lipid phases in live tissues. The small molecular antioxidant arsenal in plants consists of ascorbate, glutathione, thioredoxins, tocopherols and carotenoids while many of the phenolic compounds in plants also have considerable antioxidant activity (Pietta 2000; Hernandez et al. 2009). To maintain the redox status and to regenerate antioxidants in their active form an array of enzymes acts in the support of the antioxidative defences: Dehydroascorbate reductase (DHAR), thioredoxin (TRX) reductase, glutathione (GSH) reductase, lipoamide dehydrogenase, thiol transferase (glutaredoxin). The efficiency of antioxidant system is increased by direct enzymatic elimination of deleterious ROS by superoxide dismutases (SODs), catalases (CAT), and peroxidases (PRX). There are also a number of enzymes detoxifying LP products: glutathione S-transferases, phospholipid-hydroperoxide glutathione peroxidase and ascorbate peroxidase.

7.6.1 Low Molecular Weight Antioxidants

7.6.1.1 Glutathione

The tripeptide glutathione (GSH, glutamylcysteinylglycine) is an abundant compound in plant tissues being present in virtually all cell compartments: cytosol, ER, vacuole and mitochondria (Jimenez et al. 1998). GSH together with its oxidized form (GSSG) maintains the cellular redox balance, which is of great importance as it allows fine-tuning of the cellular redox environment under normal conditions and provides the basis for GSH stress signalling. Indeed, GSH has a role in redox regulation of gene expression (Wingate et al. 1988; Alscher 1989), and also in the regulation of the cell cycle (Sanchez-Fernandez et al. 1997). GSH functions under oxidative stress by scavenging cytotoxic H_2O_2, and reacts non-enzymatically with other ROS: singlet oxygen, superoxide radical and hydroxyl radical (Larson 1988). The central role of GSH in the antioxidative defense is due to its ability to regenerate another powerful water-soluble antioxidant, ascorbic acid, via ascorbate-glutathione cycle (Foyer and Halliwell 1976; Noctor and Foyer 1998), Fig. 7.2.

7.6.1.2 Ascorbic acid

Vitamin C is one of the most studied and powerful antioxidants (Noctor and Foyer 1998; Arrigoni and De Tullio 2000; Horemans et al. 2000). It is astonishing that the last steps in the biosynthesis of ascorbic acid (AA) have been elucidated as late as in 2000 by Smirnoff. Ascorbate has not only been detected in the majority of plant cell types and cellular organelles, but also in the apoplast. Ascorbic acid exists mostly in its reduced form (90% of the ascorbate pool) in leaves and chloroplasts (Smirnoff

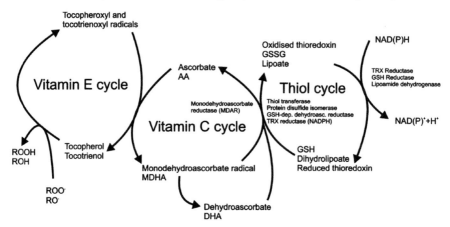

Fig. 7.2 Antioxidant turnover in plant tissues. The reducing power needed comes from NAD(P)H through the several enzymes catalysing reactions in the cycles. GSH, reduced glutathione, GSSG, oxidised glutathione. Adapted and redrawn from May et al. (1998)

2000), and its intracellular concentration can be in the high millimolar range (e.g. 20 mM in the cytosol and 20–300 mM in the chloroplast stroma (Foyer and Lelandais 1996).

AA acts in many ways in the protection of tissues against oxidative damage. It can directly scavenge superoxide, hydroxyl radicals and singlet oxygen and reduce H_2O_2 to water via ascorbate peroxidase reaction (Noctor and Foyer 1998). In chloroplasts AA acts in the xanthophyll cycle as a cofactor of violaxantin de-epoxidase thus sustaining dissipation of excess exitation energy (Smirnoff 2000). Albeit AA acts in the aqueous phase, it also protects membrane lipids by regenerating tocopherol from tocopheroxyl radicals (Thomas et al. 1992). In addition, AA carries out a number of non-antioxidant functions in the cell. It has been implicated in the regulation of the cell division, cell cycle progression from G1 to S phase (Liso et al. 1988; Smirnoff 1996) and cell elongation (de Tullio et al. 1999). Very recently dehydroascorbic acid has emerged as a signalling molecule regulating stomatal closure (Fotopoulos et al. 2008).

7.6.1.3 Tocopherol (Vitamin E)

Tocochromanols, i.e. a group of four tocopherols and four tocotrienols that collectively constitute vitamin E, is a vital nutrient in human diet only produced in photosynthetic organisms (Della Penna 2005). During the recent decade genetic engineering of the biosynthetic pathway has provided significant insights into the molecular genetics and biochemical control of tocochromanol biosynthesis in plants (Della Penna 2005). The importance of tocopherols and tocotrienols lies in the fact that they are essential components of biological membranes where they have both antioxidant and non-antioxidant functions (Kagan 1989). α-Tocopherol with its three methyl substitutes has the highest antioxidant activity of tocopherols (Kamal-Eldin and Appelqvist 1996), while β-, γ-, δ-isomers form the other three tocopherol and tocotrienols. Chloroplast membranes of higher plants contain α-tocopherol as the predominant tocopherol isomer, and are hence well protected against photooxidative damage (Fryer 1992).

The fact that makes Vitamin E especially important during the postanoxic phase in plant tissues is its chain-breaking antioxidant activity: It is able to repair oxidizing radicals directly, preventing the chain propagation step during lipid autoxidation (Serbinova and Packer 1994). The reaction between vitamin E and lipid radicals occurs in the membrane-water interphase where vitamin E donates a hydrogen ion to the lipid radical with the consequent formation of tocopheroxyl radical (TOH•) formation (Buettner 1993). Regeneration of the tocopheroxyl radical back to its reduced form can be achieved by vitamin C (ascorbate), reduced glutathione (Fryer 1992) or coenzyme Q (Kagan et al. 2000). In addition, tocopherols may act as chemical scavengers of oxygen radicals, especially singlet oxygen, and as physical deactivators of singlet oxygen by charge transfer mechanism (Fryer 1992).

It has only very recently been demonstrated for the first time that tocopherol deficiency may alter endogenous phytohormone levels, especially jasmonate, thereby reducing plant growth and triggering anthocyanin accumulation in leaves (Munné-Bosch et al. 2007). In another study on ethylene insensitive *Arabidopsis* mutants it has been concluded that tocopherol biosynthesis may be regulated by ethylene: The disruption of ethylene perception in *ein3-1*, *etr1-1* and its overproduction in *eto1-1* mutants of *Arabidopsis thaliana* correlated positively with tocopherol content (Cela et al. 2009). It was shown that a mutation in the EIN3 gene delayed the water-stress related increase in α-tocopherol and caused a reduction in the levels of this antioxidant by ca. 30% compared to the wild type. In contrast to the wild type and ein3-1 mutants, both etr1-1 and eto1-1 mutants showed a sharp (up to fivefold) increase in α-tocopherol levels during leaf aging (Cela et al. 2009).

It has also been noticed that nonenzymatic LP reprograms gene expression and activates defence markers in *Arabidopsis* tocopherol-deficient mutants (Sattler et al. 2006).

7.6.1.4 Phenolic Compounds as Antioxidants

Phenolics (flavonoids, tannins, hydroxycinnamate esters and lignin) are the largest group of secondary compounds in many plant tissues (Grace and Logan 2000). Polyphenols possess ideal structural chemistry for free radical scavenging activity, and they have been shown to be more effective antioxidants in vitro than tocopherols and ascorbate. Antioxidative properties of polyphenols arise from their high reactivity as hydrogen or electron donors, and from the ability of the polyphenol-derived radical to stabilize and delocalize the unpaired electron (chain-breaking function), as well as their ability to chelate transition metal ions (termination of the Fenton reaction) (Rice-Evans et al. 1997). Another mechanism underlying the antioxidative properties of phenolics is the ability of flavonoids to alter peroxidation kinetics by modification of the lipid packing order and to decrease fluidity of the membranes (Arora et al. 2000). These changes could sterically hinder diffusion of free radicals and restrict peroxidative reactions. Moreover, it has been shown that phenolic compounds can be involved in the hydrogen peroxide scavenging cascade in plant cells (Takahama and Oniki 1997). Induction of the biosynthesis of phenolic compounds has been shown in several articles, in connection with various stress conditions such as heavy metal stress (Sgherri et al. 2003; Mithöfer et al. 2004) and UV-B irradiation (Kondo and Kawashima 2000; Lavola et al. 2000).

7.6.2 Enzymes Participating in Quenching ROS

7.6.2.1 Superoxide Dismutase

Superoxide dismutases (SODs) are key enzymes in the antioxidative defense system of plants. SODs are present in all aerobic organisms and in all subcellular compartments susceptible of oxidative stress (Bowler et al. 1992). It is well known that enhanced formation of ROS and superoxide, $O_2^{-\bullet}$, under stress conditions may

7 Oxygen Deprivation, Metabolic Adaptations and Oxidative Stress

induce both protective responses and cellular damage. The scavenging of $O_2^{-\bullet}$ is achieved through SODs, which catalyse the dismutation of superoxide to H_2O_2. This reaction has a 10–10,000-fold faster rate than spontaneous dismutation (Bowler et al. 1992). SODs are classified by their metal cofactor into FeSOD (prokaryotic organisms, chloroplast stroma) and MnSOD (prokaryotic organisms and the mitochondrion of eukaryotes) which are structurally similar, and into the structurally unrelated Cu/ZnSOD (cytosolic and chloroplast enzyme, gram-negative bacteria).

Excessive accumulation of superoxide due to the reduced activity of SOD under flooding stress has been shown by Yan and coworkers (1996). Hence, it was to be expected that SOD activities will increase and this has indeed been shown in several experiments on plants under abiotic stress conditions. An increase in total SOD activity has been detected in wheat roots under anoxia but not under hypoxia. The degree of increase positively correlated with duration of anoxia (Biemelt et al. 2000). Induction of SOD activity under hypoxia by 40–60% in roots and leaves under hypoxia of *Hordeum vulgare* has been shown by Kalashnikov et al. (1994). Also salt stress has an effect on SOD transcription level and it seems that in different accession of *Arabidopsis* the responses can be different (Attia et al. 2008).

In *Arabidopsis* plants under various stresses including high light, ozone fumigation, and ultraviolet-B radiation that may lead to increased oxidative stress, differential regulation of seven SOD mRNAs and four SOD proteins have been noticed (Kliebenstein et al. 1998). On the whole, antioxidant defences are induced in plants under mild oxidative stress conditions (Lee et al. 2007), while a severe stress, such as anoxia, results in antioxidant depletion or slower turnover, and hence increased oxidative damage on re-oxygenation (Blokhina et al. 2000). Very recently a view on the mechanisms for the induction of SODs in plants has emerged: It has been noticed that there is a microRNA (miRNA, a class of regulatory RNAs of c. 21 nucleotides that posttranscriptionally regulate gene expression by directing mRNA cleavage or translational inhibition) in *Arabidopsis* called miR398 that targets two closely related Cu/ZnSODs (Sunkar et al. 2006). The induction of these SODs is mediated by the downregulation of miR398 but the actual molecular mechanism is still unknown (Sunkar et al. 2006). Another novel mitochondrial protein, AtMTM1, has also been implicated in the activation of MnSOD in *Arabidopsis* (Su et al. 2007). In a recent study on the significance of MnSOD (the mitochondrial SOD) with the antisense technique, it was shown that reduced MnSOD activity results in altered mitochondrial redox balance and reduced root growth while total respiratory CO_2 output did not change (Morgan et al. 2008).

7.6.2.2 Catalases, Peroxidases and Ascorbate Peroxidases

Catalases and peroxidases are important enzymes present in the intercellular spaces, where they can regulate the level of H_2O_2 (reviewed by Willekens et al. 1995). The catalase function has been explained in a review on plant antioxidants recently (Blokhina et al. 2003). In addition to catalysing the breakdown of H_2O_2 to water and dioxygen (catalase action), catalase can also work in a peroxidatic reaction and splits hydrogen peroxide and produces another strong oxidant, the hydroxyl radical (OH•; Elstner and Osswald 1994). OH• is a very strong oxidant and can initiate radical chain reactions with organic molecules, particularly with PUFA in membrane lipids.

In waterlogged citrus trees increases in both antioxidant enzyme activities (SOD, ascorbate peroxidase, catalase and glutathione reductase) as well as in ascorbate and glutathione have been detected (Arbona et al. 2008). Under anoxia a differential response of the peroxidase system has been observed in coleoptiles and roots of rice seedlings. A decrease in activities of cell wall-bound guaiacol and syringaldazine peroxidase activities has been reported, while soluble peroxidase activity was not affected in coleoptiles. In contrast, anoxia-grown roots have shown an increase in the cell wall-bound peroxidases (Lee and Lin 1995). Acclimation to anoxia has been shown to be dependent, at least partly, on peroxidases, which are up-regulated by anoxic stress in soybean cell cultures (Amor et al. 2000). In rice seedlings ADH and SOD activities responded non-significantly to submergence, while catalase activity increased upon re-oxygenation (Ushimaru et al. 1999).

7.6.2.3 Phospholipid Hydroperoxide Glutathione Peroxidase

The phospholipid hydroperoxide glutathione peroxidase (PHGPX) is an important antioxidant enzyme in the protection of biological membranes exposed to oxidative stress. It is known to be inducible under various stress conditions. This enzyme catalyses the regeneration of phospholipid hydroperoxides using the reducing power of GSH. It is localised in the cytosol and the inner membrane of mitochondria of animal cells. A cDNA clone homologous to PHGPX has been isolated from tobacco, maize, soybean, and Arabidopsis (Sugimoto et al. 1997). The PHGPX protein and its encoded gene *CSA* have been isolated and characterised in citrus. It has been shown that *CSA* is directly induced by the substrate of PHGPX under heat, cold and salt stresses, and that this induction occurs mainly via the production of ROS (Avsian-Kretchmer et al. 1999). It remains to be seen whether this gene is induced as ROS production increases after flooding or anoxia. In an earlier study, PHGPX has been suggested of being a chloroplast-localised enzyme (Mullineaux et al. 1998) preventing oxidative modification of lipids in chloroplast stroma (Baier and Dietz 1999) together with another alkyl hydroperoxide reductase, 2-Cys peroxiredoxin (Rouhier and Jacquot 2002). Recently, it has been proven to be localised in the mitochondrion (Yang et al. 2006).

7.7 Concluding Remarks

The paradoxic connection between oxygen deprivation and oxidative stress is substantiated by changes on transcriptional, translational and metabolic levels. Transcripts coding for both anaerobic proteins and ROS-detoxifying enzymes have been shown to be upregulated during hypoxia. ROS and RNS formation is tightly connected with hypoxia-induced pathways and metabolic adjustments during glycolysis, fermentation, cytoplasmic acidification, respiratory inhibition and the declining ATP level (Fig. 7.1). Many anoxia-induced enzymes have been shown to produce ROS or RNS under hypoxic conditions. In turn, ROS and RNS carry out a feedback regulation of e.g. respiration, ATP level and the TCA cycle. Their signalling functions rely on pH changes and Ca^{2+} elevation, as is the case with Rop GTPase signalling. The ability to control respiration and to preserve energy leads to manageable levels of ROS and RNS, (depending on the severity and length of oxygen deprivation), favouring cell survival. The energy preservation can be achieved via activation of glycolysis, via the PPi-dependent glycolysis (in anoxia tolerant plants) and by using of the anoxic metabolite nitrite as final electron acceptor in anoxic mitochondria. Control of the respiration rate can be realized through regulation of the glycolytic flux (to handle pyruvate levels and its entrance into the TCA cycle) and, probably, by the fine regulation of mitochondrial terminal oxidases via the dynamic equilibrium between ROS and NO. The balance between O_2 transport to hypoxic tissues, regulatory adaptations, anoxic metabolites, ROS−RNS chemistry and signalling, determine the survival of plants under waterlogging and oxygen deprivation.

References

Ahsan N, Lee D, Lee S, Lee K, Bahk J, Lee B (2007) A proteomic screen and identification of waterlogging-regulated proteins in tomato roots. Plant Soil 295:37–51

Alscher RG (1989) Biosynthesis and antioxidant function of glutathione in plants. Physiol Plant 77:457–464

Amor Y, Chevion M, Levine A (2000) Anoxia pretreatment protects soybean cells against H_2O_2-induced cell death: possible involvement of peroxidases and of alternative oxidase. FEBS Lett 477:175–180

Andreyev A, Kushnareva Y, Starkov A (2005) Mitochondrial metabolism of reactive oxygen species. Biochem (Moscow) 70:200–214

Arbona V, Hossain Z, López-Climent MF, Pérez-Clemente RM, Gómez-Cadenas A (2008) Antioxidant enzymatic activity is linked to waterlogging stress tolerance in citrus. Physiol Plant 132:452–466

Arcisio-Miranda M, Abdulkader F, Brunaldi K, Curi R, Procopio J (2009) Proton flux induced by free fatty acids across phospholipid bilayers: new evidences based on short-circuit measurements in planar lipid membranes. Arch Biochem Biophys 484:63–69

Armstrong W, Webb T, Darwent M, Beckett PM (2009) Measuring and interpreting respiratory critical oxygen pressures in roots. Ann Bot 103:281–293

Arora A, Byrem TM, Nair MG, Strasburg GM (2000) Modulation of liposomal membrane fluidity by flavonoids and isoflavonoids. Arch Biochem Biophys 373:102–109

Arrigoni O, de Tullio MC (2000) The role of ascorbic acid in cell metabolism: between gene-directed functions and unpredictable chemical reactions. J Plant Physiol 157:481–488

Attia H, Arnaud N, Karray N, Lachaal M (2008) Long-term effects of mild salt stress on growth, ion accumulation and superoxide dismutase expression of *Arabidopsis* rosette leaves. Physiol Plant 132:293–305

Avsian-Kretchmer O, Eshdat Y, Gueta-Dahan Y, Ben-Hayyim G (1999) Regulation of stress-induced phospholipid hydroperoxide glutathione peroxidase expression in citrus. Planta 209:469–477

Baier M, Dietz KJ (1999) Alkyl hydroperoxide reductases: the way out of the oxidative breakdown of lipids in chloroplasts. Trends Plant Sci 4:166–168

Bailey-Serres J, Chang R (2005) Sensing and signalling in response to oxygen deprivation in plants and other organisms. Ann Bot 96:507–518

Bailey-Serres J, Voesenek LACJ (2008) Flooding stress: acclimations and genetic diversity. Annu Rev Plant Biol 59:313–339

Baxter-Burrell A, Yang Z, Springer PS, Bailey-Serres J (2002) RopGAP4-dependent Rop GTPase rheostat control of *Arabidopsis* oxygen deprivation tolerance. Science 296:2026–2028

Benamar A, Rolletschek H, Borisjuk L, Avelange-Macherel M, Curien G, Mostefai HA, Andriant-sitohaina R, Macherel D (2008) Nitrite–nitric oxide control of mitochondrial respiration at the frontier of anoxia. Biochim Biophys Acta Bioenergetics 1777:1268–1275

Besson-Bard A, Pugin A, Wendehenne D (2008) New insights into nitric oxide signaling in plants. Annu Rev Plant Biol 59:21–39

Bethke PC, Badger MR, Jones RL (2004) Apoplastic synthesis of nitric oxide by plant tissues. Plant Cell 16:332–341

Biemelt S, Keetman U, Albrecht G (1998) Re-aeration following hypoxia or anoxia leads to activation of the antioxidative defense system in roots of wheat seedlings. Plant Physiol 116:651–658

Biemelt S, Keetman U, Mock H, Grimm B (2000) Expression and activity of isoenzymes of superoxide dismutase in wheat roots in response to hypoxia and anoxia. Plant Cell Environ 23:135–144

Blokhina O, Chirkova TV, Fagerstedt KV (2001) Anoxic stress leads to hydrogen peroxide formation in plant cells. J Exp Bot 52:1179–1190

Blokhina O, Fagerstedt KV (2006) Oxidative stress and antioxidant defences in plants. In: Quek (ed) Oxidative stress, disease and cancer, pp 151–199

Blokhina OB, Fagerstedt KV, Chirkova TV (1999) Relationships between lipid peroxidation and anoxia tolerance in a range of species during post-anoxic reaeration. Physiol Plant 105:625–632

Blokhina O, Virolainen E, Fagerstedt KV (2003) Antioxidants, oxidative damage and oxygen deprivation stress: a review. Ann Bot 91:179–194

Blokhina O, Virolainen E, Fagerstedt KV, Hoikkala A, Wähälä K, Chirkova TV (2000) Antioxidant status of anoxia-tolerant and -intolerant plant species under anoxia and reaeration. Physiol Plant 109:396–403

Bolwell GP, Bindschedler LV, Blee KA, Butt VS, Davies DR, Gardner SL, Gerrish C, Minibayeva F (2002) The apoplastic oxidative burst in response to biotic stress in plants: a three-component system. J Exp Bot 53:1367–1376

Bolwell GP, Wojtaszek P (1997) Mechanisms for the generation of reactive oxygen species in plant defence – a broad perspective. Physiol Mol Plant Pathol 51:347–366

Boo YC, Jung J (1999) Water deficit-induced oxidative stress and antioxidative defenses in rice plants. J Plant Physiol 155:255–261

Borisjuk L, Macherel D, Benamar A, Wobus U, Rolletschek H (2007) Low oxygen sensing and balancing in plant seeds: a role for nitric oxide. New Phytol 176:813–823

Borisjuk L, Rolletschek H (2009) The oxygen status of the developing seed. New Phytol 182:17–30

Bowler C, van Montagu M, Inze D (1992) Superoxide dismutase and stress tolerance. Annu Rev Plant Physiol Plant Mol Biol 43:83–116

Braidot E, Petrussa E, Vianello A, Macri F (1999) Hydrogen peroxide generation by higher plant mitochondria oxidizing complex I or complex II substrates. FEBS Lett 451:347–350

Branco-Price C, Kaiser KA, Jang CJH, Larive CK, Bailey-Serres J (2008) Selective mRNA translation coordinates energetic and metabolic adjustments to cellular oxygen deprivation and reoxygenation in *Arabidopsis thaliana*. Plant J 56:743–755

7 Oxygen Deprivation, Metabolic Adaptations and Oxidative Stress

Branco-Price C, Kawaguchi R, Ferreira RB, Bailey-Serres J (2005) Genome-wide analysis of transcript abundance and translation in Arabidopsis seedlings subjected to oxygen deprivation. Ann Bot 96:647–660

Brandalise M, Maia IG, Borecký J, AbE V, Arruda P (2003) Overexpression of plant uncoupling mitochondrial protein in transgenic tobacco increases tolerance to oxidative stress. J Bioenerg Biomembr 35:203–209

Buettner GR (1993) The pecking order of free radicals and antioxidants: lipid peroxidation, alpha-tocopherol, and ascorbate. Arch Biochem Biophys 300:535–543

Cela J, Falk J, Munné-Bosch S (2009) Ethylene signaling may be involved in the regulation of tocopherol biosynthesis in *Arabidopsis thaliana*. FEBS Lett 583:992–996

Chakraborti T, Das S, Mondal M, Roychoudhury S, Chakraborti S (1999) Oxidant, mitochondria and calcium: an overview. Cell Signal 11:77–85

Chirkova TV, Sinyutina NF, Blyudzin YA, Barsky IE, Smetannikova SV (1989) Phospholipid fatty acids of root mitochondria and microsomes from rice and wheat seedlings exposed to aeration or anaerobiosis. Plant Physiol (Russian) 36:126–134

Colmer TD (2003) Aerenchyma and an inducible barrier to radial oxygen loss facilitate root aeration in upland, paddy and deep-water rice (*Oryza sativa* L.). Ann Bot 91:301–309

Cooper CE, Davies NA, Psychoulis M, Canevari L, Bates TE, Dobbie MS, Casley CS, Sharpe MA (2003) Nitric oxide and peroxynitrite cause irreversible increases in the K-m for oxygen of mitochondrial cytochrome oxidase: in vitro and in vivo studies. Biochim Biophys Acta Bioenergetics 1607:27–34

de Oliveira HC, Wulff A, Saviani EE, Salgado I (2008) Nitric oxide degradation by potato tuber mitochondria: evidence for the involvement of external NAD(P)H dehydrogenases. Biochim Biophys Acta Bioenergetics 1777:470–476

del Rio LA, Sandalio LM, Corpas FJ, Palma JM, Barroso JB (2006) Reactive oxygen species and reactive nitrogen species in peroxisomes. Production, scavenging, and role in cell signaling. Plant Physiol 141:330–335

Della Penna D (2005) Progress in the dissection and manipulation of vitamin E synthesis. Trends Plant Sci 10:574–579

Dordas C, Hasinoff B, Rivoal J, Hill R (2004) Class-1 hemoglobins, nitrate and NO levels in anoxic maize cell-suspension cultures. Planta 219:66–72

Dordas C, Rivoal J, Hill RD (2003) Plant haemoglobins, nitric oxide and hypoxic stress. Ann Bot (Lond) 91:173–178

Drew MC (1997) Oxygen deficiency and root metabolism: injury and acclimation under hypoxia and anoxia. Annu Rev Plant Physiol Plant Mol Biol 48:223–250

Elstner EF, Osswald W (1994) Mechanisms of oxygen activation during plant stress. Proc Royal Soc Edinb 102B:31–154

Freeman BA, Baker PRS, Schopfer FJ, Woodcock SR, Napolitano A, d'Ischia M (2008) Nitro-fatty acid formation and signaling. J Biol Chem 283:15515–15519

Fotopoulos V, De Tullio MC, Barnes J, Kanellis AK (2008) Altered stomatal dynamics in ascorbate oxidase over-expressing tobacco plants suggests a role for dehydroascorbate signalling. J Exp Bot 59:729–737

Foyer CH, Halliwell B (1976) The presence of glutathione and glutathione reductase in chloroplasts: a proposed role in ascorbic acid metabolism. Planta 133:21–25

Foyer CH, Lelandais MA (1996) A comparison of the relative rates of transport of ascorbate and glucose across the thylakoid, chloroplast and plasmalemma membranes of pea leaves mesophyll cells. J Plant Physiol 148:391–398

Fryer MJ (1992) The antioxidant effects of thylakoid vitamin E (α-tocopherol). Plant Cell Environ 15:381–392

Fukao T, Bailey-Serres J (2004) Plant responses to hypoxia – is survival a balancing act? Trends Plant Sci 9:449–456

Geigenberger P (2003) Response of plant metabolism to too little oxygen. Curr Opin Plant Biol 6:247–256

Generosova IP, Vartapetian BB (2005) On the physiological role of anaerobically synthesized lipids in *Oryza sativa* seedlings. Russ J Plant Physiol 52:481–488

Gibbs J, Greenway H (2003) Mechanisms of anoxia tolerance in plants. I. Growth, survival and anaerobic catabolism. Funct Plant Biol 30:1–47

Gille L, Nohl H (2001) The ubiquinol/bc1 redox couple regulates mitochondrial oxygen radical formation. Arch Biochem Biophys 388:34–38

Godber BL, Doel JJ, Sapkota GP, Blake DR, Stevens CR, Eisenthal R, Harrison R (2000) Reduction of nitrite to nitric oxide catalyzed by xanthine oxidoreductase. J Biol Chem 275:7757–7763

Goglia F, Skulachev VP (2003) A function for novel uncoupling proteins: antioxidant defense of mitochondrial matrix by translocating fatty acid peroxides from the inner to the outer membrane leaflet. FASEB J 17:1585–1591

Grabel'nykh OI, Kolesnichenko AV, Pobezhimova TP, Zykova VV, Voinikov VK (2006) Mechanisms and functions of nonphosphorylating electron transport in respiratory chain of plant mitochondria. Russ J Plant Physiol 53:418–429

Grace S, Logan BA (2000) Energy dissipation and radical scavenging by the plant phenylpropanoid pathway. Trans R Soc Lond B 355:1499–1510

Graham JWA, Williams TCR, Morgan M, Fernie AR, Ratcliffe RG, Sweetlove LJ (2007) Glycolytic enzymes associate dynamically with mitochondria in response to respiratory demand and support substrate channeling. Plant Cell 19:3723–3738

Greenway H, Gibbs J (2003) Review: mechanisms of anoxia tolerance in plants. II. Energy requirements for maintenance and energy distribution to essential processes. Funct Plant Biol 30:999–1036

Gupta KJ, Stoimenova M, Kaiser WM (2005) In higher plants, only root mitochondria, but not leaf mitochondria reduce nitrite to NO, in vitro and in situ. J Exp Bot 56:2601–2609

Halliwell B (2006) Reactive species and antioxidants. Redox biology is a fundamental theme of aerobic life. Plant Physiol 141:312–322

Halliwell B (2009) The wanderings of a free radical. Free Radic Biol Med 46:531–542

Hamberg M, Fahlstadius P (1992) On the specificity of a fatty acid epoxygenase in broad bean (*Vicia faba* L.). Plant Physiol 99:987–995

Harrison R (2002) Structure and function of xanthine oxidoreductase: where are we now? Free Radic Biol Med 33:774–797

Hassan W, Ibrahim M, Deobald AM, Braga AL, Nogueira CW, Rocha JBT (2009) pH-Dependent Fe (II) pathophysiology and protective effect of an organoselenium compound. FEBS Lett 583:1011–1016

Hebelstrup KH, Igamberdiev AU, Hill RD (2007) Metabolic effects of hemoglobin gene expression in plants. Gene 398:86–93

Hernandez I, Alegre L, Van Breusegem F, Sergi Munne-Bosch S (2009) How relevant are flavonoids as antioxidants in plants? Trends Plant Sci 14:125–132

Horemans N, Foyer CH, Potters G, Asard H (2000) Ascorbate function and associated transport systems in plants. Plant Physiol Biochem 38:531–540

Huang S, Colmer TD, Millar AH (2008) Does anoxia tolerance involve altering the energy currency towards PPi? Trends Plant Sci 13:221–227

Huang X, von Rad U, Durner J (2002) Nitric oxide induces transcriptional activation of the nitric oxide-tolerant alternative oxidase in *Arabidopsis* suspension cells. Planta 215: 914–923

Igamberdiev AU, Baron K, Manac'h-Little N, Stoimenova M, Hill RD (2005) The haemoglobin/nitric oxide cycle: Involvement in flooding stress and effects on hormone signalling. Ann Bot 96:557–564

Igamberdiev AU, Hill RD (2004) Nitrate, NO and haemoglobin in plant adaptation to hypoxia: an alternative to classic fermentation pathways. J Exp Bot 55:2473–2482

Igamberdiev AU, Hill RD (2009) Plant mitochondrial function during anaerobiosis. Ann Bot 103:259–268

Jackson MB (2008) Ethylene-promoted elongation: an adaptation to submergence stress. Ann Bot 101:229–248

Jasid S, Simontacchi M, Bartoli CG, Puntarulo S (2006) Chloroplasts as a nitric oxide cellular source. Effect of reactive nitrogen species on chloroplastic lipids and proteins. Plant Physiol 142:1246–1255

Jimenez A, Hernandez JA, Pastori G, del Río LA, Sevilla F (1998) Role of the ascorbate-glutathione cycle of mitochondria and peroxisomes in the senescence of pea leaves. Plant Physiol 118:1327–1335

Kagan VE (1989) Tocopherol stabilizes membrane against phospholipase A, free fatty acids, and lysophospholipids. In: Diplock AT, Machlin J, Packer L, Pryor WA (Eds) Vitamin E: biochemistry and health implications. Ann New York Acad Sci 570:121–135

Kagan VE, Fabisiak JP, Quinn PJ (2000) Coenzyme Q and vitamin E need each other as antioxidants. Lipids 214:11–18

Kalashnikov JUE, Balakhnina TI, Zakrzhevsky DA (1994) Effect of soil hypoxia on activation of oxygen and the system of protection from oxidative destruction in roots and leaves of *Hordeum vulgare*. Russ J Plant Physiol 41:583–588

Kamal-Eldin A, Appelqvist L-Å (1996) The chemistry and antioxidant properties of tocopherols and tocotrienols. Lipids 31:671–701

Kliebenstein DJ, Monde RA, Last RL (1998) Superoxide dismutase in *Arabidopsis*: an eclectic enzyme family with disparate regulation and protein localization. Plant physiol 118:637–650

Klok EJ, Wilson IW, Wilson D, Chapman SC, Ewing RM, Somerville SC, Peacock WJ, Dolferus R, Dennis ES (2002) Expression profile analysis of the low-oxygen response in *Arabidopsis* root cultures. Plant Cell 14:2481–2494

Kondo N, Kawashima M (2000) Enhancement of tolerance to oxidative stress in cucumber (Cucumis sativus L.) seedlings by UV-B irradiation: possible involvement of phenolic compounds and antioxidant enzymes. J Plant Res 113:311–317

Kreuzwieser J, Hauberg J, Howell KA, Carroll A, Rennenberg H, Millar AH, Whelan J (2009) Differential response of gray poplar leaves and roots underpins stress adaptation during hypoxia. Plant Physiol 149:461–473

Lamattina L, Garcia-Mata C, Graziano M, Pagnussat G (2003) Nitric oxide: the versatility of an extensive signal molecule. Annu Rev Plant Biol 54:109–136

Lanteri ML, Laxalt AM, Lamattina L (2008) Nitric oxide triggers phosphatidic acid accumulation via phospholipase D during auxin-induced adventitious root formation in cucumber. Plant Physiol 147:188–198

Larson RA (1988) The antioxidants of higher plants. Phytochem 27:969–978

Lasanthi-Kudahettige R, Magneschi L, Loreti E, Gonzali S, Licausi F, Novi G, Beretta O, Vitulli F, Alpi A, Perata P (2007) Transcript profiling of the anoxic rice coleoptile. Plant Physiol 144:218–231

Lavola A, Julkunen-Tiitto R, DeE La Rosa TM, Lehto T, Aphalo PJ (2000) Allocation of carbon to growth and secondary metabolites in birch seedlings under UV-B radiation and CO_2 exposure. Physiol Plant 109:260–267

Lee SH, Ahsan N, Lee KW, Lee DG, Kwark SS, Kwon SY, Kim TH, Lee BH (2007) Simultaneous overexpression of both CuZn superoxide dismutase and ascorbate peroxidase in transgenic tall fescue plants confers increased tolerance to a wide range of abiotic stresses. J Plant Physiol 164:1626–1638

Lee TM, Lin YN (1995) Changes in soluble and cell wall-bound peroxidase activities with growth in anoxia-treated rice (*Oryza sativa* L.) coleoptiles and roots. Plant Sci 106:1–7

Libourel IG, van Bodegom PM, Fricker MD, Ratcliffe RG (2006) Nitrite reduces cytoplasmic acidosis under anoxia. Plant Physiol 142:1710–1717

Liso R, Innocenti AM, Bitonti MB, Arrigoni O (1988) Ascorbic acid-induced progression of quiescent centre cells from G_1 to S phase. New Phytol 110:469–471

Liu F, VanToai T, Moy LP, Bock G, Linford LD, Quackenbush J (2005) Global transcription profiling reveals comprehensive insights into hypoxic response in *Arabidopsis*. Plant Physiol 137:1115–1129

Loreti E, Poggi A, Novi G, Alpi A, Perata P (2005) A genome-wide analysis of the effects of sucrose on gene expression in *Arabidopsis* seedlings under anoxia. Plant Physiol 137:1130–1138

Maxwell DP, Wang Y, McIntosh L (1999) The alternative oxidase lowers mitochondrial reactive oxygen production in plant cells. Proc Natl Acad Sci U S A 96:8271–8276

May MJ, Vernoux T, Leaver C, Van Montagu M, Inze D (1998) Glutathione homeostasis in plants: implications for environmental sensing and plant development. J Exp Bot 49:649–667

McDonald AE, Vanlerberghe GC (2007) The organization and control of plant mitochondrial metabolism. In: Plaxton WC, McManus MT (eds) Control of primary metabolism in plants. Blackwell, Oxford, pp 290–324

Millenaar FF, Lambers H (2003) The alternative oxidase: in vivo regulation and function. Plant Biol 5:2–15

Miller G, Shulaev V, Mittler R (2008) Reactive oxygen signaling and abiotic stress. Physiol Plant 133:481–489

Mithöfer A, Schulze B, Boland W (2004) Biotic and heavy metal stress response in plants: evidence for common signals. FEBS Lett 566:1–5

Mittler R (2002) Oxidative stress, antioxidants and stress tolerance. Trends Plant Sci 7:405–410

Moller IM (2001) Plant mitochondria and oxidative stress: electron transport, NADPH turnover, and metabolism of reactive oxygen species. Annu Rev Plant Physiol Plant Mol Biol 52:561–591

Moller IM, Jensen PE, Hansson A (2007) Oxidative modifications to cellular components in plants. Annu Rev Plant Biol 58:459–481

Moller M, Botti H, Batthyany C, Rubbo H, Radi R, Denicola A (2005) Direct measurement of nitric oxide and oxygen partitioning into liposomes and low density lipoprotein. J Biol Chem 280:8850–8854

Monk LS, Fagerstedt KV, Crawford RMM (1989) Oxygen toxicity and superoxide dismutase as an antioxidant in physiological stress. Physiol Plant 76:456–459

Morgan MJ, Lehmann M, Schwarzländer M, Baxter CJ, Sienkiewicz-Porzucek A, Williams TCR, Schauer N, Fernie AR, Fricker MD, Ratcliffe RG, Sweetlove LJ, Finkemeier I (2008) Decrease in manganese superoxide dismutase leads to reduced root growth and affects tricarboxylic acid cycle flux and mitochondrial redox homeostasis. Plant Physiol 147:101–114

Mullineaux PM, Karpinski S, Jiménez A, Cleary SP, Robinson C, Creissen GP (1998) Identification of cDNAS encoding plastid-targeted glutathione peroxidase. Plant J 13:375–379

Munné-Bosch S, Weiler E, Alegre L, Müller M, Düchting P, Falk J (2007) α-Tocopherol may influence cellular signaling by modulating jasmonic acid levels in plants. Planta 225:681–691

Murphy MP (2009) How mitochondria produce reactive oxygen species. Biochem J 417:1–17

Mustroph A, Albrecht G, Hajirezaei M, Grimm B, Biemelt S (2005) Low levels of pyrophosphate in transgenic potato plants expressing *E. coli* pyrophosphatase lead to decreased vitality under oxygen deficiency. Ann Bot 96:717–726

Nakazono M, Imamura T, Tsutsumi N, Sasaki T, Hirai A (2000) Characterization of two cDNA clones encoding isozymes of the F1F0-ATPase inhibitor protein of rice mitochondria. Planta 210:188–194

Navarre DA, Wendehenne D, Durner J, Noad R, Klessig DF (2000) Nitric oxide modulates the activity of tobacco aconitase. Plant Physiol 122:573–582

Neill SJ, Desikan R, Hancock JT (2003) Nitric oxide signalling in plants. New Phytol 159:11–35

Noctor G, De Paepe R, Foyer CH (2007) Mitochondrial redox biology and homeostasis in plants. Trends Plant Sci 12:125–134

Noctor G, Foyer CH (1998) Ascorbate and glutathione: keeping active oxygen under control. Annu Rev Plant Physiol Mol Biol 49:249–279

7 Oxygen Deprivation, Metabolic Adaptations and Oxidative Stress

Nowak G, Grant DF, Moran JH (2004) Linoleic acid epoxide promotes the maintenance of mitochondrial function and active Na^+ transport following hypoxia. Toxicol Lett 147: 161–175

O'Donnell VB, Freeman BA (2001) Interactions between nitric oxide and lipid oxidation pathways: implications for vascular disease. Circ Res 88:12–21

Ohyashiki T, Nunomura M (2000) A marked stimulation of Fe^{3+}-dependent lipid peroxidation in phospholipid liposomes under acidic conditions. Biochim Biophys Acta Mol Cell Biol Lipids 1484:241–250

Pastore D, Trono D, Laus MN, Di Fonzo N, Flagella Z (2007) Possible plant mitochondria involvement in cell adaptation to drought stress: a case study: durum wheat mitochondria. J Exp Bot 58:195–210

Pavelic D, Arpagaus S, Rawyler A, Brandle R (2000) Impact of post-anoxia stress on membrane lipids of anoxia-pretreated potato cells. A re-appraisal. Plant Physiol 124:1285–1292

Pietta PG (2000) Flavonoids as antioxidants. J Nat Prod 63:1035–1042

Planchet E, Jagadis Gupta K, Sonoda M, Kaiser WM (2005) Nitric oxide emission from tobacco leaves and cell suspensions: rate limiting factors and evidence for the involvement of mitochondrial electron transport. Plant J 41:732–743

Polgreen KE, Featherstone J, Willis AC, Harris DA (1995) Primary structure and properties of the inhibitory protein of the mitochondrial ATPase (H^+-ATP synthase) from potato. Biochim Biophys Acta Bioenergetics 1229:175–180

Pryor WA, Houk KN, Foote CS, Fukuto JM, Ignarro LJ, Squadrito GL, Davies KJA (2006) Free radical biology and medicine: it's a gas, man! Am J Physiol Regul Integr Comp Physiol 291: R491–R511

Qiao W, Fan L (2008) Nitric oxide signaling in plant responses to abiotic stresses. J Integr Plant Biol 50:1238–1246

Rawyler A, Arpagaus S, Braendle R (2002) Impact of oxygen stress and energy availability on membrane stability of plant cells. Ann Bot 90:499–507

Rice-Evans CA, Miller NJ, Paganga G (1997) Antioxidant properties of phenolic compounds. Trends Plant Sci 2:152–159

Rinalducci S, Murgiano L, Zolla L (2008) Redox proteomics: basic principles and future perspectives for the detection of protein oxidation in plants. J Exp Bot 59:3781–3801

Rockel P, Strube F, Rockel A, Wildt J, Kaiser WM (2002) Regulation of nitric oxide (NO) production by plant nitrate reductase in vivo and in vitro. J Exp Bot 53:103–110

Rouhier N, Jacquot J (2002) Plant peroxiredoxins: alternative hydroperoxide scavenging enzymes. Photosynth Res 74:259–268

Rouslin W (1991) Regulation of the mitochondrial ATPase in situ in cardiac muscle: role of the inhibitor subunit. J Bioenerg Biomembr 23:873–888

Rubbo H, Radi R (2008) Protein and lipid nitration: role in redox signaling and injury. Biochim Biophys Acta 1780:1318–1324

Sairam R, Kumutha D, Ezhilmathi K, Deshmukh P, Srivastava G (2008) Physiology and biochemistry of waterlogging tolerance in plants. Biol Plant 52:401–412

Sanchez-Fernandez R, Fricker M, Corben LB, White NS, Sheard N, Leaver CJ, van Montagu M, Inzé D, May MJ (1997) Cell proliferation and hair tip growth in the Arabidopsis root are under mechanistically different forms of redox control. PNAS 94:2745–2750

Santosa I, Ram P, Boamfa E, Laarhoven L, Reuss J, Jackson M, Harren F (2007) Patterns of peroxidative ethane emission from submerged rice seedlings indicate that damage from reactive oxygen species takes place during submergence and is not necessarily a post-anoxic phenomenon. Planta 226:193–202

Sattler SE, Mene-Saffrane L, Farmer EE, Krischke M, Mueller MJ, DellaPenna D (2006) Nonenzymatic lipid peroxidation reprograms gene expression and activates defense markers in *Arabidopsis* tocopherol-deficient mutants. Plant Cell 18:3706–3720

Scebba F, Sebastiani L, Vitagliano C (1998) Changes in activity of antioxidative enzymes in wheat (*Triticum aestivum*) seedlings under cold acclimation. Physiol Plant 104:747–752

Serbinova EA, Packer L (1994) Antioxidant properties of α-tocopherol and α-tocotrienol. Methods Enzymol 234:354–366

Sgherri C, Cosi E, Navari-Izo F (2003) Phenols and antioxidative status of *Raphanus sativus* grown in copper excess. Physiol Plant 118:21–28

Skulachev VP (1998) Uncoupling: new approaches to an old problem of bioenergetics. Biochim Biophys Acta Bioenergetics 1363:100–124

Smirnoff N (1996) The function and metabolism of ascorbic acid in plants. Ann Bot 78: 661–669

Smirnoff N (2000) Ascorbic acid: metabolism and functions of a multi-facetted molecule. Curr Opin Plant Biol 3:229–235

Sorg O (2004) Oxidative stress: a theoretical model or a biological reality? C R Biol 327: 649–662

Stohr C, Stremlau S (2006) Formation and possible roles of nitric oxide in plant roots. J Exp Bot 57:463–470

Stoimenova M, Igamberdiev AU, Gupta K, Hill RD (2007) Nitrite-driven anaerobic ATP synthesis in barley and rice root mitochondria. Planta 226:465–474

Stoimenova M, Libourel IG, Ratcliffe RG, Kaiser WM (2003) The role of nitrate reduction in the anoxic metabolism of roots II. Anoxic metabolism of tobacco roots with or without nitrate reductase activity. Plant Soil 253:155–167

St-Pierre J, Brand MD, Boutilier RG (2000) Mitochondria as ATP consumers: cellular treason in anoxia. PNAS 97:8670–8674

Su Z, Chai M-F, Lu P-L, An R, Chen J, Wang X-C (2007) AtMTM1, a novel mitochondrial protein, may be involved in activation of the manganese-containing superoxide dismutase in *Arabidopsis*. Planta 226:1031–1039

Sugimoto M, Furui S, Suzuki Y (1997) Molecular cloning and characterisation of a cDNA encoding putative phospholipid hydroperoxide glutathione peroxidase from spinach. Biosci Biotech Biochem 61:1379–1381

Sunkar R, Kapoor A, Zhu J-K (2006) Posttranscriptional induction of two Cu/Zn superoxide dismutase genes in *Arabidopsis* is mediated by downregulation of miR398 and important for oxidative stress tolerance. Plant Cell 18:2051–2065

Szal B, Jolivet Y, Hasenfratz-Sauder M, Dizengremel P, Rychter AM (2003) Oxygen concentration regulates alternative oxidase expression in barley roots during hypoxia and post-hypoxia. Physiol Plant 119:494–502

Takahama U, Oniki T (1997) A peroxide/phenolics/ascorbate system can scavenge hydrogen peroxide in plant cells. Physiol Plant 101:845–852

Testerink C, Munnik T (2005) Phosphatidic acid: a multifunctional stress signaling lipid in plants. Trends Plant Sci 10:368–375

Thomas CE, McLean LR, Parker RA, Ohlweiler DF (1992) Ascorbate and phenolic antioxidant interactions in prevention of liposomal oxidation. Lipids 27:543–550

Triantaphylidès C, Havaux M (2009) Singlet oxygen in plants: production, detoxification and signaling. Trends Plant Sci 14:219–228

Turrens JF, Boveris A (1980) Generation of superoxide anion by the NADH dehydrogenase of bovine heart mitochondria. Biochem J 2:421–427

Van Breusegem F, Bailey-Serres J, Mittler R (2008) Unraveling the tapestry of networks involving reactive oxygen species in plants. Plant Physiol 147:978–984

van Dongen JT, Frohlich A, Ramirez-Aguilar SJ, Schauer N, Fernie AR, Erban A, Kopka J, Clark J, Langer A, Geigenberger P (2009) Transcript and metabolite profiling of the adaptive response to mild decreases in oxygen concentration in the roots of *Arabidopsis* plants. Ann Bot 103:269–280

van Dongen JT, Schurr U, Pfister M, Geigenberger P (2003) Phloem metabolism and function have to cope with low internal oxygen. Plant Physiol 131:1529

Vianello A, Macrì F (1999) Proton pumping pyrophosphatase from higher plant mitochondria. Physiol Plant 105:763–768

Visser EJW, Colmer TD, Blom CWPM, Voesenek LACJ (2000) Changes in growth, porosity, and radial oxygen loss from adventitious roots of selected mono- and dicotyledonous wetland species with contrasting types of aerenchyma. Plant Cell Environ 23:1237–1245

Visser EJW, Voesenek LACJ, Vartapetian BB, Jackson MB (2003) Flooding and plant growth. Ann Bot 91:107–109

Ushimaru T, Kanematsu S, Shibasaka M, Tsuji H (1999) Effect of hypoxia on the antioxidative enzymes in aerobically grown rice (*Oryza sativa*) seedlings. Physiol Plant 107:81–187

Willekens H, Inzé D, van Montagu M, van Camp W (1995) Catalase in plants. Mol Breed 1:207–228

Wingate VPM, Lawton MA, Lamb CJ (1988) Glutathione causes a massive and selective induction of plant defense genes. Plant Physiol 87:206–210

Wink DA, Mitchell JB (1998) Chemical biology of nitric oxide: insights into regulatory, cytotoxic, and cytoprotective mechanisms of nitric oxide. Free Radic Biol Med 25:434–456

Yamasaki H, Sakihama Y (2000) Simultaneous production of nitric oxide and peroxynitrite by plant nitrate reductase: in vitro evidence for the NR-dependent formation of active nitrogen species. FEBS Lett 468:89–92

Yan B, Dai Q, Liu X, Huang S, Wang Z (1996) Flooding-induced membrane damage, lipid oxidation and activated oxygen generation in corn leaves. Plant Soil 179:261–268

Yang X-D, Dong C-J, Liu J-Y (2006) A plant mitochondrial phospholipid hydroperoxide glutathione peroxidase: its precise localization and higher enzymatic activity. Plant Mol Biol 62:951–962

Yordanova R, Popova L (2007) Flooding-induced changes in photosynthesis and oxidative status in maize plants. Acta Physiol Plant 29:535–541

Yu Z, Zhang J, Wang X, Chen J (2008) Excessive copper induces the production of reactive oxygen species, which is mediated by phospholipase D, nicotinamide adenine dinucleotide phosphate oxidase and antioxidant systems. J Integr Plant Biol 50:157–167

Zabalza A, van Dongen JT, Froehlich A, Oliver SN, Faix B, Gupta KJ, Schmalzlin E, Igal M, Orcaray L, Royuela M, Geigenberger P (2009) Regulation of respiration and fermentation to control the plant internal oxygen concentration. Plant Physiol 149:1087–1098

Zsigmond L, Rigo G, Szarka A, Szekely G, Otvos K, Darula Z, Medzihradszky KF, Koncz C, Koncz Z, Szabados L (2008) *Arabidopsis* PPR40 connects abiotic stress responses to mitochondrial electron transport. Plant Physiol 146:1721–1737

Part III
Membrane Transporters in Waterlogging Tolerance

Chapter 8
Root Water Transport Under Waterlogged Conditions and the Roles of Aquaporins

Helen Bramley and Steve Tyerman

Abstract Water flow through plants roots can be affected when the soil is water-logged and oxygen deficient. For species not adapted to these conditions, water flow usually decreases within minutes to days, depending on the oxygen concentration in the root and rhizosphere. During this time, the decrease in water flow is attributed to decreased root hydraulic conductance, through an inhibition of plasma-membrane aquaporins. There is increasing evidence that aquaporins may also be involved in the transport of gases, end products of anaerobic respiration, and signalling molecules; all of which are relevant to oxygen-deficient conditions. Eventually, primary roots die if continually starved of oxygen, but may be replaced with adventitious roots that can maintain the supply of water to the shoot. Here, we review the effects of waterlogging and oxygen deficiency on root hydraulic conductance and aquaporin activity.

8.1 Introduction

Some plants wilt within hours of their roots being waterlogged unless they close their stomata. This phenomenon, called physiological drought, is believed to be caused by an insufficient supply of water from the roots (Cannell and Jackson 1981). Water uptake into roots and transport to the shoots is often inhibited when the soil is waterlogged (Kramer 1949; Kramer 1983). The abundant supply of water in direct contact with roots should reduce the drop in water potential required across the root/rhizosphere interface to induce water uptake. However, although water

---

H. Bramley
Institute of Agriculture, The University of Western Australia, 35 Stirling Highway, Crawley, Western Australia 6009, Australia

S. Tyerman (✉)
School of Agriculture, Food and Wine, The University of Adelaide (Waite Campus), Plant Research Centre, PMB 1, Glen Osmond SA 5064, Australia
e-mail: stephen.tyerman@adelaide.edu.au

S. Mancuso and S. Shabala (eds.), *Waterlogging Signalling and Tolerance in Plants*,
DOI 10.1007/978-3-642-10305-6_8, © Springer-Verlag Berlin Heidelberg 2010

moves entirely passively in response to gradients in water potential, roots are not invariable pathways for water flow (Tyree and Zimmermann 2002), so flow rate is determined by their hydraulic conductance. Hydraulic conductance of some species varies with time of day, transpiration rate and in response to biotic or abiotic perturbation (for recent reviews see Javot and Maurel 2002; Vandeleur et al. 2005; Bramley et al. 2007; Maurel et al. 2008). Water transport through the root is a spatially and temporally complex process with many physiological mechanisms that can influence it. The activity of aquaporins (density in the membrane and gating) is seen to be a component accounting for changes in root hydraulic conductance; however anatomical factors and the distribution of water flow along the root and in different parts of the root system are also important. At the plasma-membrane and in membranes surrounding intracellular compartments aquaporins may also be involved in the flux of gases (O_2, CO_2 NO), end products of anaerobic respiration (ethanol, lactic acid and other organic acids), or of molecules important in signalling (H_2O_2, ethylene). In this chapter, we review what mechanisms may be involved during root submergence and oxygen deficiency.

8.2 Variable Root Hydraulic Conductance (L_r)

Oxygen deficiency has apparently varying effects on L_r. For example, 0.5 h of anoxia reduced L_r of *Arabidopsis thaliana* by almost half and 2–4 h of flooding *Ricinus communis* reduced L_r by 65% (Else et al. 2001; Tournaire-Roux et al. 2003). In comparison, L_r was not influenced by 3 h of anoxia in *Agave deserti*, 24 h of flooding *Lycopersicon esculentum* or 10 d of flooding conifer species (Nobel et al. 1990; Reece and Riha 1991; Jackson et al. 1996). In other studies, oxygen deficiency transiently reduced L_r of *L. esculentum*, *Helianthus annuus*, *Zea mays* and *Musa* spp. (Bradford and Hsiao 1982; Everard and Drew 1989; Gibbs et al. 1998; Aguilar et al. 2003). Some of this variability may be related to differences in methodology such as concentration of oxygen or duration of oxygen deficiency. Many studies simulated oxygen-deficient conditions by rapidly applying anoxia, which likely induces "shock" responses. In their natural environment, roots experience hypoxia before anoxia, which stimulates acclimation responses and imparts greater tolerance to subsequent anoxia (Waters et al. 1991; Xia and Roberts 1996; Dennis et al. 2000; Kato-Noguchi 2000). On the other hand, studies that attempted to simulate waterlogging by inundating root systems grown in pots often imposed the treatment for weeks at a time, so by the time L_r was measured the root systems may have deteriorated or died.

The variability in L_r may also be species dependent and few studies have addressed this variability in terms of differences in root hydraulic properties. The conductance of the tissue to water and the magnitude of the driving force determine the rate of water flow through roots. Consequently, root hydraulic properties are influenced by morphology, anatomy and activity of proteinaceous water channels called aquaporins (Bramley et al. 2009). Changes induced by oxygen deficiency to each of these attributes can potentially influence L_r.

8.3 Changes in Root Morphology and Anatomy

Anatomy plays a major role in determining where water is absorbed along the root and the pathway water takes across the root. From the root surface to the xylem (radial pathway), water can travel extracellularly through the apoplast and/or intercellularly across membranes and through plasmodesmata. The contribution of these pathways depends on their hydraulic conductivity and can change with environmental conditions or whether water flow is induced by transpiration or accumulation of solutes (Steudle and Peterson 1998). Roots of many species exhibit phenotypic plasticity and long-term exposure to waterlogging can cause changes in morphology and anatomy, through cell death and/or modifications to cell walls.

8.3.1 Root Death and Adventitious Roots

The extent of injury induced by waterlogging depends on the tolerance of the species to oxygen deficiency (Drew 1992). Root tips tend to have high metabolic rates and are generally the most sensitive part of the root to oxygen deficiency (Drew 1997). In particularly sensitive species, such as *Lupinus*, root tips start to visually deteriorate within a few days of waterlogging (Bramley 2006). Root tips lose turgor pressure and the cortex deteriorates (Atwell 1991). Further from the root tip, turgor pressure of cortical cells decreases even with short exposure (0.5 h) to mild hypoxia (Bramley et al. 2010). For intolerant species, the longer the duration of waterlogging the greater the extent of root death, with progressive deterioration towards the basal region. Water absorption commences behind the root tip, so death of this region of the root may have little effect on L_r. It is not clear how root death under waterlogging affects L_r, as barriers to flow may increase or decrease, depending on which tissue dies and where the greatest limitation to L_r occurred prior to waterlogging. Plugging or collapse of xylem vessels as opposed to death of the cortex could have contrasting effects on L_r.

To investigate the role of roots in water absorption by transpiring plants Kramer (1933) killed root systems of seven different species by immersing them in hot water. Plants with dead roots still absorbed "considerable quantities" of water during the first few days after root death, so that leaves continued to transpire and remained hydrated. Eventually, transpiration declined, leaves wilted and shoots died because xylem vessels in stems plugged above the killed tissue. Applying suction to decapitated dead root systems induced water flux more than when root systems were alive, which was attributed to a decrease in the resistance to water movement in the radial pathway (Kramer 1933). In addition, L_r is commonly measured by the pressure chamber technique, but pressurising flooded roots or roots in solution culture causes tissue of some species to become infiltrated with water and/or deteriorate (Bramley 2006).

The growth of adventitious roots during waterlogging is also likely to influence L_r. These roots emerge from nodes on the base of the stem and are constitutive in Gramineae (called nodal or crown roots) and many wetland species (Barlow 1994; Kovar and Kuchenbuch 1994). However, waterlogging often stimulates their growth in species that do not develop adventitious roots under ambient conditions. Adventitious roots usually develop aerenchyma (see 8.4.2), which enables them to survive and grow in waterlogged soil, albeit with limited length (Armstrong et al. 1991). The development of adventitious roots appears to be particularly important where oxygen deficiency causes extensive mortality of primary roots (Etherington 1984; Harrington 1987; Moog and Janiesch 1990; Solaiman et al. 2007) and their importance in maintaining plant growth during waterlogging was demonstrated in pruning experiments (Etherington 1984). Adventitious roots become the main source for nutrient and water uptake, but their contribution to L_r of waterlogged root systems has not been investigated.

8.3.2 Barriers to Radial Flow

Once water has been absorbed by the root, it has to cross a series of concentric tissues to reach the vasculature. Oxygen deficiency triggers physical changes to some of these tissues, which may influence L_r. The most common change is the development of large gas-filled tubes in the cortex, called aerenchyma (Fig. 8.1). Aerenchyma forms predominantly by cell lysis in adventitious roots, but can also form in new seminal roots and some above-ground organs (Colmer 2003; Vartapetian 2006). Schizogenous aerenchyma also forms constitutively in wetland species. Often coinciding with aerenchyma in the basal part of the root is an impermeable barrier to oxygen in the outer cell layers (Fig. 8.1), but the occurrence of this feature depends on the particular species (see review in Colmer 2003). Both features are adaptations to enhance internal root aeration and both constitutively occur in many species inhabiting flooded soils (Vartapetian 2006).

Aerenchyma potentially poses a large resistance to radial flow because it dramatically reduces the surface area for liquid flow. Without septa providing an apoplastic pathway across the air spaces, water would have to cross in the form of vapour. There have been few examinations of the influence of aerenchyma on L_r under waterlogging, but a decrease in L_r of phosphorus deficient Z. mays was associated with the formation of aerenchyma (Fan et al. 2007). Other nutrient deficiencies and abiotic stresses can induce aerenchyma formation, but these may also cause other anatomically or physiologically confounding changes. To investigate the influence of aerenchyma on L_r, it would be more appropriate to study a species that does not develop a barrier to oxygen loss, such as T. aestivum (Colmer 2003). T. aestivum develops aerenchyma in nodal roots under oxygen deficiency,

8 Root Water Transport Under Waterlogged Conditions and the Roles of Aquaporins 155

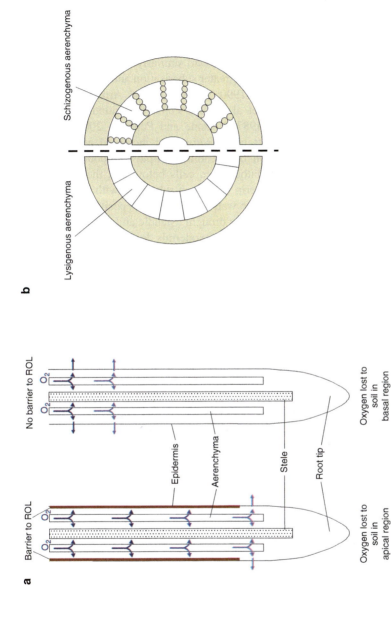

Fig. 8.1 (**a**) Diagram comparing roots with and without a barrier to radial oxygen loss (ROL). Oxygen transport occurs by diffusion from the aerial parts of the plant to the roots via aerenchyma. The barrier has a low oxygen permeability that reduces oxygen loss to the soil. A root with a barrier should have a higher concentration of oxygen in the apical part of the root, compared with an otherwise identical root without a barrier. The barrier may also block water transport via the apoplast in the outer cell layers, so water uptake may be restricted to the apical zone. The apoplast may not be blocked in a root without the barrier to ROL, but aerenchyma may still form a large resistance to radial water flow. (**b**) Diagram of a root cross-section showing lysigenous aerenchyma or schizogenous aerenchyma. Cell walls or cells form septa that could provide a pathway for water flow across aerenchyma

but has non-aerenchymatous nodal roots under ambient conditions. However, increased cellulose deposition with cell wall thickening has been observed in *T. aestivum* seedling roots exposed to 4 d hypoxia, which was speculated to provide mechanical strength where aerenchyma develops (Albrecht and Mustroph 2003).

Barriers to reduce oxygen loss may also increase the resistance to radial water flow as their low permeability to oxygen may impart similarly low permeability to water. The barrier appears to be formed by greater suberisation and/or lignification of the exodermis (Moog and Janiesch 1990; Armstrong and Armstrong 2005; Enstone and Peterson 2005; Soukup et al. 2007), whilst the endodermis often becomes less suberised (Harrington 1987; Enstone and Peterson 2005). Under stagnant conditions, adventitious roots of *O. sativa* had increased aerenchyma development and increased deposition of phenolic or lipid compounds in the hypodermis, the layer of sclerenchymatous fibre cells below and the cell layer beneath that (Insalud et al. 2006). Suberin is a waxy substance with hydrophilic properties creating an apoplastic barrier to water flow and its deposition is usually associated with the retention of water during drought (Steudle and Peterson 1998). However, whether increased suberisation of the exodermis decreases L_r is a matter of debate (Steudle and Peterson 1998).

The influence of the barrier to oxygen loss on L_r has only been studied in two species, *Hordeum marinum* and *O. sativa* (Ranathunge et al. 2003; Garthwaite et al. 2006). Adventitious roots of *H. marinum* developed a barrier to oxygen loss when grown under stagnant conditions, which did not influence L_r compared with roots grown under aerated conditions (Garthwaite et al. 2006). However, the barrier formed proximal of 60 mm from the root tip, which could be the reason it had little effect on L_r, i.e. if most water uptake occurs distall from this position. Maximal water uptake of cereal roots generally occurs in the apical region (Greacen et al. 1976), where the barrier did not form. Despite the development of a barrier in *O. sativa* roots, which reduced oxygen loss, water flow across the outer cell layers apparently occurred predominantly through the apoplast (Ranathunge et al. 2004; Ranathunge et al. 2005) and the outer part of the root was not rate limiting to water flow (Ranathunge et al. 2003). This is contrasting with Miyamoto et al. (2001) who reported that apoplastic barriers and development of the endodermis resulted in low L_r of *O. sativa* roots. *O. sativa* roots tended to have lower hydraulic conductivity per unit surface area than other herbaceous species (Miyamoto et al. 2001), which may be because water uptake occurs predominantly in the apical zone. In addition, measurements with ion-selective microelectrodes on *O. sativa* adventitious roots demonstrated that rates of NO_3^- and NH_4^+ uptake were greatest in the apical region and decreased with distance from the tip (Colmer and Bloom 1998). The decrease in ion uptake in the basal region was related to the development of sclerenchymatous fibres in the outer cortex.

Whilst the limited evidence indicates that barriers to oxygen loss and aerenchyma do not limit L_r of cereal roots under waterlogged conditions they may pose some constraint when the soil drains and there is less water available. The region of root involved in water uptake may be restricted, with little capacity for adjustment apart from growth of new roots. In addition, adventitious roots will be shallow and may be susceptible to drought and fluctuations in temperature. However, there are some suggestions that aerenchyma may have a secondary role in water storage (van der Weele et al. 1996; Kozela and Regan 2003). Plants that inhabit environments that frequently flood tend to have dimorphic root systems, where roots that penetrate the soil are thick, with few branches and have well developed aerenchyma, but surface roots are fine and heavily branched and are responsible for most nutrient uptake (Koncalová 1990).

8.3.3 Varying the Root or Root Region Involved in Water Uptake

The combined development of adventitious roots and death of primary roots may thus explain some of the transient responses in L_r. Measurements on anoxic *H. annuus* roots also indicate other mechanisms causing transient changes in L_r. Under anaerobic treatment, L_r of *H. annuus* initially declined, but after 22 h increased to values greater than aerated roots (Everard and Drew 1989). The roots were not dead and there was no evidence of xylem blockage. Anaerobic treatment for 6 d killed roots and L_r of these roots was similar to roots killed by heat immersion. Everard and Drew (1989) argued that L_r initially decreased because water flow occurs through hydrophilic pores spanning cell membranes that require metabolic energy to sustain them; a rather insightful conclusion considering these water "pores", now called aquaporins (see 8.6.1), were not discovered in plants until a few years later. The closure or reduced expression of aquaporins also probably explains the decrease in L_r of oxygen deficient *Z. mays* root segments (Birner and Steudle 1993; Gibbs et al. 1998), although it was not tested. L_r of hypoxic root segments recovered after 4–6 h of treatment, but did not recover in anoxic roots (Birner and Steudle 1993; Gibbs et al. 1998). An interesting observation from both of these studies was that L_r measured using a hydrostatic gradient was reduced more than L_r measured with an osmotic gradient, indicating different pathways for radial water flow.

Some species can vary the root or region of root for preferential water uptake, which may be particularly beneficial in stochastic and heterogeneous environments. During waterlogging, different parts of the root system may be located in wet soil if the soil is not saturated to the surface. Even if the plant is highly sensitive to oxygen deficiency and its submerged roots die, the oxygen-sufficient roots may be able to supply adequate water to maintain shoot growth. In an experiment on *Larix laricina*, where only the lower half of the root system was flooded, water uptake in the upper non-flooded part transiently increased (Reece and Riha 1991), but decreased in the flooded part. In roots totally flooded, water uptake decreased in both parts and transpiration and stomatal conductance decreased more in full-flooded plants (Reece and Riha 1991).

T. aestivum is a particularly good example of a species that can regulate L_r, as it can alter L_r of individual roots and the hydraulic conductance of different parts of an individual root (Vysotskaya et al. 2004; Bramley et al. 2009; Bramley et al. 2010). Severing roots from plants grown in solution culture resulted in L_r of the remaining root increasing, so that water supply to the shoot was maintained (Vysotskaya et al. 2004).

In an experiment on excised root segments, 0.5 h hypoxia only reduced L_r of roots shorter than 120 mm (Bramley et al. 2010). This was explained by longer roots subjected to hypoxia being able to increase the length of root over which preferential water transport occurred, so that there was then no apparent change in root conductance. When roots were re-aerated, L_r increased in all roots irrespective of length and L_r was greater than control roots. Hypoxia dramatically decreases the hydraulic conductivity of root cortical cells through decreased aquaporin activity (Zhang and Tyerman 1991; Zhang and Tyerman 1999; Bramley et al. 2010). Modelling and measurements on cells in different cell layers indicated that variation in the water permeability of the endodermis could account for the variation in L_r (Bramley et al. 2010). In an unrelated study, L_r of *T. aestivum* grown under stagnant conditions but measured under aerated conditions, was greater than plants grown under aerated conditions (Garthwaite et al. 2006). No explanation for this observation was purported, but the response may be similar to the overshoot in L_r reported by Bramley et al. (2010).

8.4 Volatile and Toxic Compounds in Anaerobic Soils

Ethylene, CO_2 and CH_4 accumulate in flooded soils because of restricted gas exchange with the atmosphere (Vartapetian 2006). Elevated levels of these gases may have an effect on L_r through their interaction with aquaporin activity, provided water flows across membranes. Ethylene is a signalling molecule that triggers aerenchyma development. In hypoxic aspen roots, ethylene enhanced L_r, which was speculated to increase phosphorylation of aquaporins (Kamaluddin and Zwiazek 2002). In contrast, CO_2 tends to decrease L_r, but the mechanism is unknown (Newman 1976; Smit and Stachowiak 1988). Anaerobic soils also accumulate reduced ions such as nitrous oxide, ferrous ions, sulphides and organic substances such as lactic acid, ethanol, acetylaldehyde and aliphatic acids (Cannell and Jackson 1981). The effect of these substances during waterlogging is often overlooked because primary injury occurs from oxygen deficiency (Drew 1997). However, their accumulation may have an antagonistic effect with oxygen deficiency on L_r.

8.5 Water Permeability of Root Cells and Aquaporins

The water permeability of root cells during waterlogging has received little attention and yet the measurements at the cell level provide direct evidence for the temporal sensitivity to oxygen deficiency, as well as transport processes that are affected. For example, turgor pressure of cortical cells near the tip in *T. aestivum* roots decreased during 0.5 h of hypoxia, which was caused by membrane depolarisation and leakage of osmotica (Zhang and Tyerman 1991; Zhang and Tyerman 1997). In another study, cortical cells in the mature root region maintained turgor pressure during similar hypoxic treatment of *T. aestivum*, but not in two *Lupinus*

8 Root Water Transport Under Waterlogged Conditions and the Roles of Aquaporins 159

species (Bramley et al. 2010). In all of these species, hypoxia (0.04–0.05 mol O_2 m^{-3}) dramatically reduced cell hydraulic conductivity (Lp_c), which we now attribute to an inhibition of aquaporin activity (Zhang and Tyerman 1991; Zhang and Tyerman 1999; Bramley et al. 2010).

Flow across cell membranes and through cells may be part of the pathway for radial water transport and as such, may impart great influence on water flow through roots. It is therefore, important to understand the mechanisms affecting cell water permeability. Since the discovery of plant aquaporins in the early 1990's (Maurel et al. 1993; Chrispeels and Agre 1994) there has been renewed interest in plant hydraulics, because these ubiquitous proteins control water flow across membranes.

8.5.1 Plant Aquaporins

The majority of aquaporin genes are predominantly expressed in roots (Bramley et al. 2007) and therefore, their location, abundance and gating, can potentially control the rate of water flow through whole plants and the fluxes of other molecules and gasses that may need compartmentation or extrusion from the cell. In this section, we review the effects of root submergence and oxygen deficiency on cell water relations and the role of aquaporins in water transport and transport of other molecules relevant to anoxia and hypoxia.

Aquaporins are members of the major intrinsic protein (MIP) group of transmembrane channels found ubiquitously in all organisms. In the last decade MIPs have also been shown to facilitate the transport of other small neutral molecules, in addition to water. This is particularly relevant to the metabolic consequences of hypoxia and anoxia, as we elaborate further below.

MIPs form tetramers in the membrane, where each monomer appears to function as an individual functioning pore. However, interactions between tetramers that affect transport may occur and the central pore in the tetramer may function as a channel (Yu et al. 2006; Bertl and Kaldenhoff 2007; Fig. 8.2). The MIP superfamily is a large gene family in plants; 35 MIP encoding genes have been identified in *Arabidopsis thaliana* (Johanson et al. 2001), 31 in *Z. mays* (Chaumont et al. 2001), 33 in *O. sativa* (Sakurai et al. 2005), 24–28 in *Vitis vinifera* (Fouquet et al. 2008; Shelden et al. 2009) and 37 in *L. esculentum* (Sade et al. 2009). *T. aestivum* appears to have an even larger set of aquaporin genes based on the identification of 35 genes just from the PIP and TIP groups (Forrest and Bhave 2008).

Plant aquaporins have been traditionally divided into four groups based on sequence homology, and two of these groups (tonoplast intrinsic proteins, TIP; and plasma-membrane intrinsic proteins, PIP) seems to match their membrane localization; though not always for some PIPs (Kirch et al. 2000; Whiteman et al. 2008). In addition to the PIPs and TIPs, there are the NOD26-like intrinsic proteins (NIPs), and the small basic intrinsic proteins (SIPs) (Johanson et al. 2001; Johanson and Gustavsson 2002). New subgroups of MIPs have been identified from the moss *Physcomitrella patens*, which have 23 MIPs (Danielson and Johanson 2008). One of these (X Intrinsic Proteins, XIPs) has been identified in other plants

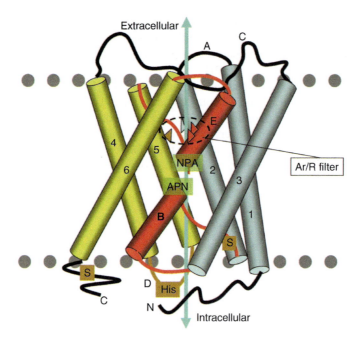

Fig. 8.2 Diagram of the structure of a PIP aquaporin monomer with six transmembrane helices (1–6) and five connecting loops (**a–e**). PIPs normally occur as a tetramer in the membrane. Shown are the highly conserved NPA motifs; histidine (H) residue, involved in cytosolic pH sensing; two phosphorylation sites [serine (S) residues] in the loop B and C-terminal tail of aquaporins of the PIP2 subgroup. The Ar/R selectivity filter is indicated. The diagram is not to scale and is only a general indicator of the structure (adapted from Luu and Maurel (2005) and Tornroth-Horsefield et al. (2006))

including *L. esculentum* (Sade et al. 2009). The NIPs have been located to both endomembranes and the plasma-membrane (Zhang and Roberts 1995; Mizutani et al. 2006; Choi and Roberts 2007). The NIPs are an important group with respect to the effects of oxygen deficiency, because of their potential to facilitate the transport of gasses and end products of anaerobic respiration.

The PIPs have been divided into two groups: the PIP1 group often show lower or no water permeability when expressed in Xenopus oocytes, in contrast with the PIP2 group that show high water permeability (Chaumont et al. 2000; Moshelion et al. 2002; Fetter et al. 2004; Vandeleur et al. 2009). The PIPs are an interesting group because: (1) They are probably the rate limiting aquaporins in transcellular water flow, because the tonoplast membrane has much higher water permeability under normal conditions (Maurel et al. 1997; Niemietz and Tyerman 1997). Though there may be exceptions given the extraordinarily high water permeability measured for *Beta vulgaris* plasma-membrane (Alleva et al. 2006); (2) They are inhibited by low pH in the physiological range (Tournaire-Roux et al. 2003; Alleva et al. 2006; Verdoucq et al. 2008); (3) They are gated by phosphorylation and cytosolic Ca^{2+} concentration (Johansson et al. 1996; Johansson et al. 1998;

Törnroth-Horsefield et al. 2006). But note that NIPs and TIPs can be regulated by phosphoryation (Maurel et al. 1995; Guenther et al. 2003; Wallace et al. 2006); (4) Some are implicated in facilitating the permeation of CO_2 (Uehlein et al. 2003; Hanba et al. 2004; Flexas et al. 2006; Uehlein et al. 2008) and there is evidence for facilitated oxygen transport by mammalian AQP1 and a link to hypoxia-inducible transcription factor (Echevarria et al. 2007).

The selectivity of aquaporins is determined by the asparagine-proline-alanine (NPA) motif at the end of the two half helices that extend into the pore (Fig. 8.2) and the aromatic-arginine (ar/R) region, which is composed of a tetrad of residues on helices 2 and 5, and two positions on loop E (Fig. 8.2). The ar/R region is on the apoplastic side of the NPA filter. The ar/R region determines both the selectivity and transport rate of MIPs. In plant MIPs, the structure of the ar/R filter has been examined for each of the four groups, revealing that PIPs have an ar/R signature typical for high water transport, TIPs have three separate conserved ar/R filters, and NIPs possess two conserved ar/R filters (Wallace and Roberts 2004). The NIPs can be subdivided into two groups that differ mainly by the substitution of an alanine (NIP subgroup II) for a tryptophan (NIP subgroup I) in the helix 2 of the filter. A subgroup I representative, NOD26, can have high water permeability and transports glycerol and formamide, while AtNIP1;6 showed no water transport but transported glycerol, formamide, and urea (Wallace and Roberts 2005). Eight *A. thaliana* NIPs (out of nine) appear to be expressed in roots, but those that have highest expression in roots, NIP1;1 and NIP5;1 represent members of subgroup I and II respectively (Wallace et al. 2006). NIP2;1, which can transport lactic acid and is upregulated, is a member of subgroup I, but does not transport water or glycerol (Choi and Roberts 2007).

8.5.2 Responses at the Cell Level Affecting Water Permeability and Potential Mechanisms

There is no known mechanism of direct oxygen sensing in plants, but indirect sensing may occur through changes in cytosolic pH, calcium concentration, reduction in ATP, or production of NO and H_2O_2. Various membrane transport processes are regulated directly or indirectly by cytosolic ATP, pH, free Ca^{2+} and reactive oxygen species (ROS) (Amtmann and Blatt 2009). It would not be surprising therefore, if MIPs were sensitive to these factors and hence, to hypoxia and anoxia. Indeed, we have an increased understanding now, of the mechanisms of how PIPs are regulated by pH, free Ca^{2+}, phosphorylation (Alleva et al. 2006; Hedfalk et al. 2006) and H_2O_2 (Boursiac et al. 2008). In addition, like ion channels, one might expect water channels to be sensitive to the gradients that generate flow, either osmotic gradients, pressure gradients or a combination of both.

8.5.2.1 Changes in Water Potential

Flooding may occur with solutions of different composition and salinity. Flooding with non-saline water would elevate the soil matrix potential to near zero and consequently, root cells adjacent to this medium would rapidly equilibrate (within several tens of seconds) to a high water potential. Saline water, on the other hand, brings with it both low osmotic potential and ion toxicity effects. The latter has been shown to depress hydraulic conductivity of roots (Azaizeh and Steudle 1991; Martinez-Ballesta et al. 2003), root cortical cells (Azaizeh et al. 1992) or protoplasts and vesicles from roots (Martínez-Ballesta et al. 2008), but not always (Tyerman et al. 1989). External calcium can ameliorate this response, probably via a combination of effects on aquaporin gating and expression (Martínez-Ballesta et al. 2008). Depression of aquaporin activity (expression and density in the membrane) has been implicated in the response to salinity (Martinez-Ballesta et al. 2003; Boursiac et al. 2008) and the internalisation of PIP aquaporins from the plasma-membrane is mediated by H_2O_2. This observation is important to keep in mind when considering the mechanism of depressed water permeability under hypoxia and anoxia.

Water potential or components thereof have been implicated in the control of aquaporins, either by phosphorylation (Johansson et al. 1996; Guenther et al. 2003), or by direct interactions of osmotica (Vandeleur et al. 2005; Ye et al. 2005). Drought (low water potential) certainly has large effects on aquaporin expression and activity in roots (Parent et al. 2009; Vandeleur et al. 2009) and ABA is implicated in these responses (Parent et al. 2009).

8.5.2.2 Decreased ATP, Implications for Transport and Interactions with Aquaporins

Although it is beyond the scope of this chapter to discuss the various pathways of carbohydrate catabolism that occur during anaerobiosis, it is prudent to consider some of the end products that can accumulate under anaerobic metabolism, because these may be substrates for transport out of the cytosol (to the vacuole or apoplast) via some MIPs (Table 8.1). For roots that do not contain aerenchyma, flooding and the hypoxia/anoxia that eventuates cause a respiratory crisis where energy production is reduced by more than 65% (Gibbs and Greenway 2003). This necessitates ATP production via fermentation with the end products of lactic acid or ethanol. Much less ATP is produced per glucose molecule catabolized (2:36–38) than when the tricarboxylic acid cycle (TCA) uses O_2 as the final electron acceptor in the mitochondrial electron transport chain. Plant cells can adapt to these conditions by increasing glycolysis (Pasteur effect) and increasing the regeneration of NAD^+ for glycolysis by induction of enzymes of fermentation pathways (ethanolic fermentation, lactic acid fermentation, and alanine fermentation) (Kennedy et al. 1992; Gibbs and Greenway 2003). Some anoxia tolerant plants, such as *O. sativa*, may also use a pyrophosphate (PPi) dependent glycolysis, which can increase the yield

8 Root Water Transport Under Waterlogged Conditions and the Roles of Aquaporins

Table 8.1 Small molecules related to anoxia, hypoxia or re-aeration after anoxia, that may or do permeate via MIPs

	Molecule	Link with MIP	pKa of acids	Reference to MIP or accumulation of molecule where relevant
Gases	O_2	HsAQP1		MIP = (Echevarria et al. 2007)
	CO_2	NtAQP1; HvPIP2;1		MIP = (Uehlein et al. 2003; Hanba et al. 2004; Flexas et al. 2006)
Signalling	NO	?		Acc = (Greenway et al. 2006)
	H_2O_2	AtTIP1;1, AtTIP1;2 AtPIP2;4 > AtPIP2;1 > AtNIP1;2 > AtTIP2;3		Acc = (Dordas 2009); Acc = (Blokhina et al. 2003); MIP = (Bienert et al. 2007; Dynowski et al. 2008)
	Ethylene	Ethylene can enhance L_r, inhibited by Hg	Neutral	Acc = (Fukao and Bailey-Serres 2008); MIP = (Kamaluddin and Zwiazek 2002)
Catabolic end products	Lactic acid	AtNIP2;1	3.1	MIP = (Choi and Roberts 2007); Acc = (Bailey-Serres and Voesenek 2008)
	Alanine	Peribacteroid membrane, Hg inhibited	Zwitterion	MIP = (Day et al. 2001); Acc = (Kato-Noguchi and Ohashi 2006)
	Ethanol	Chara plasma-membrane, ROS inhibited	Neutral	MIP = (Henzler et al. 2004)
	Acetaldehyde	?	Neutral	Acc = (Meguro et al. 2006)
	Acetic acid	?	4.8	Acc = (Meguro et al. 2006)
	γ-aminobutyric acid (GABBA)	?	Zwitterion	Acc = (Kato-Noguchi and Ohashi 2006)
	γ-hydroxybutyric acid	?	4.7	Acc = (Bailey-Serres and Voesenek 2008)
	Succinic acid	?	4.2, 5.6	Acc = (Bailey-Serres and Voesenek 2008)
	Malic acid	Once thought to permeate via NOD26	3.4, 5.1	MIP = (Ouyang et al. 1991); Acc = (Bailey-Serres and Voesenek 2008)
Lipid peroxidation	Ethane	?	Neutral	Acc = (Santosa et al. 2007)
Antioxidants	Dehydroascorbate	High affinity transporter identified	Neutral	Acc = (Blokhina et al. 2000; Garnczarska 2005); Tran = (Horemans et al. 2000)
	Ascorbic acid	? Transporter identified	4.1	Acc = (Blokhina et al. 2000; Garnczarska 2005); Tran = (Horemans et al. 2000)

For acids it is more likely the neutral form will permeate and this depends on the pKa (s). References are given for the association to MIPs (MIP=) or other transporters (Tran=), and for the accumulation of the molecules (Acc=), where relevant of ATP per glucose molecule from 2 to 5 molecules of ATP (Huang et al. 2008). Increased PPi will maintain proton pumping across the tonoplast via the H^+-PPiase. Another interesting adaptation is the potential use of nitrite as an electron acceptor. Mitochondria isolated from *O. sativa* and *Hordeum vulgare*, under anaerobic conditions, can use nitrite as an electron acceptor to oxidize cytosolic NADH/NADPH and generate ATP (Stoimenova et al. 2007). Nitric oxide (NO) is an end product of this process and may be scavenged by a non-symbiotic haemoglobin and ascorbate (Igamberdiev and Hill 2009).

To conserve energy, expression and translation of many genes are suppressed with the exception of particular genes required for acclimation. Some of the genes that are upregulated in the early stages of hypoxia/anoxia include some TIPs and NIPs. Interestingly, two of the TIPs from *A. thaliana* (AtTIP1;2 and AtTIP4;1) are also implicated in urea transport across the tonoplast and are upregulated in roots under nitrogen deficiency (Liu et al. 2003), while AtNIP2;1 transports lactic acid and is located on the plasma-membrane (Choi and Roberts 2007).

The energy crisis means that electrochemical proton gradients established across the plasma-membrane and tonoplast, generated by H^+-ATPases, may decline. This is subject to the extent of H^+-ATPase deactivation and the leakage of protons down the gradients into the cytoplasm. The latter can be via direct leakage of protons, thought to be minor (but see below), or co-transport reactions which drive the much needed active influx of reduced carbon (Felle 2005). Hypoxia and anoxia depolarize the plasma-membrane voltage (Buwalda et al. 1988; Zhang and Tyerman 1997), as would be expected if the H^+-ATPase is inhibited (Greenway and Gibbs 2003). The voltage gradient, which is large across the plasma-membrane, can be sustained at a somewhat less hyperpolarised level for a short period without the H^+-pump. This occurs by slight K^+ leakage out of the cell via K^+-outward rectifier channels, which tend to maintain a membrane voltage equal to the potassium diffusion potential, the so called K^+-state. Proton pumping across the tonoplast may be sustained by more reliance on the H^+-PPiase (Greenway and Gibbs 2003). Under hypoxia, both efflux of K^+ (Buwalda et al. 1988) and decline in net K^+ flux has been observed in seminal roots of *T. aestivum* seedlings (Kuiper et al. 1994) and the mature zone of *H. vulgare* roots of a waterlogging sensitive variety, but not in a tolerant variety (Pang et al. 2006). The mature zone response has been attributed, in part, to increased K^+ efflux via K^+-outward rectifier channels (Pang et al. 2006). The activation of K^+-outward rectifier channels was not associated with altered cell membrane hydraulic conductivity in algae (Schutz and Tyerman 1997) but remains to be tested directly in roots. There is, however, an association of mRNA levels of PIP aquaporins and K^+ transporters in *O. sativa* roots (Liu et al. 2006).

The extent of activation of anion efflux transport, for which NO_3^- and Cl^- can have high intrinsic efflux rates from roots (Britto et al. 2004; Segonzac et al. 2007) may have the effect of strong membrane voltage depolarisation (going more positive). If both K^+ and anions are effluxed by the combined action of both K^+ channels and anion channels there will be a loss of osmotica that will decrease turgor pressure. At high concentrations of nutrients, where low affinity influx mechanisms dominate, there can be energy wasting recycling via efflux from root cells (Britto and Kronzucker 2006) and this would seem to be particularly problematical under anoxia. Acid load of *A. thaliana* roots increased unidirectional NO_3^- efflux, leading to NO_3^- loss (Segonzac et al. 2007). The role of NO_3^- is particularly interesting because of its stimulatory effect on root water transport (Gloser et al. 2007), via aquaporins (Gorska et al. 2008a, b), and because of its ameliorative effects during anoxia (Libourel et al. 2006). In the study by Gorska et al. (2008a, b), anoxia treatment was used as an inhibitor of aquaporins to test the effect of nitrate on root hydraulic conductivity and to indicate a role for aquaporins in NO_3^- stimulation. Nitrate in the cytosol increased aquaporin activity, but not via increases in mRNA expression of PIP1 and PIP2. Given the ameliorative effect of NO_3^- and NO_2^- on cytosolic pH, and the proposal that stimulation of water flow by NO_3^- is to advect the ion to the roots (Gloser et al. 2007), one would predict that there may be an advantage of NO_3^- to counteract the inhibitory affect of anoxia on PIP aquaporins.

Both K^+ and Cl^- have been observed to leak from anoxia treated *T. aestivum* roots, more-so in the expansion zone (Greenway et al. 1992) where there is also concentration of water uptake (Bramley et al. 2009). It is interesting to note that when substances that block aquaporins, (which are also toxic to metabolism e.g. Hg) are applied to roots there is invariably a reduction in turgor pressure, particularly when the inhibition of water transport is severe. This reduction is not caused by a reduced reflection coefficient, which can be a consequence of large reduction in hydraulic conductivity (Schutz and Tyerman 1997), but rather a leak of K^+ and anions (Bramley and Tyerman unpublished). The link with aquaporin inhibition is interesting and it is worthwhile exploring why aquaporin inhibition, resulting in reduced cellular hydraulic conductivity, is linked to any treatment that appears to compromise the energy status of the cell.

One mechanism of post-translational regulation of aquaporin activity is reversible phosphorylation. Phosphorylation of plant MIPs can increase water permeability (Maurel et al. 1995; Johansson et al. 1998; Guenther et al. 2003). In *Spinacia oleracea* leaves, SoPIP2;1 is dephosphorylated under drought stress and therefore inactivated (Johansson et al. 1996). Dephosphorylation occurs at two highly conserved serine residues, Ser115 in cytosolic loop B and, Ser274, in the C terminus, by a Ca^{2+} dependent protein kinase (Johansson et al. 1996; Johansson et al. 1998). It is not known whether dephosphorylation of PIPs and TIPS during hypoxia and anoxia is involved in reduced water permeability of root membranes (but see 8.3. above).

8.5.2.3 Decrease in Cytosolic pH

The cytosol becomes more acidic under anoxia, depending upon species and tissue and pre-exposure to hypoxia (Vartapetian and Jackson 1997). Decreases in pH from around 7.4–7.5 to 6.5–7.3 have been recorded (Felle 2005). There has been some considerable effort to explain the origin of the excess protons. Lactate fermentation acidifies the cytoplasm and lactic acid must be effluxed from the cell, alternatively there can be a switch to ethanolic fermentation through suppression of LDH and enhancement of PDC. Ethanol also must exit the cell. There are mixed results regarding the association between the ability to switch to ethanolic fermentation and anoxia tolerance (Vartapetian and Jackson 1997; Felle 2005). Some plants show a short initial production of lactic acid, while others show more sustained production in addition to ethanol production (Vartapetian and Jackson 1997). Nucleotide triphosphate hydrolysis has also been associated with cytoplasmic acidification (Gout et al. 2001). An alternative view of the acidification of the cytosol is that it represents a new set point for regulation (Greenway and Gibbs 2003; Felle 2005). Several organic acids besides lactic acid may accumulate as end products under anaerobic conditions and these are listed in Table 8.1. The change in cytosolic pH under anoxia links directly to the molecular basis of decreased plasma-membrane water permeability via pH gating of PIP aquaporins.

Plasma-membrane water permeability is strongly inhibited by reduced cytosolic pH and one clear example is plasma-membrane vesicles from *Beta vulgaris* storage roots, which have very high water permeabilities (>500 μm s^{-1}). In this case, a one-hundred-fold inhibition was observed when the cytosolic face of the plasma-membrane was made acidic. The pH for half-maximum inhibition was pH 6.6 and only the cytosolic face of the membrane showed sensitivity (Alleva et al. 2006). Measurements with yeast expressing PIP aquaporin gave half-inhibition at pH 7.1 (Verdoucq et al. 2008) and plasma-membrane vesicles from *A. thaliana* roots had a half-inhibition of pH 7.2–7.5 (Gerbeau et al. 2002). Plasma-membrane vesicles from *T. aestivum* seedling roots had maximum water permeability at pH 8.5 and a minimum at pH 6.5, so that minor changes in pH near seven affected water permeability of the plasma-membrane (Vandeleur et al. 2005). The tonoplast water permeability also shows strong sensitivity to cytosolic pH (Sutka et al. 2005). These examples demonstrate that this effect is within physiological cytosolic pH and within the range of changes in cytosolic pH observed under anoxia.

Cytosolic and apoplastic pH are major factors in control of K$^+$ and anion/Cl$^-$ channels at the plasma-membrane of guard cells (Amtmann and Blatt 2009) and some evidence is available also for K$^+$ transport and K$^+$-channels in root cells (Hartje et al. 2000; Babourina et al. 2001). It may not be so surprising, therefore, that PIP aquaporins are also regulated by pH. However, unlike the situation with K$^+$ and Cl$^-$-channels, where there is a link between proton gradients and the energization of transport of these ions, the link between proton gradients and water transport is less obvious.

When aquaporins are expressed in Xenopus oocytes or yeast cells, cytoplasmic acidification with a weak acid can be used to examine the sensitivity of PIPs and TIPs to acidic pH (Fischer and Kaldenhoff 2008). *A. thaliana* AtTIP1;1 was insensitive while AtPIP2;1, AtPIP2;2, AtPIP2;3 were inhibited (Tournaire-Roux et al. 2003). PIP1 isoforms generally do not show water permeability in Xenopus unless co-expressed with PIP2 ioforms (Fetter et al. 2004; Vandeleur et al. 2009). Only *A. thaliana* AtPIP1;2 displayed a significant water transport activity in oocytes, which was blocked by acidifying the cytoplasm (Tournaire-Roux et al. 2003). With respect to AtPIP2;3, this isoform is located in the stele of *A. thaliana* roots (Wang and

Tyerman unpublished) and may be important in water transfer to the xylem vessels. As the stele is likely to become anoxic sooner under hypoxic conditions it would be interesting to examine this PIP's pH sensitivity in further detail.

All PIP1 and PIP2 aquaporins have conserved histidine residues and His 197 (Fig. 8.2) appears to be a major determinant of pH sensitivity for AtPIP2;2 (Tournaire-Roux et al. 2003). The decrease in cytoplasmic pH may therefore, explain the sudden reduction in hydraulic conductivity when roots were subjected to anoxic stress (Tournaire-Roux et al. 2003). A number of residues in loop D of *S. oleracea* SoPIP2;1 have been identified as being involved in gating of the channel (Törnroth-Horsefield et al. 2006). Structural studies show that loop D forms a hydrophobic gate over the pore and phosphorylation of serines in loop B and at the C terminus can open the gate (Törnroth-Horsefield et al. 2006). The situation may be more complicated than this or PIPs from different species may have different gating mechanisms, since combined effects of the serines and histidine on loop D were not supported in a study on tobacco NtPIP2;1 and NtAQP1 (Fischer and Kaldenhoff 2008). They also only observed about 50% inhibition of these PIPs under cytoplasmic acidification of living yeast cells. Complexity was also indicated in another recent study showing that serine mutation to glutamic acid (S115E and S274E) could not replicate opening the gate in SoPIP2;1 (Nyblom et al. 2009).

The question arises as to why PIPs should respond in this way under anoxic conditions, or indeed, any condition that leads to cytosolic acidification. It has been proposed that the inhibition is to divert flows from the cell-to-cell pathway to the apoplast pathway (Zhang and Tyerman 1991), which was extended to a hypothesis for diversion of flow from one part of the root system to another (Vandeleur et al. 2005), analogous to hydraulic redistribution in a root system subject to regions of dry soil. Toxic compounds accumulate in anoxic soils (Bailey-Serres and Voesenek 2008), thus if some roots were exposed to anoxic conditions it may be advantageous to divert water flow to roots in more favourable zones in the soil. One problem with this hypothesis is that all PIP aquaporins have the conserved His residues on loop D, even those that may not be present in roots and cytosolic acidification seems to inhibit water permeation even in leaves and storage organs.

One hypothesis that has not been examined, to our knowledge, is that aquaporins may disrupt the membrane pH gradient under some circumstances. Water transport must be facilitated without dissipation of proton gradients, thus protons, hydronium ions and hydroxyl ions must be excluded from the pore of aquaporins. This is accomplished by the NPA residues on loops B and E (Fig. 8.2) forming a central constriction, and the ar/R constriction in the exterior half of the pore. Mutations that remove positive charge in the ar/R constriction in AQP1 allow proton permeation (Beitz et al. 2006). Molecular dynamic simulations have shown that different exclusion mechanisms may exist for protons, hydronium ions and hydroxyl ions in the pore. Furthermore, there are differences between the periplasmic and luminal mechanisms, specifically for the hydroxyl ion (Jensen et al. 2005). There is also the possibility that the central (fifth) pore of the tetramer may be an ion conducting pathway and it is interesting that Loop D has been implicated in gating this pathway in AQP 1 (Yu et al. 2006). There is evidence for ammonia permeation through this pathway in a TIP (Bertl and Kaldenhoff 2007). It should be noted that it is probably not energetically feasible for water flow through an aquaporin to move a proton against its normal electrochemical gradient, even when the cytosol becomes acidified. Some thousands of water molecules moving down typical gradients (0.1 MPa) would be required to move a proton in the opposite direction. A discovery that links pH gradients and aquaporin function at reduced cytosolic pH would be highly significant in plant biology.

8.5.2.4 Increase in Cytosolic Free Ca^{2+}

Cytosolic free Ca^{2+} increases rapidly after the onset of anoxia and induces increased mRNA of alcohol dehydrogenase (ADH) and sucrose synthase (Subbaiah et al. 1994a, b). Ruthenium red, an inhibitor of Ca^{2+} pumps and channels, inhibited anoxia induced Ca^{2+} influx into maize roots, but external Ca^{2+} was not required for induction of ADH activity. The increase in Ca^{2+} also occurred independently of extracellular Ca^{2+} (Subbaiah et al. 1994a, b). Aequorin expressed in *A. thaliana* seedlings revealed biphasic increases in Ca^{2+} in leaves that could not be detected in roots, but upon return to air transients were observed in roots (Sedbrook et al. 1996). Calcium-sensitive dye imaging of cultured maize cells subject to anoxia showed that the increase in Ca^{2+} can be heterogeneous within a cell (sometimes around the periphery) and between cells, but was strongly associated with mitochondria (Subbaiah et al. 1998).

Ca^{2+} effects on aquaporins appear to be quite complex. Based on structural studies, divalent binding anchors loop D to the N-terminus keeping the gate in the closed state (Nyblom et al. 2009). For plasma-membrane vesicles obtained from *B. vulgaris* storage root, Ca^{2+} showed a biphasic effect on water permeability wherein a steep effect indicating co-operativity was observed at very low concentrations (half-inhibition: 4.5 nM) and a second less intense inhibitory effect was observed at higher concentrations (half-inhibition: 200 μM) (Alleva et al. 2006). Only the high concentration effect has been observed for vesicles isolated from *A. thaliana* roots (half-inhibition: 50–100 μM) (Gerbeau et al. 2002). It was proposed that Ca^{2+} could modulate both opening and closing of the channel based on the biphasic effect observed in *B. vulgaris* (Alleva et al. 2006). Calcium may also regulate PIP aquaporins by phosphorylation via a Ca^{2+}-dependent protein kinase, as has been observed for a *Tulipa gesnerina* plasma-membrane aquaporin (Azad et al. 2004) and the extensively studied SoPIP2;1 (Johansson et al. 1998). It is likely that anoxia effects on cytosolic Ca^{2+} will also affect the function of aquaporins in addition to the effect of acidification.

8.5.2.5 Increase in ROS

ROS production may occur under hypoxia, anoxia and re-aeration (Blokhina et al. 2003). H_2O_2 accumulates in the roots of *H. vulgare* and *T. aestivum* under hypoxia (Kalashnikov et al. 1994) and re-aeration in *T. aestivum* roots (Biemelt et al. 2000).

Re-aeration after anoxia can elevate ROS because of the highly reduced state of the cytoplasm, low energy charge, depleted antioxidants and membrane damage. (Blokhina et al. 2003). The iron-catalysed Fenton reaction will produce highly reactive and damaging OH^- from H_2O_2. Lipid peroxidation products (ethane) have been detected for submerged *O. sativa* seedlings with a transient peak after de-submergence (Santosa et al. 2007). Intermittent anoxia and re-aeration in *T. aestivum* caused oxidative stress in the expanded zone of the root, but there was no evidence of terminal membrane damage (Goggin and Colmer 2005). *A. thaliana* seedlings, when exposed to hypoxia, use a Rop (RHO-like small G protein) signal transduction pathway that activates NADPH oxidase, resulting in increased H_2O_2 production. A RopGAP4 (GTPase that inactivates Rop) provides negative feedback resulting in rheostatic control of H_2O_2 and induction of ADH (Baxter-Burrell et al. 2002). Both H_2O_2 and antioxidants may need to be transported to different compartments under oxidative stress, and some high affinity transporters have been identified for antioxidants, ascorbate/dehydroascorbate (Horemans et al. 1998) and glutathione (Raichaudhuri et al. 2009). Compared to the ascorbate anion ($pKa = 4.17$), dehydroascorbate is more membrane permeant since it is not charged. This is another candidate for permeation via aquaporins, though a high affinity DHA/ascorbate exchanger has been characterised on the plasma-membrane (Horemans et al. 2000).

ROS have been shown to be potent and reversible inhibitors of water transport across the plasma-membrane of algal cells, root cell and root segments (Henzler et al. 2004; Ye and Steudle 2006; Boursiac et al. 2008). This effect was originally suggested to be a direct blockade by hydroxyl radicals, but it has been subsequently shown that H_2O_2 induces a signalling pathway that internalises PIPs to intracellular membranes (Boursiac et al. 2008). Salicylic acid also inhibits water transport in *A. thaliana* roots and this is partially mediated by H_2O_2 signalling (Boursiac et al. 2008).

Besides affecting the density of aquaporins in the plasma-membrane, H_2O_2 also permeates through some aquaporins (Henzler and Steudle 2000; Bienert et al. 2007). Using a yeast screen for sensitivity to H_2O_2, several aquaporins were identified that could facilitate H_2O_2 transport and from plants these were TIP1;1 and TIP1; 2 (Bienert et al. 2007). Representatives of *A. thaliana* aquaporins were tested in yeast for permeation of H_2O_2 (AtPIP1;1, AtPIP2;1, AtPIP2;4, AtTIP2;3, AtNIP1;1, AtNIP1;2) (Dynowski et al. 2008). AtPIP2;4 and AtPIP2;1 showed the most permeation. AtPIP2;4 is relatively highly expressed in roots and shows relatively poor water permeations compared to other PIP2s (Wang and Tyerman unpublished). It is also interesting that AtNIP1;1 can transport glycerol, but appeared not to transport H_2O_2 with the yeast assay (Dynowski et al. 2008).

8.5.3 Other Changes Under Oxygen Deficiency that Could Affect Water Transport

Apart from the factors described above, there are other changes to membranes and in the cell that could affect water permeation under oxygen deficient conditions.

The cell becomes highly reduced under anoxia. Other plant transporters and channels are sensitive to redox potential (Scholz-Starke et al. 2005); for example, potassium channels in root hairs (Grabov and Bottger 1994), Ca^{2+} channels in epidermal cells (Demidchik et al. 2007) the slow vacuolar (SV) channel from *Daucus carota* (Scholz-Starke et al. 2004), and the sucrose transporter SUT1 from *Solanum tuberosum* (Krugel et al. 2008). Redox agents that react with SH groups affect water transport across plasma-membrane vesicles isolated from *Pisum sativum* root and shoot. Roots appeared to have more SH groups on the plasma-membrane and were more sensitive to oxidation (4-fold increase in osmotic water permeability), but the reducing agent Dithiothreitol only resulted in a 40% decrease in osmotic water permeability (Ampilogova et al. 2006). Lipid structure may also interact with aquaporin gating, since gating of voltage sensing K^+ channels is affected by the surrounding lipid (Swartz 2008). Under anoxia, the acute shortage of energy causes changes in lipid composition, with a decrease in desaturation of newly synthesised lipids and decrease in membrane fluidity (Rawyler et al. 2002).

Cell walls may also change in composition under anoxia and hypoxia that will affect extensibility and volumetric elasticity. Extensibility determines the expansion rate for growth while volumetric elasticity will affect the turgor/volume changes for changes in water potential. The volumetric elastic modulus may change under hypoxia (Bramley unpublished), and this could influence the rate of change of turgor and volume to water potential changes in the plant and soil medium. Changes in volumetric elastic modulus and hydraulic conductivity will equally affect the half-time for water potential equilibration. If both are reduced, this will considerably increase the half-time. Stress strain characteristics of *Z. mays* coleoptiles show classic viscoelastic hysteresis loops (Hohl and Schopfer 1995) that can affect the water relations in other systems (Tyerman 1982), and pretreatment of segments with anoxia and H_2O_2 modify the hysteresis loop consistent with wall stiffening (Hohl and Schopfer 1995). ROS production has been observed in the apoplast of the apical region of the elongation zone of water-stressed *Z. mays* roots and was proposed to enhance wall loosening for growth at reduced turgor (Zhu et al. 2007).

8.5.4 Transport of Other Molecules Besides Water Through MIPs Relevant to Flooding

MIPs have been shown to transport, or have been implicated in the transport, of specific neutral molecules relevant to mineral nutrition (Holm et al. 2005; Ma et al. 2006; Tanaka et al. 2008), toxic substances (Bienert et al. 2008), and gas exchange (Flexas et al. 2006). However, only one of them (AtNIP2;1) has been linked to flooding as a facilitator of lactic acid efflux from root cells (Choi and Roberts 2007). Lactic acid is an interesting example because it has a relatively low pKa (3.1). If this

8 Root Water Transport Under Waterlogged Conditions and the Roles of Aquaporins

acid can permeate a NIP from cells with a cytosolic pH > 6.5ish, as the neutral lactic acid, there are numerous other candidate acids that may be transported by other MIPs, and which may prevent toxic levels of the acids building up in the cytosol and assist in maintaining cytosolic pH. Table 8.1 lists some acids and amino acids that may accumulate under flooding and may be worth examining as MIP substrates. Note that some preliminary evidence exists for permeation of alanine via a MIP (Day et al. 2001). In addition, gases such as CO_2, NO and O_2 may be facilitated by MIPs, that may reduce diffusion gradients both in and out of root cells. H_2O_2 has already been shown to be transported by some MIPs, but a role in flooding has not been examined. Transport of ethanol may be enhanced by MIPs (Henzler et al. 2004). If water transporting MIPs are also involved in the transport of other molecules, such as O_2, it would seem counterintuitive for conditions of reduced O_2 to inhibit aquaporins in one way or another, as is indicated from the information above. Similarly the pH dependence of some aquaporins that may transport molecules from the cytoplasm needs to be investigated, e.g. TIP1;2 and TIP4;1.

8.6 Signalling

We stated previously that decreased root water transport is the cause of leaf wilting during waterlogging. However, the accuracy of this statement requires comment. Often there is no clear correlation between wilting, stomatal closure and changes in L_r. For example, L_r did not change in two waterlogged conifer species, but stomatal conductance and transpiration decreased (Reece and Riha 1991). Contrary to this, flooding influenced L_r in *Populus trichocarpa* and xylem water potential was more negative, but stomatal closure was not observed (Harrington 1987). *Lupinus* species wilt and close their stomata so that the shoot rehydrates (Dracup et al. 1998), but oxygen deficiency does not influence L_r, at least in the short-term (Bramley et al. 2010). In flooded *Ricinus communis* plants, gas exchange, leaf wilting and L_r also did not correlate temporally (Else et al. 2001). In waterlogged *L. esculentum,* the relationship between L_r and gas exchange depended on whether the plants were flooded starting in the evening or during the day (Bradford and Hsiao 1982). However, stomatal closure was accompanied by a decline in L_r (due to occlusion of xylem vessels) of *Pyrus* species (Andersen et al. 1984).

There appear, therefore, complex regulation of gas exchange, leaf water potential and L_r, which is species dependent. Hormonal imbalance is suspected to be involved, but as yet the signalling molecule remains elusive (Jackson 2002). Some of the molecules described above, could also be involved, and facilitated by aquaporins and/or aquaporins may be involved in the signalling pathway. Parallels may be drawn with isohydric and anisohydric behaviour of plants experiencing water deficit. Isohydric species tend to close stomata to maintain constant leaf water potential, in comparison with anisohydric species where leaf water potential becomes more negative (Tardieu and Simonneau 1998). An element of iso/anisohydry appears to be the regulation of L_r (Vandeleur et al. 2009). The same mechanisms may thus, also occur during waterlogging. There appears to be a signal from leaves to roots that allows matching of the transpirational demand to water supplied by the roots (Vandeleur et al. 2009). Thus, if anoxia tends to close stomata, this may feedback also to further reduce L_r. Hydraulic signalling has also been proposed for coordination of hydraulic conductance in root systems in response to nitrate in different regions (Subbaiah et al. 1994a, b).

8.7 Conclusion and Future Perspectives

The pathways for water flow through roots are dynamic, so hydraulic conductance is variable. During waterlogging, oxygen deficiency has spatial and temporal effects on the pathways that are not mutually exclusive. These effects influence root hydraulic conductance controlling the rate of water flow to the shoot. The spatial effects have two components; (1) depth of root submergence, i.e. which parts of the root system experience oxygen deficiency. In some species, aquaporins may facilitate increased hydraulic conductance of oxygen-sufficient roots or root regions to compensate for decreased hydraulic conductance of oxygen deficient regions. (2) changes in morphology and anatomy of root systems. Primary roots that die may be replaced by adventitious roots, but changes in anatomy to increase internal aeration may act as barriers to radial water flow. The temporal effects depend on the concentration of oxygen in the root and rhizosphere, length of exposure to oxygen deficiency and the prevailing energy crisis.

Prior to root death or changes in anatomy, decreased root hydraulic conductance may be accounted for by inhibition of aquaporin activity. PIPs are regulated by cytoplasmic pH, free Ca^{2+}, phosphorylation and H_2O_2, all of which are pertinent to cell anoxia/hypoxia. It is likely that anoxia effects on cytosolic Ca^{2+} will also affect the function of aquaporins in addition to the effect of acidification. During re-aeration the increase in ROS may inhibit aquaporins at the plasma-membrane, so recovery from anoxia may also impose a water stress. In addition, expression of aquaporins may by reduced by the energy crisis. On the other hand, at the plasma-membrane and in membranes surrounding intracellular compartments, aquaporins may be involved in the flux of gases (O_2, CO_2 NO), end products of anaerobic respiration (ethanol, lactic acid and other organic acids), or of molecules important in signalling (H_2O_2, NO, ethylene). But the role of these aquaporins during water-logging needs further investigation.

There is now a wealth of information regarding root hydraulics and aquaporins that has greatly advanced our understanding of the potential mechanisms influencing root water flow. More integrated research is required to understand these mechanisms during waterlogging and oxygen deficiency.

References

Aguilar EA, Turner DW et al (2003) Oxygen distribution and movement, respiration and nutrient loading in banana roots (*Musa* spp. L.) subjected to aerated and oxygen-depleted environments. Plant Soil 253(1):91–102

Albrecht G, Mustroph A (2003) Localization of sucrose synthase in wheat roots: increased in situ activity of sucrose synthase correlates with cell wall thickening by cellulose deposition under hypoxia. Planta 217:252–260

Alleva K, Niemietz CM et al (2006) Plasma membrane of *Beta vulgaris* storage root shows high water channel activity regulated by cytoplasmic pH and a dual range of calcium concentrations. J Exp Bot 57(3):609–621

Ampilogova Y, Zhestkova I et al (2006) Redox modulation of osmotic water permeability in plasma membranes isolated from roots and shoots of pea seedlings. Russ J Plant Physiol 53(5):622–628

Amtmann A, Blatt MR (2009) Regulation of macronutrient transport. New Phytol 181(1):35–52

Andersen P, Lombard P et al (1984) Effect of root anaerobiosis on the water relations of several *Pyrus* species. Physiol Plant 62(2):245–252

Armstrong J, Armstrong W (2005) Rice: sulfide-induced barriers to root radial oxygen loss, Fe^{2+} and water uptake, and lateral root emergence. Ann Bot 96(4):625–638

Armstrong W, Beckett PM et al (1991) Modelling and other aspects of root aeration. In: Jackson MB, Davies DD, Lambers H (eds) Plant Life Under Oxygen Stress. SPB Academic, The Hague, The Netherlands, pp 267–282

Atwell BJ (1991) Factors which affect the growth of grain legumes on a solonized brown soil. I. Genotypic responses to soil physical factors. Aust J Agric Res 42(1):95–105

Azad AK, Sawa Y et al (2004) Phosphorylation of plasma membrane aquaporin regulates temperature-dependent opening of tulip petals. Plant Cell Physiol 45(5):608–617

Azaizeh H, Gunsé B et al (1992) Effects of NaCl and $CaCl_2$ on water transport across root cells of maize (*Zea mays* L.) seedlings. Plant Physiol 99:886–894

Azaizeh H, Steudle E (1991) Effects of salinity on water transport of excised maize (*Zea mays* L.) roots. Plant Physiol 97:1136–1145

Babourina O, Hawkins B et al (2001) K^+ transport by *Arabidopsis* root hairs at low pH. Funct Plant Biol 28(7):637–643

Bailey-Serres J, Voesenek LACJ (2008) Flooding stress: acclimations and genetic diversity. Annu Rev Plant Biol 59(1):313–339

Barlow PW (1994) The origin, diversity and biology of shoot-borne roots. In: Davis TD, Haissig BE (eds) Biology of adventitious root formation, vol 62. Plenum, New York

Baxter-Burrell A, Yang Z et al (2002) RopGAP4-Dependent Rop GTPase rheostat control of *Arabidopsis* oxygen deprivation tolerance. Science 296(5575):2026–2028

Beitz E, Wu B et al (2006) Point mutations in the aromatic/arginine region in aquaporin 1 allow passage of urea, glycerol, ammonia, and protons. Proc Natl Acad Sci USA 103(2):269–274

Bertl A, Kaldenhoff R (2007) Function of a separate NH_3-pore in Aquaporin TIP2;2 from wheat. FEBS lett 581(28):5413–5417

Biemelt S, Keetman U et al (2000) Expression and activity of isoenzymes of superoxide dismutase in wheat roots in response to hypoxia and anoxia. Plant Cell Environ 23(2):135–144

Bienert G, Thorsen M et al (2008) A subgroup of plant aquaporins facilitate the bi-directional diffusion of $As(OH)_3$ and $Sb(OH)_3$ across membranes. BMC Biol 6(1):26

Bienert GP, Moller ALB et al (2007) Specific aquaporins facilitate the diffusion of hydrogen peroxide across membranes. J Biol Chem 282(2):1183–1192

Birner TP, Steudle E (1993) Effects of anaerobic conditions on water and solute relations, and on active transport in roots of maize (*Zea mays* L.). Planta 190:474–483

Blokhina O, Virolainen E et al (2003) Antioxidants, oxidative damage and oxygen deprivation stress: a review. Ann Bot 91(2):179–194

Blokhina OB, Virolainen E et al (2000) Antioxidant status of anoxia-tolerant and -intolerant plant species under anoxia and reaeration. Physiol Plant 109(4):396–403

Boursiac Y, Boudet J et al (2008) Stimulus-induced down regulation of root water transport involves reactive oxygen species-activated cell signalling and plasma membrane intrinsic protein internalization. Plant J 56(2):207–218

Bradford KJ, Hsiao TC (1982) Stomatal behaviour and water relations of waterlogged tomato plants. Plant Physiol 70:1508–1513

Bramley H (2006) Water flow in the roots of three crop species: the influence of root structure, aquaporin activity and waterlogging. PhD thesis, Plant Biology, The University of Western Australia, Crawley, WA, Australia

Bramley H, Turner DW et al (2007) Water flow in the roots of crop species: The influence of root structure, aquaporin activity, and waterlogging. Adv Agron 96:133–196

Bramley H, Turner NC et al (2009) Roles of morphology, anatomy and aquaporins in determining contrasting hydraulic behavior of roots. Plant Physiol 150:1–17

Bramley H, Turner NC et al (2010) Contrasting influence of mild hypoxia on hydraulic properties of cells and roots of wheat and lupin. Funct Plant Biol (in press)

Britto D, Kronzucker H (2006) Futile cycling at the plasma membrane: a hallmark of low-affinity nutrient transport. Trends Plant Sci 11:529–534

Britto D, Ruth T et al (2004) Cellular and whole-plant chloride dynamics in barley: insights into chloride-nitrogen interactions and salinity responses. Planta 218:615–622

Buwalda F, Thomson CJ et al (1988) Hypoxia induces membrane depolarization and potassium loss from wheat roots but does not increase their permeability to sorbitol. J Exp Bot 39 (206):1169–1183

Cannell RQ, Jackson MB (1981) Alleviating aeration stresses. In: Arkin GF, Taylor HM (eds) Modifying the root environment to reduce crop stress. American Society of Agricultural Engineers, St. Joseph, Michigan, pp 141–192

Chaumont F, Barrieu F et al (2000) Plasma membrane intrinsic proteins from maize cluster in two sequence subgroups with differential aquaporin activity. Plant Physiol 122:1025–1034

Chaumont F, Barrieu F et al (2001) Aquaporins constitute a large and highly divergent protein family in maize. Plant Physiol 125(3):1206–1215

Choi W-G, Roberts DM (2007) *Arabidopsis* NIP2;1, a major intrinsic protein transporter of lactic acid induced by anoxic stress. J Biol Chem 282(33):24209–24218

Chrispeels MJ, Agre P (1994) Aquaporins: water channel proteins of plant and animal cells. Trends Biochem Sci 19:421–425

Colmer TD (2003) Long-distance transport of gases in plants: a perspective on internal aeration and radial oxygen loss from roots. Plant Cell Environ 26:17–36

Colmer TD, Bloom AJ (1998) A comparison of NH_4^+ and NO_3^- net fluxes along roots of rice and maize. Plant Cell Environ 21(2):240–246

Danielson J, Johanson U (2008) Unexpected complexity of the Aquaporin gene family in the moss *Physcomitrella patens*. BMC Plant Biol 8(1):45

Day DA, Poole PS et al (2001) Ammonia and amino acid transport across symbiotic membranes in nitrogen-fixing legume nodules. Cell Mol Life Sci 58(1):61–71

Demidchik V, Shabala SN et al (2007) Spatial variation in H_2O_2 response of *Arabidopsis thaliana* root epidermal Ca^{2+} flux and plasma membrane Ca^{2+} channels. Plant J 49(3):377–386

Dennis ES, Dolferus R et al (2000) Molecular strategies for improving waterlogging tolerance in plants. J Exp Bot 51(342):89–97

Dordas C (2009) Nonsymbiotic hemoglobins and stress tolerance in plants. Plant Sci 176 (4):433–440

Dracup M, Turner NC et al (1998) Responses to abiotic stress. In: Gladstones JS, Atkins CA, Hamblin J (eds) Lupins As Crop Plants: Biology, Production and Utilisation. CAB International, Oxon, pp 227–262

Drew MC (1992) Soil aeration and plant root metabolism. Soil Sci 154(4):259–268

Drew MC (1997) Oxygen deficiency and root metabolism: Injury and acclimation under hypoxia and anoxia. Annu Rev Plant Physiol Plant Mol Biol 48:223–250

Dynowski M, Schaaf G et al (2008) Plant plasma membrane water channels conduct the signalling molecule H_2O_2. Biochem J 414(1):53–61

Echevarria M, Munoz-Cabello AM et al (2007) Development of cytosolic hypoxia and HIF stabilization are facilitated by aquaporin 1 expression. J Biol Chem 282(41):30207–30215

Else MA, Coupland D et al (2001) Decreased root hydraulic conductivity reduces leaf water potential, initiates stomatal closure and slows leaf expansion in flooded plants of castor oil (*Ricinus communis*) despite diminished delivery of ABA from the roots to shoots in xylem sap. Physiol Plant 111:46–54

Enstone DE, Peterson CA (2005) Suberin lamella development in maize seedling roots grown in aerated and stagnant conditions. Plant Cell Environ 28:444–455

Etherington JR (1984) Comparative studies of plant growth and distribution in relation to water-logging: X. Differential formation of adventitious roots and their experimental excision in *Epilobium Hirsutum* and *Chamerion Angustifolium*. J Ecol 72(2):389–404

Everard JD, Drew MC (1989) Mechanisms controlling changes in water movement through the roots of *Helianthus annuus* L. during continuous exposure to oxygen deficiency. J Exp Bot 40(210):95–104

Fan M, Bai R et al (2007) Aerenchyma formed under phosphorus deficiency contributes to the reduced root hydraulic conductivity in maize roots. J Integr Biol 49(5):598–604

Felle HH (2005) pH regulation in anoxic plants. Ann Bot 96(4):519–532

Fetter K, Van Wilder V et al (2004) Interactions between plasma membrane aquaporins modulate their water channel activity. Plant Cell 16(1):215–228

Fischer M, Kaldenhoff R (2008) On the pH regulation of plant aquaporins. J Biol Chem 283 (49):33889–33892

Flexas J, Ribas-Carbo M et al (2006) Tobacco aquaporin NtAQP1 is involved in mesophyll conductance to CO_2 in vivo. Plant J 48(3):427–439

Forrest K, Bhave M (2008) The PIP and TIP aquaporins in wheat form a large and diverse family with unique gene structures and functionally important features. Funct Integr Genomics 8(2):115–133

Fouquet R, Léon C et al (2008) Identification of grapevine aquaporins and expression analysis in developing berries. Plant Cell Rep 27(9):1541–1550

Fukao T, Bailey-Serres J (2008) Ethylene – a key regulator of submergence responses in rice. Plant Sci 175(1–2):43–51

Garnczarska M (2005) Response of the ascorbate-glutathione cycle to re-aeration following hypoxia in lupine roots. Plant Physiol Biochem 43(6):583–590

Garthwaite AJ, Steudle E et al (2006) Water uptake by roots of *Hordeum marinum*: formation of a barrier to radial O_2 loss does not affect root hydraulic conductivity. J Exp Bot 57(3):655–664

Gerbeau P, Amodeo G et al (2002) The water permeability of *Arabidopsis* plasma membrane is regulated by divalent cations and pH. Plant J 30(1):71–81

Gibbs J, Greenway H (2003) Mechanisms of anoxia tolerance in plants. I. Growth, survival and anaerobic catabolism. Funct Plant Biol 30:1–47

Gibbs J, Turner DW et al (1998) Response to oxygen deficiency in primary maize roots. II. Development of oxygen deficiency in the stele has limited short-term impact on radial hydraulic conductivity. Aust J Plant Physiol 25:759–763

Gloser V, Zwieniecki MA et al (2007) Dynamic changes in root hydraulic properties in response to nitrate availability. J Exp Bot 58(10):2409–2415

Goggin DE, Colmer TD (2005) Intermittent anoxia induces oxidative stress in wheat seminal roots: assessment of the antioxidant defence system, lipid peroxidation and tissue solutes. Funct Plant Biol 32(6):495–506

Gorska A, Ye Q et al (2008a) Nitrate control of root hydraulic properties in plants: translating local information to whole plant response. Plant Physiol 148(2):1159–1167

Gorska A, Zwieniecka A et al (2008b) Nitrate induction of root hydraulic conductivity in maize is not correlated with aquaporin expression. Planta 228(6):989–998

Gout E, Boisson AM et al (2001) Origin of the cytoplasmic pH changes during anaerobic stress in higher plant cells. Carbon-13 and phosphorous-31 nuclear magnetic resonance studies. Plant Physiol 125(2):912–925

Grabov A, Bottger M (1994) Are redox reactions involved in regulation of K^+ channels in the plasma membrane of *Limnobium stoloniferum* root hairs? Plant Physiol 105(3):927–935

Greacen EL, Ponsana P et al (1976) Resistance to water flow in the roots of cereals. In: Lange OL, Kappen L, Schulze E-D (eds) Water and plant life, vol 19. Springer, Berlin, pp 86–100

Greenway H, Armstrong W et al (2006) Conditions leading to high CO_2 (>5 kPa) in waterlogged-flooded soils and possible effects on root growth and metabolism. Ann Bot 98(1):9–32

Greenway H, Gibbs J (2003) Mechanisms of anoxia tolerance in plants. II. Energy requirements for maintenance and energy distribution to essential processes. Funct Plant Biol 30:999

Greenway H, Waters I et al (1992) Effects of anoxia on uptake and loss of solutes in roots of wheat. Aust J Plant Physiol 19:233–247

Guenther JF, Chanmanivone N et al (2003) Phosphorylation of soybean nodulin 26 on serine 262 enhances water permeability and is regulated developmentally and by osmotic signals. Plant Cell 15(4):981–991

Hanba YT, Shibasaka M et al (2004) Overexpression of the barley aquaporin HvPIP2;1 increases internal CO_2 conductance and CO_2 assimillation in the leaves of transgenic rice plants. Plant Cell Physiol 45(5):521–529

Harrington CA (1987) Responses of red alder and black cottonwood seedlings to flooding. Physiol Plant 69(1):35–48

Hartje S, Zimmermann S et al (2000) Functional characterisation of LKT1, a K^+ uptake channel from tomato root hairs, and comparison with the closely related potato inwardly rectifying K^+ channel SKT1 after expression in *Xenopus* oocytes. Planta 210(5):723–731

Hedfalk K, Tornroth-Horsefield S et al (2006) Aquaporin gating. Curr Opin Struct Biol 16(4): 447–456

Henzler T, Steudle E (2000) Transport and metabolic degradation of hydrogen peroxide in *Chara corallina*: model calculations and measurements with the pressure probe suggest transport of H_2O_2 across water channels. J Exp Bot 51(353):2053–2066

Henzler T, Ye Q et al (2004) Oxidative gating of water channels (aquaporins) in *Chara* by hydroxyl radicals. Plant Cell Environ 27:1184–1195

Hohl M, Schopfer P (1995) Rheological analysis of viscoelastic cell wall changes in maize coleoptiles as affected by auxin and osmotic stress. Physiol Plant 94(3):499–505

Holm LM, Jahn TP et al (2005) NH_3 and NH_4^+ permeability in aquaporin-expressing Xenopus oocytes. Pflugers Archiv 450(6):415–428

Horemans N, Asard H et al (1998) Carrier mediated uptake of dehydroascorbate into higher plant plasma membrane vesicles shows trans-stimulation. FEBS lett 421(1):41–44

Horemans N, Foyer CH et al (2000) Transport and action of ascorbate at the plant plasma membrane. Trends Plant Sci 5(6):263–267

Huang S, Colmer TD et al (2008) Does anoxia tolerance involve altering the energy currency towards PPi? Trends Plant Sci 13(5):221–227

Igamberdiev AU, Hill RD (2009) Plant mitochondrial function during anaerobiosis. Ann Bot 103 (2):259–268

Insalud N, Bell RW et al (2006) Morphological and physiological responses of rice (*Oryza sativa*) to limited phosphorus supply in aerated and stagnant solution culture. Ann Bot 98(5):995–1004

Jackson MB (2002) Long-distance signalling from roots to shoots assessed: the flooding story. J Exp Bot 53(367):175–181

Jackson MB, Davies WJ et al (1996) Pressure-flow relationships, xylem solutes and root hydraulic conductance in flooded tomato plants. Ann Bot 77:17–24

Javot H, Maurel C (2002) The role of aquaporins in root water uptake. Ann Bot 90(3):301–313

8 Root Water Transport Under Waterlogged Conditions and the Roles of Aquaporins 177

Jensen MX, Rvthlisberger U et al (2005) Hydroxide and proton migration in aquaporins. Biophys J 89(3):1744–1759

Johanson U, Gustavsson S (2002) A new subfamily of major intrinsic proteins in plants. Mol Biol Evol 19(4):456–461

Johanson U, Karlsson M et al (2001) The complete set of genes encoding major intrinsic proteins in Arabidopsis provides a framework for a new nomenclature for major intrinsic proteins in plants. Plant Physiol 126(4):1358–1369

Johansson I, Karlsson M et al (1998) Water transport activity of the plasma membrane aquaporin PM28A is regulated by phosphorylation. Plant Cell 10(3):451–459

Johansson I, Larsson C et al (1996) The major integral proteins of spinach leaf plasma membranes are putative aquaporins and are phosphorylated in response to Ca^{2+} and apoplastic water potential. Plant Cell 8:1181–1191

Kalashnikov Y, Balakhnina T et al (1994) The effect of soil hypoxia on oxygen activation and the enzyme-system protecting barley roots and leaves against oxygen destruction. Russ J Plant Physiol 41:512–516

Kamaluddin M, Zwiazek JJ (2002) Ethylene enhances water transport in hypoxic aspen. Plant Physiol 128:962–969

Kato-Noguchi H (2000) Abscisic acid and hypoxic induction of anoxia tolerance in roots of lettuce seedlings. J Exp Bot 51(352):1939–1944

Kato-Noguchi H, Ohashi C (2006) Effects of anoxia on amino acid levels in rice coleoptiles. Plant Prod Sci 9(4):383–387

Kennedy RA, Rumpho ME et al (1992) Anaerobic metabolism in plants. Plant Physiol 100(1):1–6

Kirch H-H, Vera-Estrella R et al (2000) Expression of water channel proteins in *Mesembryanthemum crystallinum*. Plant Physiol 123(1):111–124

Koncalová H (1990) Anatomical adaptations to waterlogging in roots of wetland graminoids: limitations and drawbacks. Aquat Bot 38(1):127–134

Kovar JL, Kuchenbuch RO (1994) Commercial importance of adventitious rooting to agronomy. In: Davis TD, Haissig BE (eds) Biology of aventitious root formation, vol 62. Plenum, New York

Kozela C, Regan S (2003) How plants make tubes. Trends Plant Sci 8(4):159–164

Kramer PJ (1933) The intake of water through dead root systems and its relation to the problem of absorption by transpiring plants. Am J Bot 20(7):481–492

Kramer PJ (1949) Plant and soil water relationships. McGraw-Hill, New York

Kramer PJ (1983) Water relations of plants. Academic, New York

Krugel U, Veenhoff LM et al (2008) Transport and sorting of the *Solanum tuberosum* sucrose transporter SUT1 is affected by posttranslational modification. Plant Cell 20(9):2497–2513

Kuiper P, Walton C et al (1994) Effect of hypoxia on ion uptake by nodal and seminal wheat roots. Plant Physiol Biochem 32:267–276

Libourel IGL, van Bodegom PM et al (2006) Nitrite reduces cytoplasmic acidosis under anoxia. Plant Physiol 142(4):1710–1717

Liu H-Y, Sun W-N et al (2006) Co-regulation of water channels and potassium channels in rice. Physiol Plant 128(1):58–69

Liu L-H, Ludewig U et al (2003) Urea transport by nitrogen-regulated tonoplast intrinsic proteins in Arabidopsis. Plant Physiol 133(3):1220–1228

Luu DT, Maurel C (2005) Aquaporins in a challenging environment: molecular gears for adjusting plant water status. Plant Cell Environ 28(1):85–96

Ma JF, Tamai K et al (2006) A silicon transporter in rice. Nature 440(7084):688–691

Martínez-Ballesta M, Cabañero F et al (2008) Two different effects of calcium on aquaporins in salinity-stressed pepper plants. Planta 228(1):15–25

Martinez-Ballesta MC, Aparicio F et al (2003) Influence of saline stress on root hydraulic conductance and PIP expression in *Arabidopsis*. J Plant Physiol 160(6):689–697

Maurel C, Kado RT et al (1995) Phosphorylation regulates the water channel activity of the seed-specific aquaporin alpha-TIP. Embo J 14(13):3028–3035

Maurel C, Reizer J et al (1993) The vacuolar membrane-protein gamma-TIP creates water specific channels in *Xenopus* oocytes. Embo J 12(6):2241–2247

Maurel C, Tacnet F et al (1997) Purified vesicles of tobacco cell vacuolar and plasma membranes exhibit dramatically different water permeability and water channel activity. Proc Natl Acad Sci USA 94(13):7103–7108

Maurel C, Verdoucq L et al (2008) Plant aquaporins: membrane channels with multiple integrated functions. Annu Rev Plant Biol 59(1):595–624

Meguro N, Tsuji H et al (2006) Analysis of expression of genes for mitochondrial aldehyde dehydrogenase in maize during submergence and following re-aeration. Breed Sci 56(4): 365–370

Miyamoto N, Steudle E et al (2001) Hydraulic conductivity of rice roots. J Exp Bot 52(362):1835–1846

Mizutani M, Watanabe S et al (2006) Aquaporin NIP2;1 is mainly localized to the ER membrane and shows root-specific accumulation in *Arabidopsis thaliana*. Plant Cell Physiol 47 (10):1420–1426

Moog PR, Janiesch P (1990) Root growth and morphology of *Carex* species as influenced by oxygen deficiency. Funct Ecol 4:201–208

Moshelion M, Becker D et al (2002) Plasma membrane aquaporins in the motor cells of *Samanea saman*: Diurnal and circadian regulation. Plant Cell 14(3):727–739

Newman EI (1976) Water movement through root systems. Phil Trans R Soc B 273:463–478

Niemietz CM, Tyerman SD (1997) Characterisation of water channels in wheat root membrane vesicles. Plant Physiol 115:561–567

Nobel PS, Schulte PJ et al (1990) Water influx characteristics and hydraulic conductivity for roots of *Agave deserti* Engelm. J Exp Bot 41(225):409–415

Nyblom M, Frick A et al (2009) Structural and functional analysis of SoPIP2;1 mutants adds insight into plant aquaporin gating. J Mol Biol 387(3):653–668

Ouyang L-J, Whelan J et al (1991) Protein phosphorylation stimulates the rate of malate uptake across the peribacteroid membrane of soybean nodules. FEBS Lett 293(1–2):188–190

Pang JY, Newman I et al (2006) Microelectrode ion and O_2 fluxes measurements reveal differential sensitivity of barley root tissues to hypoxia. Plant Cell Environ 29(6):1107–1121

Parent B, Hachez C et al (2009) Drought and abscisic acid effects on aquaporin content translate into changes in hydraulic conductivity and leaf growth rate: a trans-scale approach. Plant Physiol 149(4):2000–2012

Raichaudhuri A, Peng M et al (2009) Plant vacuolar ATP-binding cassette transporters that translocate folates and antifolates *in vitro* and contribute to antifolate tolerance *in vivo*. J Biol Chem 284(13):8449–8460

Ranathunge K, Kotula L et al (2004) Water permeability and reflection coefficient of the outer part of young rice roots are differently affected by closure of water channels (aquaporins) or blockage of apoplastic pores. J Exp Bot 55(396):433–447

Ranathunge K, Steudle E et al (2003) Control of water uptake by rice (*Oryza sativa* L.): role of the outer part of the root. Planta 217:193–205

Ranathunge K, Steudle E et al (2005) A new precipitation technique provides evidence for the permeability of Casparian bands to ions in young roots of corn (*Zea mays* L.) and rice (*Oryza sativa* L.). Plant Cell Environ 28(11):1450–1462

Rawyler A, Arpagaus S et al (2002) Impact of oxygen stress and energy availability on membrane stability of plant cells. Ann Bot 90(4):499–507

Reece CF, Riha SJ (1991) Role of root systems of eastern larch and white spruce in response to flooding. Plant Cell Environ 14(2):229–234

Sade N, Vinocur BJ et al (2009) Improving plant stress tolerance and yield production: is the tonoplast aquaporin SlTIP2;2 a key to isohydric to anisohydric conversion? New Phytol 181(3):651–661

Sakurai J, Ishikawa F et al (2005) Identification of 33 rice aquaporin genes and analysis of their expression and function. Plant Cell Physiol 46(9):1568–1577

Santosa I, Ram P et al (2007) Patterns of peroxidative ethane emission from submerged rice seedlings indicate that damage from reactive oxygen species takes place during submergence and is not necessarily a post-anoxic phenomenon. Planta 226(1):193–202

Scholz-Starke J, Angeli AD et al (2004) Redox-dependent modulation of the carrot SV channel by cytosolic pH. FEBS lett 576(3):449–454

Scholz-Starke J, Gambale F et al (2005) Modulation of plant ion channels by oxidizing and reducing agents. Arch Biochem Biophys 434(1):43–50

Schutz K, Tyerman SD (1997) Water channels in *Chara corallina*. J Exp Bot 48:1511–1518

Sedbrook JC, Kronebusch PJ et al (1996) Transgenic AEQUORIN reveals organ-specific cytosolic Ca^{2+} responses to anoxia in *Arabidopsis thaliana* seedlings. Plant Physiol 111(1):243–257

Segonzac C, Boyer J et al (2007) Nitrate efflux at the root plasma membrane: Identification of an *Arabidopsis* excretion transporter. Plant Cell 19:3760–3777

Shelden MC, Howitt SM et al (2009) Identification and functional characterisation of aquaporins in the grapevine, *Vitis vinifera*. Funct Plant Biol 36:1–14

Smit B, Stachowiak M (1988) Effects of hypoxia and elevated carbon dioxide concentration on water flux through *Populus* roots. Tree Physiol 4(2):153–165

Solaiman Z, Colmer TD et al (2007) Growth responses of cool-season grain legumes to transient waterlogging. Aust J Agric Res 58(5):406–412

Soukup A, Armstrong W et al (2007) Apoplastic barriers to radial oxygen loss and solute penetration: a chemical and functional comparison of the exodermis of two wetland species, *Phragmites australis* and *Glyceria maxima*. New Phytol 173(2):264–278

Steudle E, Peterson CA (1998) How does water get through roots? J Exp Bot 49(322):775–788

Stoimenova M, Igamberdiev A et al (2007) Nitrite-driven anaerobic ATP synthesis in barley and rice root mitochondria. Planta 226(2):465–474

Subbaiah CC, Bush DS et al (1994a) Elevation of cytosolic calcium precedes anoxic gene expression in maize suspension-cultured cells. Plant Cell 6(12):1747–1762

Subbaiah CC, Bush DS et al (1998) Mitochondrial contribution to the anoxic Ca^{2+} signal in maize suspension-cultured cells. Plant Physiol 118(3):759–771

Subbaiah CC, Zhang J et al (1994b) Involvement of intracellular calcium in anaerobic gene expression and survival of maize seedlings. Plant Physiol 105(1):369–376

Sutka M, Alleva K et al (2005) Tonoplast vesicles of *Beta vulgaris* storage root show functional aquaporins regulated by protons. Biol Cell 97(11):837–846

Swartz KJ (2008) Sensing voltage across lipid membranes. Nature 456(7224):891–897

Tanaka M, Wallace IS et al (2008) NIP6;1 is a boric acid channel for preferential transport of boron to growing shoot tissues in *Arabidopsis*. Plant Cell 20(10):2860–2875

Tardieu F, Simonneau T (1998) Variability among species of stomatal control under fluctuating soil water status and evaporative demand: modelling isohydric and anisohydric behaviours. J Exp Bot 49(suppl 1):419–432

Törnroth-Horsefield S, Wang Y et al (2006) Structural mechanism of plant aquaporin gating. Nature 439(7077):688–694

Tournaire-Roux C, Sutka M et al (2003) Cytosolic pH regulates root water transport during anoxic stress through gating of aquaporins. Nature 425:393–397

Tyerman SD (1982) Water relations of seagrasses – stationary volumetric elastic-modulus and osmotic-pressure of the leaf-cells of halophila-ovalis, zostera-capricorni, and posidonia-australis. Plant Physiol 69(4):957–965

Tyerman SD, Oats P et al (1989) Turgor-volume regulation and cellular water relations in *Nicotiana tabacum* roots grown in high salinities. Aust J Plant Physiol 16:517–531

Tyree MT, Zimmermann MH (2002) Xylem structure and the ascent of sap. Springer, New York

Uehlein N, Lovisolo C et al (2003) The tobacco aquaporin NtAQP1 is a membrane CO_2 pore with physiological functions. Nature 425:734–737

Uehlein N, Otto B et al (2008) Function of *Nicotiana tabacum* aquaporins as chloroplast gas pores challenges the concept of membrane CO_2 permeability. Plant Cell 20:648–657 van der Weele CM, Canny MJ et al (1996) Water in aerenchyma spaces in roots. A fast diffusion path for solutes. Plant Soil 184(1):131–141

Vandeleur R, Niemietz C et al (2005) Roles of aquaporins in root responses to irrigation. Plant Soil 274(1–2):141–161

Vandeleur RK, Mayo G et al (2009) The role of plasma membrane intrinsic protein aquaporins in water transport through roots: Diurnal and drought stress responses reveal different strategies between isohydric and anisohydric cultivars of grapevine. Plant Physiol 149(1):445–460

Vartapetian BB (2006) Plant anaerobic stress as a novel trend in ecological physiology, biochemistry, and molecular biology: 2. Further development of the problem. Russ J Plant Physiol 53(6):711–738

Vartapetian BB, Jackson MB (1997) Plant adaptations to anaerobic stress. Annals of Botany 79 (suppl 1):3–20

Verdoucq L, Grondin A et al (2008) Structure-function analysis of plant aquaporin $AtPIP2;1$ gating by divalent cations and protons. Biochem J 415(3):409–416

Vysotskaya LB, Arkhipova TN et al (2004) Effect of partial root excision on transpiration, root hydraulic conductance and leaf growth in wheat seedlings. Plant Physiol Biochem 42:251–255

Wallace IS, Choi WG et al (2006) The structure, function and regulation of the nodulin 26-like intrinsic protein family of plant aquaglyceroporins. Biochim Biophys Acta 1758(8):1165–1175

Wallace IS, Roberts DM (2004) Homology modeling of representative subfamilies of *Arabidopsis* major intrinsic proteins. Classification based on the aromatic/arginine selectivity filter. Plant Physiol 135(2):1059–1068

Wallace IS, Roberts DM (2005) Distinct transport selectivity of two structural subclasses of the nodulin-like intrinsic protein family of plant aquaglyceroporin channels. Biochemistry 44(51):16826–16834

Waters I, Morrell S et al (1991) Effects of anoxia on wheat seedlings. II. Influence of O_2 supply prior to anoxia on tolerance to anoxia, alcoholic fermentation, and sugar levels. J Exp Bot 42(244):1437–1447

Whiteman SA, Nuhse TS et al (2008) A proteomic and phosphoproteomic analysis of *Oryza sativa* plasma membrane and vacuolar membrane. Plant J 56(1):146–156

Xia J-H, Roberts JKM (1996) Regulation of H^+ extrusion and cytoplasmic pH in maize root tips acclimated to a low-oxygen environment. Plant Physiol 111:227–233

Ye Q, Muhr J et al (2005) A cohesion/tension model for the gating of aquaporins allows estimation of water channel pore volumes in *Chara*. Plant Cell Environ 28(4):525–535

Ye Q, Steudle E (2006) Oxidative gating of water channels (aquaporins) in corn roots. Plant Cell Environ 29(4):459–470

Yu J, Yool AJ et al (2006) Mechanism of gating and ion conductivity of a possible tetrameric pore in Aquaporin-1. Structure 14:1411–1423

Zhang W-H, Tyerman SD (1991) Effect of low O_2 concentration and azide on hydraulic conductivity and osmotic volume of the cortical cells of wheat roots. Aust J Plant Physiol 18:603–13

Zhang W-H, Tyerman SD (1997) Effect of low oxygen concentration on the electrical properties of cortical cells of wheat roots. J Plant Physiol 150:567–572

Zhang W-H, Tyerman SD (1999) Inhibition of water channels by $HgCl_2$ in intact wheat root cells. Plant Physiol 120:849–857

Zhang YX, Roberts DM (1995) Expression of soybean Nodulin-26 in transgenic tobacco – targeting to the vacuolar membrane and effects on floral and seed development. Mol Biol Cell 6:109–117

Zhu J, Alvarez S et al (2007) Cell wall proteome in the maize primary root elongation zone. II. Region-specific changes in water soluble and lightly ionically bound proteins under water deficit. Plant Physiol 145(4):1533–1548

Chapter 9
Root Oxygen Deprivation and Leaf Biochemistry in Trees

Laura Arru and Silvia Fornaciari

Abstract Plants are aerobic organisms, that is, they depend on oxygen for their life. Therefore, oxygen deficiency impacts on the biochemical and molecular processes of the plant cell. However, plant cells have evolved inducible strategies to cope with low oxygen stress conditions. When O_2 is reduced, energy production in the form of ATP is reduced too. Cells respond to this energy crisis by switching to fermentative metabolism, producing ATP and regenerating NAD^+ through the glycolytic and fermentative pathways.

Roots are the organs most easily subject to low O_2 stress, but changes in fermentative enzymatic activities are also seen in leaves. Nevertheless, leaves already possess a constitutive expression of these enzymes. Since leaves are the plant organs less likely exposed to low O_2 conditions, they should have evolved in addition an alternative role for the enzymes usually related to fermentative metabolism. Leaves seem to have the ability to take advantage of the enzymes of a metabolic pathway commonly useful in parts of the plant which can undergo anoxia or hypoxia stress: they make use of fermentative metabolism in a different way, to limit the damage that stress condition imposes to the whole plant.

Abbreviations

ABA	Abscisic acid
ADH	Alcohol dehydrogenase
ALDH	Aldehyde dehydrogenase
ANPs	Anaerobiosis related proteins

L. Arru (✉) and S. Fornaciari
Department of Agricultural and Food Sciences, University of Modena and Reggio Emilia, via Amendola 2–pad Besta, 42100 Reggio Emilia, Italy
e-mail: laura.arru@unimore.it

S. Mancuso and S. Shabala (eds.), *Waterlogging Signalling and Tolerance in Plants*, 181
DOI 10.1007/978-3-642-10305-6_9, © Springer-Verlag Berlin Heidelberg 2010

GAPDH	Glyceraldehyde-3-phosphate dehydrogenase
LDH	Lactate dehydrogenase
PDC	Pyruvate decarboxylase
PPi	Inorganic pyrophosphate
Suc	Sucrose
SuSy	Sucrose synthase

9.1 Introduction

There were a few reasons in the past that drove the curiosity to investigate what happens in the leaf of a tree when the plant undergoes an oxygen deficiency stress. Trees are a difficult subject to investigate: they differ from herbaceous plants not only for size or longevity, but they also possess single features that make them unique. However, waterlogging soils can determine a heavy stress also for trees, in a longer temporal scale with respect to a herbaceous plant, and survival in prolonged hypoxic or anoxic conditions can be a severe challenge also for them. What happens in a tree leaf, when the plant experiences root oxygen deficiency?

When roots are in hypoxic or anoxic conditions, the leaf physiology may be affected by significant changes, the severity of which depends on different factors such as the duration of the stress, the plant species, or the growth stage. At the macroscopic level, reduced leaf area (Else et al. 1996) and transpiration rate (Armstrong et al. 1994), reduced rate of photosynthesis and stomatal conductance (Terazawa et al. 1992; Atkinson et al. 2008), chlorosis (Drew and Sisworo 1977), reduced photosynthetic assimilation of CO_2 (Kalashnikov et al. 1994; Yordanova and Popova 2007), reduced cell-wall extensibility (Smit et al. 1989), changes in leaf water content (Jackson et al. 1996), are all symptoms shown in response to flooding.

At biochemical/molecular level, since one of the damages imposed by waterlogging is the onset of an oxidative stress with generation of reactive oxygen species (ROS) (Yan et al. 1996; Biemelt et al. 2000; Blokhina et al. 2002; Garnczarska and Bednarski 2004; Wang and Jiang 2007; Balakhnina et al 2009), an early rise in antioxidant enzyme activities can generally be measured. For example, in short-term flooding, superoxide dismutase (SOD), ascorbate peroxidase (AP), and gluta-thione reductase (GR) activities are enhanced in maize (*Zea mays* L.) (Yan et al. 1996), and mungbean (*Vigna radiata* L.) (Ahmed et al. 2002) leaves, but with the persisting of waterlogging conditions, these enzyme activities are shown to decrease. These results, in which two different stages of behaviour can be distinguished, are generally diffused, but they are not absolute. In the flooding-sensitive pea leaves, SOD activity decreases (Zakrzhevsky et al. 1995), suggesting a possible correlation between early activation of SOD and plant adaptation to flooding stress, at least in herbaceous plants (Balakhnina et al. 2009). Otherwise, in faba bean (*Vicia faba*) leaves, even if SOD activity is enhanced at the onset of hypoxic root stress, the activity of the important antioxidant enzyme, GR, shows an equally early decrease (Balakhnina et al. 2009).

Leaves are the plant organs that should be less likely to be exposed to anoxia or hypoxia, because they live in an atmosphere rich in oxygen. Virtually they should not ever suffer O_2 deficiency. Nonetheless, it has been found that leaves themselves are rich in the enzymes necessary for fermentation, even if there should not be any apparent need for the presence of enzymes for anaerobic metabolism. On the contrary, not only alcohol dehydrogenase (ADH) and pyruvate decarboxylase (PDC) have been found in plant leaves, but leaves of many trees possess a basal constitutive expression of enzymes usually related to fermentative metabolism (Kimmerer and MacDonald 1987).

In effect, tree leaves have been shown to produce aerobic ethanol, the product of ADH enzyme, upon exposure to atmospheric pollutants such as sulphur dioxide, or nitrogen oxides (Kimmerer and Kozlowski 1982; MacDonald et al. 1989). But, since air pollution is one of the last damaging insults caused by human beings to the Earth, this should not be the reason why trees have evolved the ability to maintain constitutive activities of ADH in their leaves.

Hence, there must be a reason why the leaf is able to modulate the activity of an enzyme already constitutively expressed, the presence of which apparently seems to be unnecessary in that organ. This phenomenon that occurs in the leaf however could be related to oxygen deficiency stress that occurs in other organs of the plant. The stress can be caused by external conditions at root level – such as an excess rainfall –, or by anatomical ones – such as limited permeability of oxygen into some internal tissues –, or by metabolic factors – such as high oxygen request by actively dividing cells.

9.2   Root O_2 Deprivation

In roots, under sub-optimal oxygen concentration, genes involved in lactic and alcoholic fermentation pathways are activated, and so are the genes implicated in ROS detoxification that could also have a role later, when the normal oxygen conditions are restored (Klok et al. 2002). Only after long periods of anaerobiosis, do morphological and anatomical changes also take part (Perata and Voesenek 2007). The first genes to be up-regulated encode for the so-called ANPs (Anaerobiosis Related Proteins), enzymes involved in fermentative pathways such as PDC, ADH, and lactate dehydrogenase (LDH) (Huang et al. 2005).

The oxygen is the terminal electron acceptor in the oxidative phosphorylation pathway. This pathway not only provides the vast majority of ATP in a plant cell but also regenerates NAD^+ from NADH. NAD^+ is of fundamental importance for glycolysis to continue: it is necessary for the enzyme GAPDH (glyceraldehyde-3-phosphate dehydrogenase) in the first step of the energy-conserving phase of glycolysis. Under oxygen deprivation, oxidative phosphorylation pathway is inhibited, all steps of the oxidative chain are electrons saturated and there is NADH

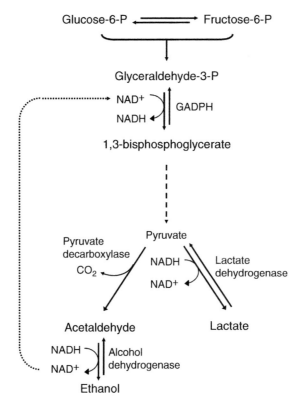

Fig. 9.1 *Schematic view of plant glycolysis and fermentation.* In the glycolysis, carbohydrates from different sources (starch, sucrose, glucose, fructose etc.) are converted to triose phosphates, and then to pyruvate. In the first reaction of the so called "energy-conserving phase of glycolysis", NAD^+ is reduced to NADH by glyceraldehyde-3-phosphate dehydrogenase, but it can be regenerated during fermentation by alcohol dehydrogenase, or lactate dehydrogenase

accumulation. The GAPDH, lacking its cofactor, comes to a halt: glycolysis thus cannot continue (Fig. 9.1). To overcome these problems, the cell metabolism then switches to one or more forms of fermentative metabolism (Davies et al. 1974; Albert et al. 1983; Ricard et al. 1994). This impairs cellular metabolism: the efficiency of energy production in the form of ATP is strongly reduced, intracellular pH is lowered, and there is an accumulation of lactate and ethanol (Drew 1997). The plant cell starts to ferment pyruvate (formed from glycolysis) to lactate, through LDH. This fermentation is often transient, since it might results in a dangerous acidification of intracellular pH by lactic acid. This lowering in pH quickly leads to a switch from lactic to ethanolic fermentation, helped by different pH optima of the involved enzymes. Indeed, at acidic pH LDH is inhibited while PDC is activated, leading first to the production of acetaldehyde, and then to ethanol through the subsequent action of ADH.

9.2.1 Root O₂ Deprivation: Effects on Leaves

In roots, *Adh* gene expression increases under oxygen deprivation stress. In leaves, *Adh* gene seems to be constitutively expressed, at least in poplars (Kreuzwieser et al. 1999) and in some other tree species (Harry and Kimmerer 1991). It's also known that in the leaf there's a response to root flooding which includes enzymatic activities, and ethanol and possibly acetaldehyde leak. But under oxygen deprivation *Quercus palustris*, a flood-tolerant tree, and *Quercus rubra*, flood-intolerant species, produce about the same amount of ethanol; also, the flood-tolerant *Betula nigra* produces less ethanol than the flood-intolerant *Betula lenta* (Kimmerer and MacDonald 1987). Little or no ethanol can be measured from the leaves of herbaceous plants upon exposure to root hypoxia (Freeling and Bennett 1985; Kimmerer 1987; Kimmerer and MacDonald 1987).

So, which kind of relation exists between ADH and ethanol, and between them and root flooding stress? The thing becomes clearer if we consider the hypothesis that there must be also an alternative physiological role of ADH in the leaves of plants, maybe related to aerobic conditions. Whereas in hypoxic tissues such as roots, germinating seeds, or even stems, ADH (gene and enzyme) is needed in order to guarantee the continuity of glycolysis producing ethanol and oxidizing NADH in the process (Tadege et al. 1999), in the aerobic leaves its expression is constitutive and independent of the condition of the roots (Kreuzwieser et al. 2001).

Moreover, focusing on *Adh* gene, if foliar *Adh* plays an exclusive role in flooding stress, its expression should be modulated. However, in the leaves of many of herbaceous species, such as soybean or maize seedlings, *Adh* gene seems to be not even inducible at all.

9.3 The Role of ADH

It is also true that ethanol produced in the roots when soil is depleted of molecular O_2 can exit from the roots and enter the xylem transpiration stream. In this way, ethanol can reach the foliar mesophyll. However, if the ADH role is to metabolize this xylem-translocated fermentation product, a higher activity in leaves from flooded plants or from flood-tolerant plants would be expected (Kimmerer 1987).

On the other hand, ADH not only mediates the acetaldheyde reduction in a two-step pathway in cooperation with PDC but can also operate in the reverse direction, promoting the ethanol oxidation and thereby producing acetaldehyde. In effect, both acetaldehyde and ethanol have been found in the transpiration stream of trees (Crawford and Finegan 1989; MacDonald and Kimmerer 1991). ADH might thus operate in a sequential action with aldehyde dehydrogenase (ALDH) that converts acetaldehyde into acetate (Fig. 9.2).

The mechanism is similar in animals where ethanol is converted to acetaldehyde through ADH, and then to acetate through ALDH (Williamson and Tischler 1979).

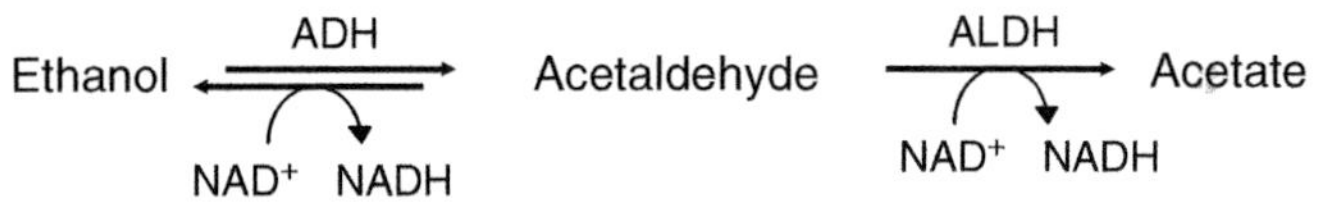

$$\text{Ethanol} \underset{\text{NAD}^+ \; \text{NADH}}{\overset{\text{ADH}}{\rightleftharpoons}} \quad \text{Acetaldehyde} \quad \overset{\text{ALDH}}{\underset{\text{NAD}^+ \; \text{NADH}}{\longrightarrow}} \quad \text{Acetate}$$

Fig. 9.2 *Oxidative pathway of ethanol metabolism.* In leaves, ethanol can be converted back to acetaldehyde by alcohol dehydrogenase, and further oxidized to acetate by aldehyde dehydrogenase

But the real analogy is between ethanol/acetate production in plants, and the lactate metabolism in animals. When muscles are under stress, lactate is produced, and enters the bloodstream. After the physical activity, in the recovery period, with O_2 disponibility, lactate is converted back to glucose by gluconeogenesis, and then respired aerobically (Lehninger 1982). In other words, similar mechanisms belong to organisms of both Plant and Animal Kingdoms. Animals possess the ability to recover the anaerobic metabolism product (lactate) and convert it into a molecule involved in the aerobic metabolism (glucose) when the stress is over. Plants, similarly, possess the ability to recover the anaerobic metabolism product (ethanol) and convert it into a molecule involved in the aerobic metabolism (acetate). Moreover, differently from animals, plants operate when the stress is not still over, and in organs not directly affected (the leaves).

The physiological significance of these reactions may be a recycling of the root-derived ethanol produced under conditions of oxygen deficiency (Kreuzwieser et al. 2001). Acetate, activated as acetyl-CoA, can enter then into many pathways of the general aerobic metabolism (MacDonald and Kimmerer 1993; Bode et al. 1997; Kesselmeier and Staudt 1999).

Not only the leaves, but also young stems possess the ability to metabolize ethanol produced in hypoxic tissues, preventing loss of carbon and helping to maintain an efficient respiration process during the stress. In effect, very high ADH activity was also measured in the vascular cambium of trees, and tree stems (Kimmerer and Stringer 1988; MacDonald and Kimmerer 1991). So, in flooded trees fermentation occurs in the roots, and possibly it may occur also along the stems (Eklund 1990; Harry and Kimmerer 1991). Ethanol that reaches the leaves carried by the transpiration stream can be metabolized by ADH before it can escape from the plant (MacDonald et al. 1989; MacDonald and Kimmerer 1990, 1991). In other words, constitutive expression of *Adh* gene in leaves of trees may be a strategy to avoid ethanol escape, and safeguard the carbon and energy invested in ethanol.

9.4 Carbon Recovery

Many studies have demonstrated that acetaldheyde is effectively emitted by leaves of trees such as spruce (Hahn et al. 1991), oak, pine (Kesselmeier et al. 1997) and poplar (Kreuzwieser et al. 1999). In general, there are significant variations in acetaldehyde concentrations in forests during the day (Enders et al. 1992; Steinbrecher et al. 1997), and flooded plants show higher acetaldheyde emission rates during the day and lower at night (Kreuzwieser et al. 2001). It has been proposed that this acetaldehyde is the result of the oxidation in leaves of the ethanol produced in flooded roots, and then transported to the leaves via the transpiration stream.

In order to confirm this hypothesis, excised poplar leaves were incubated in a buffer plus ethanol, with the addition of inhibitors of the enzymes responsible of acetaldehyde metabolism: ADH and ALDH (Kreuzwieser et al. 2001). The inhibition of ADH led to a collapse of acetaldehyde emission from the leaves. This means that ADH also generates the conversion of ethanol into acetaldehyde in the leaves, originating acetaldehyde emissions. A little enhanced ethanol emission is also noted, probably due to an ethanol accumulation as a consequence of inhibited ethanol oxidation. By contrast, treatment of excised poplar leaves with ethanol and the ALDH inhibitor caused significantly increased emission both of ethanol and acetaldehyde. Acetaldehyde oxidation was in fact inhibited, and this led to acetaldehyde and ethanol accumulation in the leaf tissue (Kreuzwieser et al. 2001). The loss of carbon in the form of ethanol or acetaldehyde therefore, should be considered as a gap in the metabolization of fermentation products coming from the roots.

Even if ADH is responsible for acetaldehyde production from ethanol, other factors seem to be related to a higher diurnal acetaldehyde emission from the leaves of flooded trees. There might be a further link between oxygen deficiency and ADH and acetaldehyde emission. It could be supposed that during the day, the higher acetaldehyde emission measured may be promoted by the opening of stomata. In fact, acetaldehyde is not an easily volatilizable molecule because of its polar nature; it may only be emitted from leaves via the stomata and not via the cuticle.

However, it has also been demonstrated that stomatal conductance does not directly influence acetaldehyde emission. Manipulating stomatal conductance by light intensity or ABA (two known factors inducing stomatal closure) did not result in altered acetaldehyde emission. Therefore, only acetaldehyde concentration can directly influence acetaldehyde emission. When is acetaldehyde concentration higher? In hypoxic/anoxic soil condition? During flooding stress, more ethanol than in aerobic condition reaches the leaves, where it is oxidized generating acetaldehyde as an intermediate in the pathway to acetate conversion. It is also true that leaf ethanol concentration increases depending on the transpiration rate. Since the transpiration is controlled by stomata, stomatal conductance may indirectly influence acetaldehyde emission from leaves by controlling ethanol transport (Kreuzwieser et al. 2001).

Moreover, acetaldehyde is a high phytotoxic compound. At the beginning it was also thought that ethanol was the toxic compound, however, later, studies have not proven its toxicity (Jackson et al. 1982; Atkinson et al. 2008). Recent works now highlight how acetaldehyde rather than ethanol possibly causes the observed anoxia-related injuries on growth and development of flooded plants (Ap Rees 1980; Perata and Alpi 1991). Therefore acetaldehyde accumulation must be avoided, or it should be further metabolized, to prevent self-poisoning. Acetaldehyde loss from the leaves has thus to be considered a problem of efficiency between production and metabolization of this molecule (Kreuzwieser et al. 1999).

9.5 Differential mRNA Translation

There is another point that deserves a little more attention. Why is ADH activity enhanced in leaves of flooded trees, while the *Adh* gene expression seems to be not altered?

As discussed above, an enhanced ADH activity may be required in order to manage all the ethanol transported into the xylem flux. This role is confirmed by the greater concentration of the ethanol-derived oxidation product, acetaldehyde, and by the increased ethanol concentrations in the xylem sap (up to 200 fold after 24 h of flooding, Kreuzwieser et al. 2009). However, microarray experiments on poplar do not report any other isoform of *Adh* that could be induced in leaves in response to anoxic/hypoxic root conditions (Kreuzwieser et al. 2009). Moreover, it should be kept in mind that a lack in the gene induction does not necessarily imply a lack of activity in the corresponding enzyme. In fact, there is more than one level in the regulation of gene expression. Besides the differential transcriptional induction of genes in response to a stress, the translation can also be selectively regulated (Branco-Price et al. 2008; Kreuzwieser et al. 2009).

Transcript and metabolite profiling was performed to characterize the different molecular and physiological responses under conditions of oxygen deprivation in flooding tolerant and non tolerant plants (Branco-Price et al. 2005, 2008; Loreti et al. 2005; Lasanthi-Kudahettige et al. 2007; Van Dongen et al. 2008; Kreuzwieser et al. 2009). The first result that appears confirms that the genes involved in changes in transcript abundance are different between roots and leaves. That means that the most significant variation in gene expression belongs to the roots, whereas changes in the metabolite levels are found in both organs. Regarding root behaviour, as soon as a root senses the oxygen deficiency, it begins to modify its internal metabolism in order to maintain energy availability, and to activate the pathways that permit an adequate supply of ATP. In Arabidopsis roots for example, the peak in *Adh* transcript abundance was observed 2–4 h after hypoxia onset (Liu et al. 2005). In roots of poplar trees, the consistent induction of *Adh* is maintained for the whole flooding period. Once again, even if *Pdc* transcript level in poplar roots is 15 times higher after 5 h of hypoxia, PDC activity has only a threefold increase (Kreuzwieser et al. 2009). This lack of proportional correlation between gene expression and enzyme activity may suggest a reduced translation efficiency (Branco-Price et al. 2008), or some post-transcriptional or post-translational regulation, as suggested by Kreuzwieser et al. 2009.

This alteration of gene expression in response to the stress is a way to adapt internal cell metabolism to the new temporary condition of low energy disponibility. Naturally, the first way to preserve energy dissipation is to reduce the transcription of those mRNAs that are not strictly necessary to the survival of the cell, such as those needed in DNA synthesis or cell division (Gibbs and Greenway 2003). But recently it has been demonstrated that this is not the only way (Branco-Price et al. 2005, 2008; Kreuzwieser et al. 2009). Changes have been observed in the association between mRNA and polysomes. Polysomes (polyribosomes) are large cytoplasmic assemblies made up of several ribosomes. During low oxygen stress, it has demonstrated a decrease in polysome complexes inside the cells, which reflect a decrease in initiation of protein synthesis (Branco-Price et al. 2008). In addition, not all the mRNAs have the same affinity to associate with polysomes: this capability depends on numerous factors such as the length of the $5'$-UTR, the ability to form secondary structures, etc. (Kawaguchi and Bailey-Serres 2005). So, a little sub-set of mRNAs still maintains association with polysomes under stress conditions. This led to the conclusion that the regulation in protein synthesis observed can also occur at translational level, even in absence of variation in mRNA transcription. mRNA translation is in fact an extremely energy requiring step, since it needs multiple molecules of ATP and one of GTP in removing secondary mRNA structure and in the recognition of the initiation codon. It seems advantageous a regulation of protein synthesis at level of the initiation phase. Moreover, the fact that protein synthesis can be regulated without the request for a decrease in transcript abundance permits the quick re-recruitment of these not down-regulated mRNA to polysomes, as soon as oxygen comes back (Branco-Price et al. 2008).

9.6 Effects on Cell Metabolism

A comparison of the response of a flooding-sensitive plant with that of a flooding-tolerant one reveals important differences in the number of genes selectively expressed during oxygen deprivation stress. Over 5,000 genes show significantly altered expression in the hypoxic stressed poplar tissue, the flooding-tolerant, whereas similar conditions alter much less the transcript abundance of Arabidopsis, the flooding-sensitive, where only approximately 150 genes change (Klok et al. 2002; Liu et al. 2005). In spite of this difference, transcriptional factors represent a high percentage of altered genes in both species. This differential expression of transcriptional factors suggests a regulation of the transcriptional network, aimed at a stress adaptation in response to hypoxia (Kreuzwieser et al. 2009).

It seems that the stimulation of glycolysis and the supply of sugars can be important for tolerating anoxia, as suggested by Drew in 1997. On the other hand, also the inhibition of energy consuming processes such as biosynthesis of cellulose, hemicellulose, and cell wall proteins may be of some help. The synthesis of secondary cell wall seems to be not a priority in hypoxic or anoxic conditions, when the life of the plant is at risk, and can be postponed if the plant has survived, at the end of the stress.

But as the situation of oxygen deprivation continues, in contrast to the early activation response, genes in roots start to be down-regulated. This behaviour is a common response to survival, found both in herbaceous flooding-tolerant species such as rice (*Oryza sativa*) (Lasanthi-Kudahettige et al. 2007), and in flooding-tolerant trees such as grey poplar (Kreuzwieser et al. 2009).

When poorly drained soil becomes waterlogged, the roots of plants experience a deficiency in O_2 disponibility, because of slow diffusion of O_2 in the water-filled pore space of the soil. When oxygen becomes unavailable, glycolysis remains the primary source of energy for the cell, while fermentation recycles the NADH produced by glycolysis. The efficiency of fermentation, measured as net synthesis of ATP, is very low with respect to the full respiration of carbohydrates in the mitochondrion, since most of the energy available in the sugars remains inside the not fully oxidized products ethanol and/or lactate. In fact, fermentation yields only 2–3 mol ATP mol^{-1} glucose (Mancuso and Marras 2006), compared with 24–36 mol ATP mol^{-1} glucose in O_2 presence (Gibbs and Greenway 2003). Because of this, it may happen to verify an increased rate of glycolysis, in order to balance the lower energy yield of alcoholic fermentation. This is known as the Pasteur effect (Gibbs et al. 2000). As a result, under prolonged hypoxic conditions the activation of glycolytic enzymes can be observed (Setter et al. 1997; Liu et al. 2005; Loreti et al. 2005), and the consequent high rate of fermentation increases the demand for carbohydrates.

The hypothesis that carbohydrate supplies for an enhanced glycolysis rate becomes critical for survival in oxygen deficiency stress is supported by some studies that show an improved sugar supply to flooded roots, precisely to compensate the higher request for oxidable substrate under these conditions. This behaviour might make one think that an optimal carbohydrate supply could be important for survival under prolonged hypoxic conditions. Indeed, many flood-sensitive species show a reduced carbohydrate concentration in their roots (Vu and Yelenosky 1991), but their survival can be improved by providing them with exogenous sugars (Waters et al. 1991; Loreti et al. 2005).

However, even if maintaining adequate levels of fermentable sugars can help for long term survival, there are various exceptions to this trend. Some herbaceous plants as well as trees show a build-up of sugars in roots (Albrecht et al. 1993; Castonguay et al. 1993; Huang and Johnson 1995), possibly due to a stop in root growth in order to save energy, with a consequent reduced carbon demand (Angelov et al. 1996). This is another survival strategy (metabolic depression strategy), as seen in *Vitis*, in order to slow down metabolic activity, and to depress the rate of ATP use (Mancuso and Marras 2006). Furthermore in some species, an ATP deficiency seems to delay glycolysis, and sugars again accumulate rather than decline (Fukao and Bailey-Serres 2004).

To assure cell survival, the roots of flooded plants can require a higher carbohydrate supply to sustain the increased rate of glycolysis necessary for ATP production. In this case the need for sugars can be satisfied by an increase in transport of sucrose (Suc) via phloem from leaves to roots. There is experimental evidence of an increase in Suc concentration in the phloem sap of hypoxically treated plants, accompanied by a decrease in Suc concentration in leaves (Kreuzwieser et al. 2009). Some species such as peas, pumpkin, or several herbaceous plants, have shown that they need a continuous supply of sugars from shoot to root to survive under flooding (Jackson and Drew 1984; Saglio et al. 1980; Webb and Armstrong 1983).

On the other hand, there are also cases in which flooding decreased carbohydrate translocation, as in alfalfa (*Medicago sativa* L.) and *Lotus corniculatus* L. (Barta

1987). This trend may be due to a request reduction for carbohydrate by roots (Wample and Davis 1983), but also to an impaired phloem translocation (Saglio 1985; Topa and Cheeseman 1992a). The reduction of carbohydrate request is probably caused by the slowed root metabolism under hypoxia which in turn generates a decreased demand for sugars from leaves (Hsu et al. 1999; Wample and Davis 1983) and therefore a reduced rate of carbohydrate translocation (Barta 1987). This could be one reason why under flooding conditions starch accumulates in leaves, although the photosynthetic rate declines (Vu and Yelenosky 1991; Wample and Davis 1983). In flooding treatment, carbohydrate content (including the contribution of starch) can thus increase in leaves, as for sunflower (Wample and Davis 1983), purple flower alfalfa (Barta 1988), sweet orange (Vu and Yelenosky 1991), pine (Topa and Cheeseman 1992a, b), and bitter melon (Liao and Lin 1994), *H. annuus* (Wample and Davis 1983), *M. charantia* (Liao and Lin 1994), or *S. samarangense* (Hsu et al. 1999).

9.7 Conclusions

Much evidence suggests that leaf biochemistry in trees is severely affected by soil waterlogging, causing a noticeable decline in the rate of photosynthesis and transpiration (Kreuzwieser et al. 2004). Phloem transport can also be affected, and soluble sugars accumulate in the leaves instead of providing the roots with carbohydrates. On the other hand, roots need adequate levels of fermentable sugars for long term survival. When sugars are not coming from the leaves, some roots can utilize the starch accumulated there. In some species in fact, starch can be degraded and converted into fermentable substrates for anaerobic metabolism (Perata et al. 1992; Liao and Lin 2001).

As oxygen falls under normoxic conditions, there is a widespread decrease in ATP-consuming processes, mostly in the ones involved in biosynthesis and growth. Other pathways which cannot be suppressed are by-passed by alternative ones, but less energy-consuming. This is the case of sucrose degradation. Sucrose is degraded by invertase and hexokinase to generate fructose-6-phosphate, the first intermediate of glycolysis. This reaction requires two molecules of ATP. When oxygen concentrations are low, this pathway is replaced by sucrose synthase (SuSy) and UDP-glucose pyrophosphorylase, which require only one molecule of inorganic pyrophosphate (PPi). The cost is even lower if one thinks that PPi is generated by a wide range of reactions as a waste product (Geigenberger 2003; Magneschi and Perata 2009).

There is still much to learn about the molecular and biochemical adaptations in a plant to low oxygen condition, and the way each tissue responds to it. It is evident that a coordinate and cooperative response is required for the tolerance to the stress. It involves changes in numerous pathways and metabolic processes, starting from the oxygen sensing system(s) and the consequent signal transduction pathway(s) that still remain unknown.

References

Ahmed S, Nawata E, Hosokawa M, Domae Y, Sakuratani T (2002) Alterations in photosynthesis and some antioxidant enzymatic activities of mungbean subjected to waterlogging. Plant Sci 163:117–123

Albert B, Bray D, Lewis J, Raff M, Roberts K, Watson J (1983) Molecular biology of the cells. Garland Publishing, New York, pp 67–80

Albrecht G, Kammerer S, Praznik W, Wiedenroth EM (1993) Fructan content of wheat seedlings (*Triticum aestivum* L) under hypoxia and following re-aeration. New Phytol 123:471–476

Angelov MN, Sung SJS, Doong RL, Harms WR, Kormanik PP, Black CC Jr (1996) Long- and short-termflooding effects on survival and sink-source relationships of swamp-adapted tree species. Tree Physiol 16:477–484

Ap Rees T (1980) Assessment of the contributions of metabolic pathways to plant respiration. In: Stumpf PK, Conn EE (eds) The biochemistry of plants, a comprehensive treatise, 2nd edn. Academic Press, New York, pp 1–29

Armstrong W, Brandle R, Jackson MB (1994) Mechanisms of flood tolerance in plants. Acta Bot Neerl 43:307–358

Atkinson CJ, Harrison-Murraya RS, Taylora JM (2008) Rapid flood-induced stomatal closure accompanies xylem sap transportation of root-derived acetaldehyde and ethanol in Forsythia. Env Exp Bot 64:196–205

Balakhnina TI, Bennicelli RP, Stępniewska Z, Stępniewski W, Fomina IR (2009) Oxidative damage and antioxidant defense system in leaves of *Vicia faba* major L. cv. Bartom during soil flooding and subsequent drainage 2009. Plant Soil. doi:10.1007/s11104-009-0054-6

Barta AL (1987) Supply and partitioning of assimilates to roots of *Medicago sativa* L. and *Lotus corniculatus* L. under anoxia. Plant Cell Environ 10:151–156

Barta AL (1988) Response of field grown alfalfa to root water-logging and shoot removal. I. Plant injury and carbohydrate and mineral content of roots. Agron J 88:889–892

Biemelt S, Keetman U, Mock HP, Grimm B (2000) Expression and activity of isoenzymes of superoxide dismutase in wheat roots in response to hypoxia and anoxia. Plant Cell Environ 23:135–144

Blokhina O, Virolainen E, Fagestedt KV (2002) Antioxidants, oxidative damage, and oxygen deprivation stress: a review. Ann Bot (Lond) 91:179–194

Bode K, Helas G, Kesselmeier J (1997) Biogenic contribution to atmospheric organic acids. In: Helas G, Slanina J, Steinbrecher R (eds) Biogenic volatile organic compounds in the atmosphere. SPB Academic Publishers, Amsterdam, pp 157–170

Branco-Price C, Kaiser KA, Jang CJH, Larive CK, Bailey-Serres J (2008) Selective mRNA translation coordinates energetic and metabolic adjustments to cellular oxygen deprivation and reoxygenation in Arabidopsis thaliana. Plant J 56:743–755

Branco-Price C, Kawaguchi R, Ferreira RB, Bailey-Serres J (2005) Genomewide analysis of transcript abundance and translation in *Arabidopsis* seedlings subjected to oxygen deprivation. Ann Bot (Lond) 96:647–660

Castonguay Y, Nadeau P, Simard RR (1993) Effects of flooding on carbohydrate and ABA levels in roots and shoots of alfalfa. Plant Cell Environ 16:695–702

Crawford RMM, Finegan DM (1989) Removal of ethanol from lodgepole pine roots. Tree Physiol 5:53–61

Davies DD, Grego S, Kenworthy P (1974) The control of the production of lactate and ethanol by higher plants. Planta 118:297–310

Drew MC, Sisworo EJ (1977) Early effects of flooding on nitrogen deficiency and leaf chlorosis in barley. New Phytol 79:567–571

Drew MC (1997) Oxygen deficiency and root metabolism: injury and acclimation under hypoxia and anoxia. Annu Rev Plant Physiol Plant Mol Biol 48:223–250

9 Root Oxygen Deprivation and Leaf Biochemistry in Trees

Eklund L (1990) Endogenous levels of oxygen, carbon dioxide and ethylene in stems of Norway spruce trees during one growing season. Trees 4:150–154

Else MA, Tiekstra AE, Croker SJ, Davies WJ, Jackson MB (1996) Stomatal closure in flooded tomato plants involves abscisic acid and a chemically unidentified anti-transpirant in xylem sap. Plant Physiol 112:239–247

Enders G, Dlugi R, Steinbrecher R, Clement B, Daiber R, Van Eijk J, Gab S, Haziza M, Helas G, Herrmann U, Kessel K, Kesselmeier J, Kotzias D, Kourtidis K, Kurth HH, McMillan RT, Roider G, Schurmann W, Teichmann U, Torres L (1992) Biosphere/atmosphere interactions: integrated research in a European coniferous forest ecosystem. Atmos Environ 26:171–189

Freeling M, Bennett DC (1985) Maize Adh1. Annu Rev Genet 19:297–323

Fukao T, Bailey-Serres J (2004) Plant responses to hypoxia – is survival a balancing act? Trends Plant Sci 9:449–456

Garnczarska M, Bednarski W (2004) Effect of a short-term hypoxia treatment followed by re-aeration on free radicals level and antioxidant enzymes in lupine roots. Plant Physiol Biochem 42:233–240

Geigenberger P (2003) Response of plant metabolism to too little oxygen. Curr Opin Plant Biol 6:247–256

Gibbs J, Greenway H (2003) Mechanisms of anoxia tolerance in plants. I. Growth, survival and anaerobic catabolism. Funct Plant Biol 30:1–47

Gibbs J, Morrell S, Valdez A, Setter TL, Greenway T (2000) Regulation of alcoholic fermentation in coleoptiles of two rice cultivars differing in tolerance to anoxia. J Exp Bot 51:785–796

Hahn J, Steinbrecher R, Slemr J. 1991. Study of the emission of low molecular-weight organic compounds by various plants. EUROTRAC Annu Rep Part 4. BIATEX., pp 230–235

Harry DE, Kimmerer TW (1991) Molecular genetics and physiology of alcohol dehydrogenase in woody plants. For Ecol Manage 43:251–272

Hsu YM, Tseng MJ, Lin CH (1999) The fluctuation of carbohydrates and nitrogen compounds in flooded wax-apple trees. Bot Bull Acad Sin 40:193–198

Huang B, Johnson JW (1995) Root respiration and carbohydrate status of two wheat genotypes in response to hypoxia. Ann Bot (Lond) 75:427–432

Huang S, Greenway H, Colmer TD, Millar H (2005) Protein synthesis by rice coleoptiles during prolonged anoxia: implication for glycolysis, growth and energy utilization. Ann Bot (Lond) 96:703–715

Jackson MB, Davies WJ, Else MA (1996) Pressure-flow relationships, xylem solutes, and root hydraulic conductance in flooded tomato. Ann Bot 77:17–24

Jackson MB, Drew MC (1984) Effects of flooding on growth and metabolism of herbaceous plants. In: Kozlowski T (ed) Flooding and plant growth. Academic Press, New York, pp 47–128

Jackson MB, Herman B, Goodenough A (1982) An examination of the importance of ethanol in causing injury to flooded plants. Plant Cell Environ 5:163–172

Kalashnikov YUE, Zakrzhevsky DA, Balakhnina TI (1994) Effect of soil hypoxia on activation of oxygen and the system of protection from oxidative damage in roots and leaves of *Hordeum vulgare* L. Russ J Plant Physiol 41:583–588

Kawaguchi R, Bailey-Serres J (2005) mRNA sequence features responsible for translational regulation in *Arabidopsis*. Nucleic Acids Res 33:955–965

Kesselmeier J, Bode K, Hofmann U, Muller H, Schafer L, Wolf A, Ciccioli P, Brancaleoni E, Cecinato A, Frattoni M, Foster P, Ferrari C, Jacob V, Fugit JL, Dutaur L, Simon V, Torres L (1997) Emission of short chained organic acids, aldehydes and monoterpenes from *Quercus ilex* L. and *Pinus pinea* L. in relation to physiological activities, carbon budget and emission algorithms. Atmos Environ 31:119–133

Kesselmeier J, Staudt M (1999) Biogenic volatile organic compounds (VOC): an overview on emission, physiology and ecology. J Atmos Chem 33:23–88

Kimmerer TW, Kozlowski TT (1982) Ethylene, ethane, acetaldehyde and ethanol production by plants under stress. Plant Physiol 69:840–847

Kimmerer TW, MacDonald RC (1987) Acetaldehyde and ethanol biosynthesis in leaves of plants. Plant Physiol 84:1204–1209

Kimmerer TW, Stringer MA (1988) Alcohol dehydrogenase and ethanol in the stems of trees. Plant Physiol 87:693–697

Kimmerer TW (1987) Alcohol dehydrogenase and pyruvate decarboxylase activity in leaves and roots of eastem cottonwood (*Populus deltoides* Bartr.) and soybean (*Glycine max* L.). Plant Physiol 84:1210–1213

Klok EJ, Wilson IW, Wilson D, Chapman SC, Ewing RM, Somerville SC, Peacock WJ, Dolferus R, Dennis ES (2002) Expression profile analysis of the low-oxygen response in *Arabidopsis* root cultures. Plant Cell 14:2481–2494

Kreuzwieser J, Harren FJM, Laarhoven LJ, Boamfa I, Lintel-Hekkert S, Scheerer U, Huglin C, Rennenberg H (2001) Acetaldehyde emission by the leaves of trees: correlation with physiological and environmental parameters. Physiol Plant 113:41–49

Kreuzwieser J, Hauberg J, Howell KA, Carroll A, Rennenberg H, Millar AH, Whelan J (2009) Differential response of gray poplar leaves and roots underpins stress adaptation during hypoxia. Plant Physiol 149:461–473

Kreuzwieser J, Papadopoulou E, Rennenberg H (2004) Interaction of flooding with carbon metabolism of forest trees. Plant Biol 6:299–306

Kreuzwieser J, Scheerer U, Rennenberg H (1999) Metabolic origin of acetaldehyde emitted by trees. J Exp Bot 50:757–765

Lasanthi-Kudahettige R, Magneschi L, Loreti E, Gonzali S, Licausi F, Novi G, Beretta O, Vitulli F, Alpi A, Perata P (2007) Transcript profiling of the anoxic rice coleoptile. Plant Physiol 144:218–231

Lehninger AL (1982) Principles of biochemistry. Worth Publishers, New York

Liao CT, Lin CH (1994) Effect of flooding stress on photosynthetic activities of *Momordica charantia*. Plant Physiol Biochem 32:479–485

Liao CT, Lin CH (2001) Physiological adaptation of crop plants to flooding stress. Proc Natl Sci Counc 25:148–157

Liu F, VanToai T, Moy LP, Bock G, Linford LD, Quackenbush J (2005) Global transcription profiling reveals comprehensive insights into hypoxic response in *Arabidopsis*. Plant Physiol 137:1115–1129

Loreti E, Poggi A, Novi G, Alpi A, Perata P (2005) A genome-wide analysis of the effects of sucrose on gene expression in *Arabidopsis* seedlings under anoxia. Plant Physiol 137:1130–1138

MacDonald RC, Kimmerer TW, Razzaghi M (1989) Aerobic ethanol production by leaves: evidence for air pollution stress in tress of the Ohio River Valley, USA. Environ Pollut 62:337–351

MacDonald RC, Kimmerer TW (1990) Remetabolism of transpired ethanol by *Populus deltoides* (abstract No. 658). Plant Physiol 93:S112

MacDonald RC, Kimmerer TW (1991) Ethanol in the stems of trees. Physiol Plant 82:582–588

MacDonald RC, Kimmerer TW (1993) Metabolism of transpired ethanol by eastern cottonwood (*Populus deltoides*-Bartr). Plant Physiol 102:173–179

Magneschi L, Perata P (2009) Rice germination and seedling growth in the absence of oxygen. Ann Bot (Lond) 103:181–196

Mancuso S, Marras AM (2006) Adaptive response of Vitis root to anoxia. Plant Cell Physiol 47:401–409

Perata P, Alpi A (1991) Ethanol-induced injuries carrot cells. The role of acetaldehyde. Plant Physiol 95:748–752

Perata P, Pozueta-Romero J, Akazawa T, Yamaguchi J (1992) Effect of anoxia on starch breakdown in rice and wheat seeds. Planta 188:611–618

Perata P, Voesenek LA (2007) Submergence tolerance in rice requires Sub1A, an ethylene-response-factor-like gene. Trends Plant Sci 12:43–46

Ricard B, Couée I, Raymond P, Saglio PH, Saint-Ges V, Pradet A (1994) Plant metabolism under hypoxia and anoxia. Plant Physiol Biochem 32:1–10

Saglio PH, Raymond P, Pradet A (1980) Metabolic activity and energy charge of excised maize root tips under anoxia. Plant Physiol 66:1053–1057

Saglio PH (1985) Effect of path or sink anoxia on sugar translocation in roots of maize seedlings. Plant Physiol 77:285–290

Setter TL, Ellis M, Laureles EV, Ella ES, Senadhira D, Mishra SB, Sarkarung S, Datta S (1997) Physiology and genetics of submergence tolerance in rice. Ann Bot (Lond) 79:67–77

Smit B, Stachowiak M, Van Volkenburgh E (1989) Cellular processes limiting leaf growth in plants under hypoxic root stress. J Exp Bot 40:89–94

Steinbrecher R, Hahn J, Stahl K, Eichstadter G, Lederle K, Rabong R, Schreiner AM, Slemr J (1997) Investigations on emissions of low molecular weight compounds (C2-C10) from vegetation. In: Slanina S (ed) Biosphere-atmosphere exchange of pollutants and trace substances. Springer, Berlin, pp 342–351 ISBN 3-540-61711-6

Tadege M, Dupuis I, Kuhlemeier C (1999) Ethanolic fermentation: new functions for an old pathway. Trends Plant Sci 4:320–325

Terazawa K, Maruyama Y, Morikawa Y (1992) Photosynthetic and stomatal responses of Larix kaempferi seedlings to short-term waterlogging. Ecol Res 7:193–197

Topa MA, Cheeseman JM (1992a) Carbon and phosphorus partitioning in *Pinus serotina* seedlings growing under hypoxic and low-phosphorous conditions. Tree Physiol 10:195–207

Topa MA, Cheeseman JM (1992b) Effects of root hypoxia and a low P supply on relative growth, carbon dioxide exchange rates and carbon partitioning in *Pinus serotina* seedlings. Physiol Plant 86:136–144

Van Dongen JT, Frohlich A, Ramirez-Aguilar SJ, Schauer N, Fernie AR, Erban A, Kopka J, Clark J, Langer A, Geigenberger P (2008) Transcript and metabolite profiling of the adaptive response to mild decreases in oxygen concentration in the roots of *Arabidopsis* plants. Ann Bot (Lond). doi:doi/10.1093/aob/mcn126

Vu CV, Yelenosky G (1991) Photosynthetic responses of citrus trees to soil flooding. Physiol Plant 81:7–14

Wample RL, Davis RW (1983) Effect of flooding on starch accumulation in chloroplasts of sunflower (*Helianthus annuus* L.). Plant Physiol 73:195–198

Wang KH, Jiang YW (2007) Antioxidant responses of creeping bentgrass roots to waterlogging. Crop Sci 47:232–238

Waters I, Morrell S, Greenway H, Colmer TD (1991) Effects of anoxia on wheat seedlings. 2. Influence of O2 supply prior to anoxia on tolerance to anoxia, alcoholic fermentation, and sugar levels. J Exp Bot 42:832–841

Webb T, Armstrong W (1983) The effects of anoxia and carbohydrates on the growth and viability of rice, pea and pumpkin roots. J Exp Bot 34:579–603

Williamson JR, Tischler M (1979) Ethanol metabolism in perfused liver and isolated hepatocytes with associated methodologies. In: Majchrowikz E, Noble EP (eds) Biochemistry and pharmacology of ethanol, vol 1. Plenum Press, New York, pp 167–189

Yan B, Dai Q, Liu X, Huang S, Wang Z (1996) Flooding-induced membrane damage, lipid oxidation and activated oxygen generation in com leaves. Plant Soil 179:261–268

Yordanova RY, Popova LP (2007) Flooding-induced changes in photosynthesis and oxidative status in maize plants. Acta Physiol Plant 29:535–541

Zakrzhevsky DA, Balakhnina TI, Stepniewski W, Stepniewska S, Bennicelli RP, Lipiec J (1995) Oxidation and growth processes in roots and leaves of higher plants at different oxygen availability in soil. Rus J Plant Physiol 42:242–248

Chapter 10
Membrane Transporters and Waterlogging Tolerance

Jiayin Pang and Sergey Shabala

Abstract Oxygen deprivation and the subsequent accumulation of toxic secondary metabolites in soil are two major adverse processes linked with soil waterlogging stress. A plant's ability to control ion transport across cellular membranes in response to waterlogging stress is important for its waterlogging tolerance. This chapter summarises the impact of oxygen deprivation on membrane transport activity in plant cells and discusses the nature of putative "oxygen sensors" in living cells. The adverse effects of secondary metabolites produced under waterlogged conditions on the key membrane transporters mediating plant nutrient acquisition, are also discussed. It is suggested that a plant's ability to control the activity of key membrane transporters and to maintain K^+ uptake in response to oxygen deprivation and secondary metabolite toxicity should be considered as useful traits in future breeding programmes. It is suggested that urgent attention should be paid to elucidate the molecular identity and control modes of key membrane transporters mediating plant adaptive responses to waterlogging.

Abbreviations

ADH	Alcohol dehydrogenase
$[Ca]_i$	Cytosolic calcium
DACC	Depolarization-activated Ca^{2+} channel
KIR	Potassium inward-rectifying channel
KOR	Potassium outward-rectifying channel

J. Pang
School of Plant Biology, The University of Western Australia, 35 Stirling Highway, Crawley, WA 6009, Australia

S. Shabala (✉)
School of Agricultural Science, University of Tasmania, Private Bag 54, Hobart, Tas 7001, Australia
e-mail: sergey.shabala@utas.edu.au

S. Mancuso and S. Shabala (eds.), *Waterlogging Signalling and Tolerance in Plants*,
DOI 10.1007/978-3-642-10305-6_10, © Springer-Verlag Berlin Heidelberg 2010

MIFE	Microelectrode ion flux measurement
NSCC	Non-selective cation channel
PM	Plasma membrane
ROL	Radial oxygen loss
WL	Waterlogging

10.1 Introduction

Waterlogging stress imposes numerous physiological constraints on plants. First, the concentration of dissolved oxygen in the soil water solution drops dramatically, from 230 mmol m^{-3} in well-drained soil to below 50 mmol m^{-3} O_2 under waterlogged conditions (Grichko and Glick 2001). This lack of oxygen effectively blocks aerobic respiration and ATP synthesis in root mitochondria (Greenway and Gibbs 2003; Pradet and Bomsel 1978), causing a rapid decline in energy availability. The latter has immediate implications for the activity of major membrane transporters (both at the plasma and organelles membranes), affecting cell metabolism and the overall plant nutritional status. Also, a wide range of toxic secondary metabolites are produced under waterlogged conditions, directly affecting the activity of membrane transporters in root cells (Pang et al. 2007a). Given the central role of membrane-transport processes in environmental signalling and plant responses to stress (Zimmermann et al. 1999; Shabala et al. 2006), the causal link between a plant's ability to control ion transport across major cellular membranes and its waterlogging tolerance must exist. Surprisingly, most of the reported evidence is essentially phenomenological in nature and does not address this issue at the cellular/molecular level. Filling this gap is one of the purposes of this chapter. In addition, physiological studies on waterlogging tolerance thus far have focused predominantly on improving root oxygen supply by either minimising radial oxygen losses (ROL) or increasing oxygen availability in roots through the formation of aerenchyma in root cortex (Armstrong et al. 2000; Colmer 2003; Jackson and Armstrong 1999). In this chapter, an additional (and essentially unexplored) approach towards breeding plants with superior toxicity tolerance is advocated.

10.2 Waterlogging and Plant Nutrient Acquisition

10.2.1 Root Ion Uptake

The adverse effects of waterlogging on root nutrient acquisition and plant nutrition are discussed in more detail in a chapter by Elzenga in this volume. In this section, only a brief summary is given, with major emphasis on the specific membrane transporters mediating these effects.

Ion transport in roots is highly sensitive to oxygen supply, and marked changes in the concentration of ions in the soil solution take place during flooding (Drew 1988). The link between oxygen supply and ion transport is primarily through respiration and the generation of ATP to drive transport. Anaerobic conditions does not maintain energy metabolism at a level that will drive primary active transport via the H^+-translocating ATPase in the plasma membrane (Armstrong and Drew 2002). It has been speculated that cells direct their limited amounts of energy to the transport of solutes involved in pH regulation as well as to transport sugars required for anaerobic carbohydrate catabolism (Greenway and Gibbs 2003). As a result, root uptake of ions such as N, P, and K, and transport of ions to the shoots is substantially reduced in waterlogged soil and hypoxic solutions (Boem et al. 1996; Buwalda et al. 1988; Drew 1988; Pang et al. 2007b; Singh et al. 2002; Smethurst et al. 2005). Mg and Ca contents in the shoot are usually affected to a lesser extent (Cannell et al. 1980; Drew and Sisworo 1979; Pang et al. 2007b; Trought and Drew 1980), most likely reflecting the fact that the transport of these ions from the external solution is less closely linked to energy metabolism (Stieger and Feller 1994).

10.2.2 Transport Between Roots and Shoots

Both solute transport to the shoot (Gibbs et al. 1998) and phloem transport of photosynthate to the roots (Colmer 2003) become impaired under oxygen-deficient conditions. Indeed, stomatal closure is observed within 2–4 h of soil flooding (Else et al. 1996), resulting in reduced transpiration and consequently nutrient delivery to the shoot (Colinbelgrand et al. 1991; Gansert 2003; Gibbs et al. 1998; Morard et al. 2004). However, the effects of hypoxia on the ion content per se in the xylem are rather controversial. While some authors report a dramatic decline in xylem sap ion composition (e.g. K^+ or Cl^-; Gibbs et al. 1998), others suggest that oxygen deprivation may induce a greater concentration of ions (e.g. K^+, Mg^{2+}, PO_4^{3-} and SO_4^{2-}) in the xylem sap of non-aerated compared to aerated treatment (Morard et al. 2004). The reasons for such controversy may be found in the complexity of the process of xylem loading (as discussed below).

Nutrients could also be translocated from the shoots to the roots, given that extra nutrition is supplied to the shoots under waterlogging conditions. A significant increase in root N and K content was found in waterlogged barley plants after foliar nutrient application (Pang et al. 2007b), suggesting the translocation of foliar absorbed nutrients to the roots via the phloem. Such translocation was able to compensate for the reduced capacity of roots to supply the essential nutrients required for root growth and development. In addition, it was suggested to be responsible for the significant increase in adventitious root production, which is an important adaptive response to waterlogging (Pang et al. 2007b). Waterlogging stress also significantly enhances nutrient recycling and the redistribution of nutrients from older leaves to new growth (Drew et al. 1979; Drew 1988; Drew and Sisworo 1979; Boem et al. 1996).

10.2.3 Ionic Mechanisms Mediating Xylem Loading

Given the fact that even the mechanisms for loading of nutrients into the xylem are still under debate (De Boer and Volkov 2003), it comes as no surprise that little is known about the impact of waterlogging on this process. The so-called "*leakage hypothesis*" was postulated by Crafts and Broyer as early as in 1938 (Crafts and Broyer 1938). This hypothesis proposes that xylem loading in the root is a passive process because of the lack of oxygen in the core of root tissue. Indeed, oxygen concentration is lowest in the stele tissue (Gibbs et al. 1998; Aguilar et al. 2003), and both mathematical modelling (Armstrong and Beckett 1987) and direct measurements (Armstrong et al. 1994) have proved that anoxia will develop in the stele but not in the cortex of the roots exposed to a low oxygen supply. However, rather than advocating a simple "leak", the modern view is that xylem-loading mechanisms are much more sophisticated.

Being classified as "thermodynamically passive", the process of xylem loading is mediated by numerous cation and anion channels shown by both electrophysiological and molecular data to be present in the xylem parenchyma (Wegner and Raschke 1994; Roberts and Tester 1995; Wegner and de Boer 1997; Gaymard et al. 1998; Köhler and Raschke 2000; Köhler et al. 2002). It is believed that major bulk of K^+ is loaded into the xylem via depolarisation-activated outwardly rectifying K^+ channels (termed SKOR; Wegner and Raschke 1994; Wegner and De Boer 1997; Roberts and Tester 1995). Such SKOR channels are highly selective between K^+ and Na^+ but do facilitate the passage of Ca^{2+} (Wegner and De Boer 1997; De Boer and Volkov 2003). Also, several types of anion channels are present at the xylem/parenchyma boundary (Köhler and Raschke 2000; White and Broadley 2001; Köhler et al. 2002). As for Na^+, its loading into the xylem is most likely to be mediated by a non-selective outward-rectifying (NORC) channel (Wegner and Raschke 1994; Wegner and De Boer 1997). This channel is permeable to both cations ($P_{Na} \approx P_K$) and anions, and shows strong voltage dependence (open at zero potential). Its molecular identity is as yet unknown, and a possibility exists that this channel might represent a glutamate receptor, GLR (Kim et al. 2001). In Arabidopsis, 20 genes are reported to encode putative glutamate receptor subunits (Lacombe et al. 2000).

The thermodynamically passive xylem loading suggests a strict control over the channel activity, and a large number of factors that control the process of xylem loading are reported. This includes apoplastic ion concentrations, cytosolic Ca^{2+}, pH, plant hormonal status (e.g. auxin and ABA), etc. (reviewed in De Boer and Volkov 2003). Most of these characteristics undergo significant changes under hypoxic conditions. Nonetheless, the control of xylem loading under waterlogged conditions essentially remains a completely unexplored area, although a number of possible mechanisms are proposed and these are discussed below.

10.2.4 Control of Xylem Ion Loading Under Hypoxia

Xylem sap collected from the flooded plants is significantly different from that in well-drained plants, with up to 0.5 more alkaline pH reported 2 h after flooding (Else et al. 1996). The effects of this change in xylem sap pH are twofold: (1) modifying the driving force for antiport and symport and (2) acting as a regulator of ion transporters or signalling molecules. Both these issues are discussed in more detail in other chapters in this book.

Another important factor potentially contributing to the control of xylem ion loading is ABA. Several major ion channels are ABA sensitive (Roberts and Snowman 2000), and external application of ABA to excised roots decreases the net efflux of K^+ from stellar cells to the xylem vessels (Cram and Pitman 1972; Behl and Jeschke, 1981). Nonetheless, most authors unanimously propose that the early reduction of stomatal conductance and reduced rates of CO_2 assimilation under waterlogged conditions are *not* related to ABA (Keles and Unyayar 2004; Ahmed et al. 2006). However, delivery rates of ABA from roots to shoots are strongly suppressed by flooding (Else et al. 1996, 2001; Janowiak et al. 2002). Thus, it cannot be excluded that such a reduction in ABA transport rate from the root may cause its overaccumulation in the basal part of the xylem. This may affect ABA-sensitive ion channels in the xylem parenchyma and modify the rate of ions loading into the xylem. This issue needs to be addressed in detail.

10.3 Oxygen Sensing in Mammalian Systems

10.3.1 Diversity and Functions of Ion Channels as Oxygen Sensors

The ability to sense and react to changes in environmental oxygen levels is crucial for the survival of all aerobic life forms (Kemp and Peers 2007). In mammals, specialised tissues have evolved to sense and rapidly respond to an acute reduction in oxygen. Central to this ability is a dynamic modulation of ion channels to hypoxia (Lopez-Barneo et al. 1988; Kemp 2006; Szigligeti et al. 2006; Kemp and Peers 2007).

The first direct demonstration of acute oxygen sensing by ion channels was carried out by Lopez-Barneo et al. (1988) using rabbit carotid body glomus cells. Using patch-clamp experiments, they provided explicit evidence for K^+ channel inhibition by hypoxia. This inhibition was suggested to cause membrane depolarisation, leading to voltage-dependent Ca^{2+} influx, neurotransmitter release and, ultimately, increased ventilatory response. To date, the activity of specific members of almost all ion channel families have been shown experimentally to be regulated by acute changes in the partial pressure of oxygen (pO_2) (Kemp and Peers 2007). Of these, the most studied are K^+ channels, although oxygen-sensitive Ca^{2+} and Na^+ channel families has also been shown (e.g. Wang et al. 2009). Moreover, such channels have been found in a variety of cell types such as carotid body, neuroepithelial bodies, H146 cells, and pulmonary circulation (Kemp 2006). Such oxygen sensitivity plays a significant role in the modulation of excitability in several cellular compartments in the mammalian nervous system, and oxygen-sensing K^+ channels have also been implicated in diverse processes in the control of catecholamine release from the faetal adrenal medulla and in T-lymphocyte activation (Kemp and Peers 2007). A short summary of major types of oxygen-sensing K^+ channels in mammalian tissues is given in Table 10.1.

Table 10.1 A summary of oxygen-sensitive K^+ channels in mammalian tissues (based on Kemp and Peers 2007)

Family	Type	Tissue expressed	References
Shaker channels	KCNA1 (Kv1.1)	Carotid body; petrosal ganglion cells; tractus solitarius	Kline et al. (2005)
	KCNA2 (Kv1.2)	PC12 cells	Conforti et al. (2000)
	KCNA5 (Kv1.5)	Pulmonary arteriolar smooth muscle	Hartness et al. (2001)
	KCNA3 (Kv1.3)	T-lymphocyte	Szigligeti et al. (2006)
Shab channels	KCNB1 (Kv2.1)	Pulmonary smooth muscle	Archer et al. (2004); Hartness et al. (2001)
Shaw channels	KCNC1 (Kv3.1b)	Pulmonary arteriolar smooth muscle	Coppock and Tamkun (2001); Osipenko et al. (2000)
	KCNC3 (Kv3.3b)	Neuroepithelial bodies, carotid body	Wang et al (1996) Perez-Garcia et al. (2004)
Shal channels	KCND1 (Kv4.1)	Carotid body	Sanchez et al. (2002)
	KCND3 (Kv4.3)	Glomus cells	Sanchez et al. (2002)
Tandem P domain channels	KNCK3 (TASK1)	HEK293 cells, carotid body	Lewis et al. (2001) Buckler et al. (2000)
	KCNK9 (TASK3)	Neuroepithelial body (H146 cells)	Hartness et al. (2001)
	KCNK2 (TREK-1)		Hartness et al. (2001)
	KCNK13 (THIK-1)	Carotid body	Campanucci et al. (2003)
Maxi-K channels	KCNMA1	Carotid body; glomus cells	Lewis et al. (2002)
Other K^+ channels	KCN2 (SK2)	Prenatal adrenal medulla	Keating et al. (2001)
	KCNJx (KIR)	Coronary arterial smooth muscle	Park et al. (2005)

10.3.2 Mechanisms of Hypoxic Channel Inhibition

Given the absence of any direct experiments on mechanisms of hypoxic channel inhibition in plants, all the current knowledge comes from the animal literature. This has recently been explicitly summarised in a comprehensive review by Kemp and Peers (2007) and is reiterated below. In brief, four different components are proposed to act as oxygen sensors (Kemp 2006): (1) NADPH oxidase; (2) mitochondria; (3) AMP kinase; (4) haemaoxygenase-2.

1. The suggestion that NADPH oxidase acts as an oxygen sensor is based on the findings that redox modulation is key to K^+ channel inhibition by hypoxia (Wang et al. 1996; O'Kelly et al. 2000). The main idea behind this model is that ROS levels vary in response to oxygen levels, thereby altering channel activity either directly, or via changes in redox-sensitive molecules such as glutathione (Kemp and Peers 2007). However, there appears to be some controversy concerning ROS levels at hypoxic conditions, with arguments for both a *decrease* (Michelakis et al. 2004) and an *increase* (Guzy and Schumacker 2006) in hypoxia-induced ROS production. Key to this controversy may be a tissue specificity of the observed effects. Moreover, it is becoming increasingly evident that there may be a number of oxygen-sensing systems within each cell type.
2. Another major source of ROS production is the mitochondria, and the idea that ROS generation is increased under hypoxia (in mitochondria) is gaining momentum (Kemp 2006). Although the specific details remain unknown, such sensing may involve electron transfer at the Rieske iron–sulphur centre (Chandel and Schumacker 2000).
3. The AMP-activated protein kinase is another potential oxygen sensor. It is suggested that hypoxia may evoke an increase in the AMP/ATP ratio of sufficient magnitude to augment AMP-activated protein kinase activity via phosphorylation. In turn, AMP kinase activation elicits an increase in intracellular Ca^{2+}, thereby regulating the channel's activity.
4. Another mechanism may be a coupling between a channel and hemeoxygenase-2 (such as in KCNMA1 channels). In the presence of NADPH and molecular oxygen, this enzyme breaks down cellular haem to produce carbon monoxide (CO) (Kemp and Peers 2007). The latter is a potent activator of certain types of ion channels in various mammalian tissues (Jaggar et al. 2002; Riesco-fagundo et al. 2001; Williams et al. 2004). Thus, as oxygen availability decreases CO production wanes and channels close (Williams et al. 2004).

10.3.3 The Molecular Mechanisms of Oxygen Sensing in Plant Systems Remain Elusive

The search for an acute oxygen sensor has become the Holy Grail for many animal researchers (Kemp and Peers 2007). Surprisingly, not a single attempt has been made to address this issue in plant systems. Given the wide range of such oxygen-sensing channels in mammalian systems (Table 10.1), there is a dire need for experiments to reveal the identity of similar systems in plants. As shown in the following sections, some recent data obtained at the cellular level, suggests that such oxygen-sensing channels may indeed be present in plant roots. However, direct patch-clamp experiments are needed to reveal the molecular mechanism(s) underpinning the response of these channels to acute hypoxia in plants. Given the importance of regulation of membrane transporters activity in plant breeding for waterlogging tolerance (discussed in Sect. 10.6), such work should be given a high priority in future.

10.4 Impact of Anoxia and Hypoxia on Membrane Transport Activity in Plant Cells

10.4.1 Oxygen Deficiency and Cell Energy Balance

Plant tissues use molecular oxygen for a number of biosynthetic or degradative processes, a major one being mitochondrial respiration. Hypoxia metabolism is characterised by both limited respiration (aerobic metabolism) and some degree of fermentation (anaerobic metabolism) (Ricard et al. 1994). Under anoxic conditions when cytochrome oxidase activity becomes oxygen limited, ATP formation through oxidative phosphorylation is also inhibited and ATP has to be produced by fermentation (Ricard et al. 1994; Geigenberger 2003). This impairs cellular metabolism and function because the efficiency of ATP formation is sharply reduced. The respiration of one molecule of hexose equivalent produces up to 39 molecules of ATP, whereas fermentation of such a molecule provides a maximum of just three molecules of ATP (Beevers 1961). The ATP required in anaerobic tissues is generated by glycolytic processes, mainly ethanolic and lactic acid fermentation (Armstrong et al. 1994). Deprivation of oxygen leads to disturbances in the ionic balance of plant cells, reflecting energy depletion and membrane depolarisation (Buwalda et al. 1988). All these changes have a direct impact on the activity of membrane transporters.

10.4.2 H^+ and Ca^{2+} Pumps

Cytoplasmic acidosis is viewed as a determinant of cell death in plant cells, and it has been demonstrated in a variety of plant cells that hypoxia rapidly decreases cytosolic pH by 0.5 to 0.6 pH units (reviewed by Felle 2001; see also his Chapter in this volume). Such acidification results from proton release through ATP hydrolysis as well as from the low ATP concentration reducing the activity of the proton pump

(Gout et al. 2001; Ricard et al. 1994). Thus, dealing with the accumulation of protons is a particularly critical task.

Nonetheless, Xia and Roberts (1996) showed that the plasma membrane H^+-ATPase can operate under anoxia. Furthermore, the H^+ pumps are more activated under anoxic conditions and this partly offsets acidosis, thereby contributing to cytoplasmic pH regulation and tolerance of anoxia. In barley seedlings, hypoxia shifted net H^+ fluxes towards net efflux in elongation and meristematic regions (Pang et al. 2006). This also suggests an up-regulation of H^+ pumping that will counter the hypoxia-induced cytosol acidification. This is further supported by pharmacological evidence showing that in roots pre-treatment with vanadate, a known inhibitor of plasma membrane (PM) H^+-ATPase, net H^+ influx in hypoxia-affected roots was significantly increased. Hence, the effect of hypoxia on H^+-ATPase activity appears to be not as straightforward as initially believed.

It has also been suggested that pyrophosphate (PPi) can substitute for ATP as an energy source, and this would represent an important acclimation mechanism in anoxia-tolerant tissues. Carystinos et al. (1995) showed that vacuolar H^+-pyrophosphatase (V-PPase) activity increased 75-fold after 6 days of anoxia in rice seedlings. Anoxia also induced an increase in activity of PPi: Fru-6-P 1 phosphotransferase, which substitutes for ATP-dependent phosphofructokinase in glycolysis in rice seedlings (Mertens et al. 1990).

Induction of Ca^{2+}-ATPase has also been shown in the adaptation of oxygen stress. Subbaiah and Sachs (2000) identified a Ca^{2+}-ATPase in maize roots (CAP1). They found that the abundance of both CAP1 transcript and protein remained at low levels in aerobic condition, whereas CAP1 mRNA was markedly induced under anoxia stress (two- to threefold increase in the transcript level after 2 h of anoxia). Consistent with this report were findings that several specific CaM isoforms were strongly up-regulated in various oak tissues affected by hypoxia (Folzer et al. 2005). The differential expression pattern observed in these experiments were suggested to reflect a specific role for each oak CaM isoform in fine-tuning the Ca^{2+} signal during flooding responses. Overall, it appears that the up-regulation of the Ca^{2+} efflux systems could play an important role in a plant's adaptation to oxygen deprivation, restoring the basal level of intracellular Ca^{2+} disturbed by waterlogging stress signalling (as discussed in the next section).

10.4.3 Ca^{2+}-Permeable Channels

Several studies have showed that gene expression and physiological changes in response to oxygen deprivation are preceded and signalled by an elevation in cytosolic free Ca^{2+} $[Ca^{2+}]_{cyt}$ (Subbaiah et al. 1994a, b). These authors found that ruthenium red (RR), a specific blocker of endomembrane Ca^{2+} channels, repressed the activation of anoxia-inducible genes, blocked the anoxia-induced $[Ca^{2+}]_{cyt}$ elevation and impaired the post-anoxic survival of maize seedlings and cells. With the mitochondrion being the primary site of oxygen consumption as well as being an important target of RR action, it is thought it might additionally serve as a Ca^{2+} store, mobilised in response to anoxia in plant cells (Subbaiah and Sachs 2003). At the same time, Ca^{2+} uptake from the external medium appears to be critical for maintaining shoot elongation under anaerobic conditions, as revealed in a series of pharmacological experiments using PM Ca^{2+} channel blockers such as verapamil, diltiazem and La^{3+} (Tamura et al. 2001). Thus, it appears that both the plasma and endomembrane Ca^{2+} channels are involved in adaptive response to waterlogging. However, as both the sensing and signalling pathways may be triggered by changes in cellular homeostasis, and not necessarily by the direct sensing of the oxygen concentration (Bailey-Serres and Chang 2005), direct experiments are required to reveal whether any of the above types of Ca^{2+} channels serve as a primary sensor for oxygen deficiency in plant cells.

10.4.4 K^+-Permeable Channels

Potassium is one of the most important nutrients for plant functions, and K^+ uptake is disturbed by oxygen stress. Buwalda et al. (1988) found that both 6-d-old seedlings and 26-d-old wheat plants lost K^+ from the roots following their transfer from aerated to hypoxia nutrient solutions. They suggested that the K^+ loss during the early stages of hypoxia is due to membrane depolarisation as opposed to an increase in the membrane permeability to K^+ (Buwalda et al. 1988). In contrast, Colmer et al. (2001) reported a 16- to 18-fold lower permeability for K^+ efflux 19 h after the onset of stress in rice. This effect was attributed to rapid restoration of the membrane potential to values more negative than the K^+ diffusion potential, resulting in closure of KOR channels (Greenway and Gibbs 2003); cAMP-triggered KOR closure has also been advocated under anoxic conditions (Reggiani 1997). Pang et al. (2006) reported that hypoxia significantly inhibited K^+ uptake in the mature zone of the sensitive barley variety Naso Nijo, but not of the tolerant cultivar, TX9425, while hypoxia reduced K^+ efflux in the elongation and meristem root zones. They suggest that hypoxia-induced K^+ flux responses are mediated by both KIR and NSCC channels in the elongation zone, while in the mature zone KOR channels are likely to be the key players based on pharmacological data in barley.

10.5 Secondary Metabolites Toxicity and Membrane Transport Activity in Plant Cells

10.5.1 Waterlogging and Production of Secondary Metabolites

Excess water causes a sharp decrease in soil redox potential. As a result, significant accumulation of various toxic substances accumulate in waterlogged soil; the result of anaerobic metabolism in plants and microbes (Armstrong and Armstrong 1999; Lynch 1977; Tanaka et al. 1990). The specific nature of the metabolites produced is a direct function of redox potentials, so correlates with the severity of waterlogging. As soon as the free oxygen is depleted, nitrate (NO_3^-) is used by soil microorganisms as an alternative electron acceptor in respiration. Nitrate is reduced to nitrite (NO_2^-), various nitrous oxides (e.g. N_2O, NO), and molecular nitrogen (N_2) during the process of denitrification (Gambrell et al. 1991). This occurs around 225 mV redox potential (corrected to pH 7) (Gambrell et al. 1991). Manganese oxides [mainly Mn(VI)] are the next electron acceptors when the redox potential is approximately 200 mV. In acid soils that are high in manganese oxides and organic matter, but low in NO_3^-, very high levels of water-soluble and exchangeable Mn^{2+} can build up within a few days (Marschner 1995). Ferric iron is reduced to the more soluble and mobile ferrous form just above 100 mV redox potential. Iron reduction is also associated with a marked increase in soil pH (Kirk et al. 1990). Sulphate is reduced to sulphide at -150 mV. Due to the formation of sparingly soluble sulphides, the reduction of sulphate to hydrogen sulphide (H_2S) in submerged soils may decrease the solubility of iron, zinc, copper and cadmium (Ponnamperuma 1972). Finally, methane formation, from the reduction of carbon dioxide and certain organic acids, is initiated at -200 mV (Gambrell et al. 1991).

The type and amount of organic acid produced depends upon the fermentive character of the microflora, the type and amount of organic materials added, and on the prevailing soil conditions (Rao and Mikkelsen 1977). Various products of microbial carbon metabolism, such as ethylene, also accumulate in waterlogged soils (Grichko and Glick 2001; Setter and Waters 2003). During prolonged waterlogging, volatile fatty acids and phenolics accumulate in soils that are high in readily decomposable organic matter (i.e. after application of green manure or straw) (Armstrong and Armstrong 1999; Lynch 1978; Tanaka et al. 1990; Wang et al. 1967). Each of these (potentially toxic) substances may significantly affect the activity of membrane transporters and plant nutrient acquisition.

10.5.2 Secondary Metabolite Production and Plant Nutrient Acquisition

Root nutrient uptake is significantly impaired by secondary metabolites. Early reports by Glass (Glass 1973; 1974) showed that various benzoic compounds caused a rapid and substantial inhibition of phosphate (PO_4) and K^+ absorption in excised barley roots. In cucumber, 3 h treatment with 200 μM ferulic acid inhibited NO_3^- uptake and promoted K^+ efflux (Booker et al. 1992). Cinnamic acid greatly inhibited the uptake of NO_3^-, SO_4^{2-}, K^+, Ca^{2+}, Mg^{2+} in soybean and cucumber (Baziramakenga et al. 1995; Yu and Matsui 1997). Inhibition of ion uptake by phenolic acids correlates with their lipophilicity and is largely attributed to the ability of phenolic acids to cause membrane depolarisation (Glass 1973;

Yu and Matsui 1997). The majority of such studies have dealt with the analysis of the overall changes in ion content in plant tissues, or with monitoring the kinetics of nutrient depletion in a growth solution (Glass 1973; 1974; Jackson and St. John 1980).

Due to methodological limitations, early investigations failed to provide answers regarding the specific ionic mechanisms involved. For instance, radiotracers (e.g. [86]Rb; Glass 1974) allowed the measurements of unidirectional K^+ uptake, but could not show the extent to which particular K^+ efflux systems are affected. As a result, many hypotheses proposed in the literature were not substantiated by actual data. For example, it was suggested that changes in membrane lipid composition might be responsible for the observed leak of K^+ and Ca^{2+} from roots treated with monocarboxylic acids (Jackson and St. John 1980; Jackson and Taylor 1970). However, using the microelectrode ion flux measurement (MIFE) technique with high spatial (a few micrometers) and temporal resolution (several seconds) to directly measure net ion fluxes across cellular membranes, it was shown that this was not the case. Rapid (within 1 min) changes in net K^+ and Ca^{2+} fluxes were measured from barley roots in response to application of physiologically relevant concentrations of various phenolic and monocarboxilic acids (Pang et al. 2007a). However, non-specific changes in membrane permeability are usually associated with a change in membrane lipid components (Glass 1974; Jackson and St. John 1980; Jackson and Taylor 1970), and this process operates at a much slower time scale than that needed for the rapid changes in ion flux described above. Thus, it is highly unlikely that the observed rapid ion flux responses could originate from such changes in membrane lipid composition in response to secondary metabolites. Importantly, the above changes in membrane permeability are believed to be non-specific (Glass 1974). However, Pang et al. (2007a) observed that the K^+ leak gradually increased with time, while Ca^{2+} efflux was short-lived and returned back to control values within 10–15 min after treatment. This suggests that fluxes of these two ions are mediated by different transport systems, and so cannot be attributed to a general change in membrane permeability.

To date, the mechanisms behind the impaired ion uptake resulting from the presence of secondary metabolites are still not clear and needs further investigation.

10.6 Secondary Metabolites and Activity of Key Membrane Transporters

10.6.1 Pumps

Once inside the cell, permeated phenolic acids dissociate and acidify the cytosol (Ehness et al. 1997; Guern et al. 1986). This activates the plasma membrane H^+-ATPase, thereby increasing H^+ extrusion (Beffagna and Romani 1991; Felle et al. 1991; Frachisse et al. 1988). Indeed, a 40% higher ATP hydrolytic activity compared with the control is reported in PM vesicles from barley roots treated with 200 μM of 2-hydroxybenzoic acid for 30 min (Pang et al. 2007a). While net H^+ fluxes were not significantly different from the control 24 h after treatment with three phenolic acids, membrane potential values were even more negative (hyper-polarised) than control roots, suggesting higher ATP hydrolytic activity in phenolic-treated roots (Pang et al. 2007a). Conversely, other reports show that ATPase is deactivated in response to secondary metabolites (Erdei et al. 1994; Ye et al. 2006). The reasons for such discrepancy are yet to be unveiled.

In microbes, the uncoupling theory has been proposed to explain the inhibitory effect of weak acids on cell growth (Russell 1992; Palmqvist and Hahn-Hagerdal 2000). According to this theory, when weak acids flow into the cell the resultant decrease in intracellular pH is neutralised by the action of the plasma membrane ATPase, which pumps protons out of the cell at the expense of ATP hydrolysis (Verduyn et al. 1992). However, at high acid concentrations, the proton pumping capacity of the cell is exhausted. This results in a depletion of the ATP content, dissipation of the proton motive force, and acidification of the cytoplasm (Imai and Ohno 1995).

10.6.2 Carriers

It is traditionally believed that most phenolic acids cross the cell membrane in an undissociated form via passive diffusion. However, recent cloning and functional characterisation of the monocarboxylic acid transporter (MCT) family suggests that uptake of both monocarboxylic acids and benzoic acid occur via a H^+-coupled co-transport mechanism in animal systems (Kido et al. 2000; Tamai et al. 1999). Moreover, the involvement of MCT in transporting other secondary metabolites, such as m-coumaric acid, m-hydroxyphenylpropionic acid, p-coumaric acid, and ferulic acid, has also been widely reported in animal models (Konishi and Kobayashi 2004; Konishi et al. 2003; Konishi and Shimizu 2003; Watanabe et al. 2006). To date, more than 10 isoforms of MCT have been identified in animals (Halestrap and Meredith 2004); but, to the best of our knowledge, no reports on the involvement of MCT in transporting secondary metabolites in plants have been presented. A plausible model explaining the effects of secondary metabolites on kinetics of ion fluxes in plant roots has been put forward (Fig. 10.1; Pang et al. 2007a). Its validation remains an important task for the future.

10.6.3 Channels

The identification of ion channels that mediate plant adaptive responses to secondary metabolites remains elusive, and the molecular mechanisms by which these

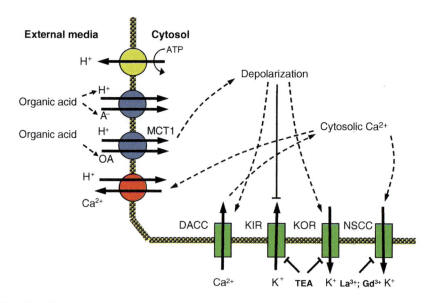

Fig. 10.1 Signal transduction pathways mediating short-term effects of secondary metabolites on membrane transport activity. *KIR* inward-rectifying K^+ channel, *KOR* outward-rectifying K^+ channel, *DACC* depolarization-activated Ca^{2+} channel. Modified from Pang et al. (2007)

compounds control ion transport across the plasma membrane are largely unknown. On the basis of the fact that removal of phenolics causes a rapid recovery of K^+ absorption, Glass (1974) suggested a direct effect on cell membranes; but, no specific details were provided. It is reported that the addition of fusicoccin counteracts the inhibitory effect of ferulic acid on net K^+ uptake, while a concurrent treatment of cucumber seedlings with ferulic acid and tetraethylammonium (a known blocker of K^+-permeable channels) reduces the average K^+ efflux by 66% (Booker et al. 1992). By combining non-invasive ion flux measurements with a pharmacological approach, Pang et al. (2007a) have recently showed that voltage-gated Shaker-type K^+ channels do play a key role in mediating K^+ transport after exposure to secondary metabolites. In addition, involvement of non-selective cation channels (NSCC) is also likely, as both Gd^{3+} and La^{3+} (two known NSCC blockers; Demidchik et al. 2002) were efficient in preventing 2-hydroxybenzoic acid- and acetic acid-induced K^+ loss.

Barley roots treated with phenolic acids also show a gradual and prolonged increase in net Ca^{2+} uptake (Pang et al. 2007a). As such treatment also results in a significant plasma membrane depolarisation, it is suggested that depolarisation-activated Ca^{2+} channels (DACC) may contribute to the observed Ca^{2+} influx after phenolic application. Such increased Ca^{2+} uptake would provide a positive feedback mechanism to further depolarise the membrane potential, thus amplifying the effect of phenolics on K^+ transport. Both these hypothesis have to be tested in direct patch-clamp experiments.

10.7 Breeding for Waterlogging Tolerance by Targeting Key Membrane Transporters

10.7.1 General Trends in Breeding Plants for Waterlogging Tolerance

Waterlogging tolerance is a complex trait conferred by many different physiological mechanisms and complicated by confounding factors such as temperature, plant development stage, nutrient availability, severity of waterlogging stress, soil physical properties, etc (Setter and Waters 2003; Setter et al. 2009). Different traits have been used as indirect selection indices for waterlogging tolerance; the majority using a straightforward phenotyping approach. A "damage index" (expressed as a percentage of yield under non-waterlogged conditions) was used to screen 4,572 barley lines (Qiu and Ke 1991). However, direct selection on grain yield has low effectiveness since the heritability of yield after waterlogging is rather low (Collaku and Harrison 2005). Leaf chlorosis after waterlogging is another major indice used in other crops such as wheat (Boru et al. 2001) and soybean (Reyna et al. 2003) as well as barley (Li et al. 2008). Even though the heritability is relatively high for leaf chlorosis and early generation selection could be efficient, well-controlled waterlogging conditions are still crucial for its precise evaluation (Li et al. 2008).

Although field experiments are, and will always remain, the ultimate test for waterlogging tolerance, multiple confounding environmental factors make targeted screening for a particular waterlogging trait unrealistic. To date, physiologists have focused predominantly on preventing oxygen loss or improving its transport to, or storage, in roots (Jackson and Armstrong 1999). As for breeders, the phenotyping approach remains the main tool at present. Given the multiple physiological mechanisms contributing to plant adaptation to waterlogging and confounding environmental factors, this approach can hardly be successful. Therefore, specific physiological traits have to be targeted in a search for the "waterlogging tolerance gene(s)". Meanwhile, other physiological mechanisms such as plant tolerance to secondary metabolites may be equally important. Such traits have never been used as selection criteria in breeding programmes. We believe that targeting these characteristics would aid the selection of waterlogging-tolerant genotypes for waterlogging breeding programmes. Some of the possible strategies and background information are provided and discussed below.

10.7.2 Improving Membrane Transporters Efficiency Under Hypoxic Conditions

To the best of our knowledge, membrane transporters have never been a *direct* target in any breeding programme for waterlogging tolerance. However, Mancuso and Boselli (2002) found that under hypoxia stress, tolerant *Vitis* species decreased the oxygen influx at a rate much slower than sensitive species. Similarly, a much higher oxygen influx was measured for the waterlogging-tolerant TX9425 barley variety compared with waterlogging-sensitive Naso Nijo (Pang et al. 2006). As noted in Sect. 10.4, a higher oxygen uptake rate in the mature zone of tolerant species would permit the maintenance of ATP thereby precluding the adverse effects of hypoxia on cytosolic metabolism. This could result in an increased ability of tolerant species to maintain a relatively stable K^+ uptake. Indeed, sensitive species did show a significant decline in net K^+ influx in the mature zone in response to hypoxia (Pang et al. 2006). Given the importance of K^+ for cell metabolism (Leigh 2001; Shabala 2003), an improved ability for K^+ uptake in roots may therefore be critical to the overall plant performance under hypoxia.

The paper by Pang et al. (2006) described above is the only direct attempt to reveal the difference in root ion flux patterns between waterlogging-sensitive and waterlogging-tolerant barley cultivars in response to oxygen deprivation. Therefore, before this parameter is incorporated into any breeding programme, the contribution to waterlogging tolerance of this ion flux response under oxygen stress needs to be addressed. It must be shown that it is not merely a consequence of the difference in some other physiological or anatomical (e.g. Pang et al. 2004) mechanisms that mediate plant's adaptation to waterlogging.

10.7.3 Reducing Sensitivity to Toxic Secondary Metabolites

Secondary metabolites associated with anaerobic soil conditions could cause detrimental effects on plant growth because of the negative effects on ion uptake for growth maintenance. As discussed in Sect. 10.5, a pronounced shift towards K^+ efflux was measured in response to both phenolic and monocarboxlic acids in barley roots, even *in the presence* of molecular oxygen. Given the tight link between net K^+ efflux and cytosolic free K^+ concentration (Shabala et al. 2006), such a K^+ loss may be detrimental to the activity of a large number of K^+-dependent enzymes (over 50; Marschner 1995). Also, a decrease in cytosolic K^+ pool can trigger programmed cell death (PCD) by activating caspase-like proteins (Hughes and Cidlowski 1999; Shabala 2009). Moreover, the magnitude of stress-induced K^+ loss is often used as a measure of stress tolerance (e.g. salinity – Chen et al. 2005, 2007; oxidative stress – Cuin and Shabala 2007). Recently, Pang et al. (2007a) has presented evidence that the same conclusion may be applicable to plant waterlogging tolerance. Indeed, in the waterlogging-sensitive barley variety Naso Nijo, all the three lower monocarboxylic acids (formic, acetic and propionic acids) and the three phenolic acids (benzoic, 2-hydroxybenzoic, 4-hydroxybenzoic acids) caused a substantial shift towards steady K^+ efflux, but such a detrimental effect on root cell K^+ homeostasis was absent in the waterlogging-tolerant TX variety (Pang et al. 2007a). It was suggested that waterlogging tolerance in barley is conferred not only by the difference in the root anatomy (high percentage of aerenchyma in TX genotype) (Pang et al. 2004) but may, to a large extent, be determined by the superior ability of tolerant genotypes to reduce the detrimental effects of secondary metabolites on membrane-transport processes in roots (specifically, improving K^+ retention) (Pang et al. 2007a). On the basis of these results, we suggest that plant tolerance to these secondary metabolites should be considered as a useful trait in breeding programmes. Similar experiments are required in a wider range of plant cultivars with contrasting waterlogging tolerance to further validate the theory.

Acknowledgements We are thankful to Dr Tracey Cuin for her technical assistance and valuable comments during the preparation of this MS.

References

Aguilar EA, Turner DW, Gibbs DJ, Armstrong W, Sivasithamparam K (2003) Oxygen distribution and movement, respiration and nutrient loading in banana roots (*Musa* spp. L.) subjected to aerated and oxygen-depleted environments. Plant Soil 253:91–102

Ahmed S, Nawata E, Sakuratani T (2006) Changes of endogenous ABA and ACC, and their correlations to photosynthesis and water relations in mungbean (*Vigna radiata* (L.) Wilczak cv. KPS1) during waterlogging. Environ Exp Bot 57:278–284

Archer SL, Wu XC, Thebaud B, Nsair A, Bonnet S, Tyrrell B, McMurtry MS, Hashimoto K, Harry G, Michelakis ED (2004) Preferential expression and function of voltage-gated, O_2-sensitive K^+ channels in resistance pulmonary arteries explains regional heterogeneity in hypoxic pulmonary vasoconstriction: ionic diversity in smooth muscle cells. Circ Res 95:308–318

Armstrong J, Armstrong W (1999) *Phragmites* die-back: toxic effects of propionic, butyric and caproic acids in relation to pH. New Phytol 142:201–217

Armstrong W, Beckett PM (1987) Internal aeration and the development of stelar anoxia in submerged roots: a multishelled mathematical model combining axial diffusion of oxygen in the cortex with radial losses to the stele, the wall layers and the rhizosphere. New Phytol 105:221–245

Armstrong W, Brandle R, Jackson MB (1994) Mechanisms of flood tolerance in plants. Acta Bot Neerl 43:307–358

Armstrong W, Cousins D, Armstrong J, Turner DW, Beckett PM (2000) Oxygen distribution in wetland plant roots and permeability barriers to gas-exchange with the rhizosphere: a microelectrode and modelling study with *Phragmites australis*. Ann Bot 86:687–703

Armstrong W, Drew MC (2002) Root growth and metabolism under oxygen deficiency. In: Waisel Y, Eshel A, Kafkafi U (eds) Plant roots: the hidden half. Marcel Dekker, New York, pp 729–761

Bailey-Serres J, Chang R (2005) Sensing and signalling in response to oxygen deprivation in plants and other organisms. Ann Bot 96:507–518

Baziramakenga R, Leroux GD, Simard RR (1995) Effects of benzoic and cinnamic acids on membrane permeability of soybean roots. J Chem Ecol 21:1271–1285

Beevers H (1961) Respiratory metabolism in plants. Row, Peterson and Company, Evanston

Beffagna N, Romani G (1991) Modulation of the plasmalemma proton pump activity by intracellular pH in *Elodea densa* leaves: correlation between acid load and H^+ pumping activity. Plant Physiol Biochem 29:471–480

Behl R, Jeschke WD (1981) Influence of abscisic acid on unidirectional fluxes and intracellular compartmentation of K^+ and Na^+ in excised barley root segments. Physiol Plant 53:95–100

Boem FHG, Lavado RS, Porcelli CA (1996) Note on the effects of winter and spring waterlogging on growth, chemical composition and yield of rapeseed. Field Crops Res 47:175–179

Booker FL, Blum U, Fiscus EL (1992) Short-term effects of ferulic acid on ion uptake and water relations in cucumber seedlings. J Exp Bot 43:649–655

Boru G, van Ginkel M, Kronstad WE, Boersma L (2001) Expression and inheritance of tolerance to waterlogging stress in wheat. Euphytica 117:91–98

Buckler KJ, Williams BA, Honore E (2000) An oxygen-, acid- and anaesthetic-sensitive TASK-like background potassium channel in rat arterial chemoreceptor cells. J Physiol 525:135–142

Buwalda F, Thomson CJ, Steigner W, Barrett-Lennard EG, Gibbs J, Greenway H (1988) Hypoxia induces membrane depolarization and potassium-loss from wheat roots but does not increase their permeability to sorbitol. J Exp Bot 39:1169–1183

Campanucci VA, Fearon IM, Nurse CA (2003) A novel O_2-sensing mechanism in rat glossopharyngeal neurones mediated by a halothane-inhibitable background K^+ conductance. J Physiol 548:731–743

Cannell RQ, Belford RK, Gales K, Dennis CW, Prew RD (1980) Effects of waterlogging at different stages of development on the growth and yield of winter wheat. J Sci Food Agric 31:117–132

Carystinos GD, Macdonald HR, Monroy AF, Dhindsa RS, Poole RJ (1995) Vacuolar H^+-translocating pyrophosphatase is induced by anoxia or chilling in seedlings of rice. Plant Physiol 108:641–649

Chandel NS, Schumacker PT (2000) Cellular oxygen sensing by mitochondria: old questions, new insight. J Appl Physiol 88:1880–1889

Chen Z, Newman I, Zhou M, Mendham N, Zhang G, Shabala S (2005) Screening plants for salt tolerance by measuring K^+ flux: a case study for barley. Plant Cell Environ 28:1230–1246

Chen Z, Zhou M, Newman I, Mendham N, Zhang G, Shabala S (2007) Potassium and sodium relations in salinised barley tissues as a basis of differential salt tolerance. Funct Plant Biol 34:150–162

Colinbelgrand M, Dreyer E, Biron P (1991) Sensitivity of seedlings from different oak species to waterlogging – effects on root-growth and mineral-nutrition. Ann Sci Forest 48:193–204

Collaku A, Harrison SA (2005) Heritability of waterlogging tolerance in wheat. Crop Sci 45: 722–727

Colmer TD (2003) Long-distance transport of gases in plants: a perspective on internal aeration and radial oxygen loss from roots. Plant Cell Environ 26:17–36

Colmer TD, Huang SB, Greenway H (2001) Evidence for down-regulation of ethanolic fermentation and K^+ effluxes in the coleoptile of rice seedlings during prolonged anoxia. J Exp Bot 52:1507–1517

Conforti L, Bodi I, Nisbet JW, Millhorn DE (2000) O_2-sensitive K^+ channels: role of the Kv1.2 subunit in mediating the hypoxic response. J Physiol 524:783–793

Coppock EA, Tamkun MM (2001) Differential expression of Kv channel α- and β-subunits in the bovine pulmonary arterial circulation. Am J Physiol Lung Cell Mol Physiol 281:L1350–L1360

Crafts AS, Broyer TC (1938) Migration of salts and water into xylem of the roots of higher plants. Am J Bot 25:529–535

Cram WJ, Pitman MG (1972) Action of abscisic acid on ion uptake and water flow in plant roots. Aust J Biol Sci 25:1125–1132

Cuin T, Shabala S (2007) Amino acids regulate salinity-induced potassium efflux in barley root epidermis. Planta 225:753–761

De Boer AH, Volkov V (2003) Logistics of water and salt transport through the plant: structure and functioning of the xylem. Plant Cell Environ 26:87–101

Demidchik V, Davenport RJ, Tester M (2002) Nonselective cation channels in plants. Annu Rev Plant Biol 53:67–107

Drew MC (1988) Effects of flooding and oxygen deficiency on plant mineral nutrition. In: Lauchli A, Tinker PB (eds) Advances in plant nutrition. Praeger, New York, pp 115–159

Drew MC, Sisworo EJ (1979) The development of waterlogging damage in young barley plants in relation to plant nutrient status and changes in soil properties. New Phytol 82:301–314

Drew MC, Sisworo EJ, Saker LR (1979) Alleviation of waterlogging damage to young barley plants by application of nitrate and a synthetic cytokinin, and comparison between the effects of waterlogging, nitrogen deficiency and root excision. New Phytol 82:315–329

Ehness R, Ecker M, Godt DE, Roitsch T (1997) Glucose and stress independently regulate source and sink metabolism and defense mechanisms via signal transduction pathways involving protein phosphorylation. Plant Cell 9:1825–1841

Else MA, Coupland D, Dutton L, Jackson MB (2001) Decreased root hydraulic conductivity reduces leaf water potential, initiates stomatal closure and slows leaf expansion in flooded plants of castor oil (*Ricinus communis*) despite diminished delivery of ABA from the roots to shoots in xylem sap. Physiol Plant 111:46–54

Else MA, Tiekstra AE, Croker SJ, Davies WJ, Jackson MB (1996) Stomatal closure in flooded tomato plants involves abscisic acid and a chemically unidentified anti-transpirant in xylem sap. Plant Physiol 112:239–247

Erdei L, Szabonagy A, Laszlavik M (1994) Effects of tannin and phenolics on the H^+-ATPase activity in plant plasma membrane. J Plant Physiol 144:49–52

Felle HH (2001) pH: signal and messenger in plant cells. Plant Biol 3:577–591

Felle HH, Peters W, Palme K (1991) The electrical response of maize to auxins. Biochim Biophys Acta 1064:199–204

Folzer H, Capelli N, Dat J, Badot PM (2005) Molecular cloning and characterization of calmodulin genes in young oak seedlings (*Quercus petraea* L.) during early flooding stress. BBA Gene Struct Exp 1727:213–219

Frachisse JM, Johannes E, Felle H (1988) The use of weak acids as physiological tools: a study of the effects of fatty acids on intracellular pH and electrical plasmalemma properties of *Riccia fluitans* rhizoid cells. BBA Biomembr 938:199–210

Gambrell RP, DeLaune RD, Patrick WH (1991) Redox processes in soils following oxygen depletion. In: Jackson MB, Davies DD, Lambers H (eds) Plant life under oxygen deprivation: ecology, physiology and biochemistry. SPB Academic, The Hague

Gansert D (2003) Xylem sap flow as a major pathway for oxygen supply to the sapwood of birch (*Betula pubescens* Ehr.). Plant Cell Environ 26:1803–1814

Gaymard F, Pilot G, Lacombe B, Bouchez D, Bruneau D, Boucherez J, Michaux-Ferriere N, Thibaud JB, Sentenac H (1998) Identification and disruption of a plant shaker-like outward channel involved in K^+ release into the xylem sap. Cell 94:647–655

Geigenberger P (2003) Response of plant metabolism to too little oxygen. Curr Opin Plant Biol 6:247–256

Gibbs J, Turner DW, Armstrong W, Darwent MJ, Greenway H (1998) Response to oxygen deficiency in primary maize roots. I. Development of oxygen deficiency in the stele reduces radial solute transport to the xylem. Aust J Plant Physiol 25:745–758

Glass ADM (1973) Influence of phenolic acids on ion uptake. I. Inhibition of phosphate uptake. Plant Physiol 51:1037–1041

Glass ADM (1974) Influence of phenolic acids upon ion uptake. III. Inhibition of potassium absorption. J Exp Bot 25:1104–1113

Gout E, Boisson AM, Aubert S, Douce R, Bligny R (2001) Origin of the cytoplasmic pH changes during anaerobic stress in higher plant cells. Carbon-13 and phosphorous-31 nuclear magnetic resonance studies. Plant Physiol 125:912–925

Greenway H, Gibbs J (2003) Mechanisms of anoxia tolerance in plants. II. Energy requirements for maintenance and energy distribution to essential processes. Funct Plant Biol 30:999–1036

Grichko VP, Glick BR (2001) Ethylene and flooding stress in plants. Plant Physiol Biochem 39:1–9

Guern J, Mathieu Y, Pean M, Pasquier C, Beloeil JC, Lallemand JY (1986) Cytoplasmic pH regulation in *Acer pseudoplatanus* cells. I. A ^{31}P NMR description of acid-load effects. Plant Physiol 82:840–845

Guzy RD, Schumacker PT (2006) Oxygen sensing by mitochondria at complex III: the paradox of increased reactive oxygen species during hypoxia. Exp Physiol 91:807–819

Halestrap AP, Meredith D (2004) The SLC16 gene family – from monocarboxylate transporters (MCTs) to aromatic amino acid transporters and beyond. Pflug Arch Eur J Phy 447:619–628

Hartness ME, Lewis A, Searle GJ, O'Kelly I, Peers C, Kemp PJ (2001) Combined antisense and pharmacological approaches implicate hTASK as an airway O_2 sensing K^+ channel. J Biol Chem 276:26499–26508

Hughes FM, Cidlowski JA (1999) Potassium is a critical regulator of apoptotic enzymes in vitro and in vivo. Adv Enzyme Regul 39:157–171

Imai T, Ohno T (1995) The Relationship between viability and intracellular pH in the yeast *Saccharomyces cerevisiae*. Appl Environ Microbiol 61:3604–3608

Jackson MB, Armstrong W (1999) Formation of aerenchyma and the processes of plant ventilation in relation to soil flooding and submergence. Plant Biol 1:274–287

Jackson PC, St. John JB (1980) Changes in membrane lipids of roots associated with changes in permeability. Plant Physiol 66:801–804

Jackson PC, Taylor JM (1970) Effects of organic acids on ion uptake and retention in barley roots. Plant Physiol 46:538–542

Jaggar JH, Leffler CW, Cheranov SY, Tcheranova D, Cheng X (2002) Carbon monoxide dilates cerebral arterioles by enhancing the coupling of Ca^{2+} sparks to Ca^{2+}-activated K^+ channels. Circ Res 91:610–617

Janowiak F, Maas B, Dorffling K (2002) Importance of abscisic acid for chilling tolerance of maize seedlings. J Plant Physiol 159:635–643

Keating DJ, Rychkov GY, Roberts ML (2001) Oxygen sensitivity in the sheep adrenal medulla: role of SK channels. Am J Physiol Cell Physiol 281:C1434–C1441

Keles Y, Unyayar S (2004) Responses of antioxidant defense system of *Helianthus annuus* to abscisic acid treatment under drought and waterlogging. Acta Physiol Plant 26:149–156

Kemp PJ (2006) Detecting acute changes in oxygen: will the real sensor please stand up? Exp Physiol 91:829–834

Kemp PJ, Peers C (2007) Oxygen sensing by ion channels. Oxygen sensing and hypoxia-induced responses. Essays Biochem 43:77–90

Kido Y, Tamai I, Okamoto M, Suzuki F, Tsuji A (2000) Functional clarification of MCT1-mediated transport of monocarboxylic acids at the blood-brain barrier using in vitro cultured cells and in vivo BUI studies. Pharm Res 17:55–62

Kim SA, Kwak JM, Jae SK, Wang MH, Nam HG (2001) Overexpression of the AtGluR2 gene encoding an *Arabidopsis* homolog of mammalian glutamate receptors impairs calcium utilization and sensitivity to ionic stress in transgenic plants. Plant Cell Physiol 42:74–84

Kirk GJD, Ahmad AR, Nye PH (1990) Coupled diffusion and oxidation of ferrous iron in soils. II. A model of the diffusion and reaction of O_2, Fe^{2+}, H^+ and HCO_3^- in soils and a sensitivity analysis of the model. J Soil Sci 41:411–431

Kline DD, Buniel MC, Glazebrook P, Peng YJ, Ramirez-Navarro A, Prabhakar NR, Kunze DL (2005) Kv1.1 deletion augments the afferent hypoxic chemosensory pathway and respiration. J Neurosci 25:3389–3399

Köhler B, Raschke K (2000) The delivery of salts to the xylem. Three types of anion conductance in the plasmalemma of the xylem parenchyma of roots of barley. Plant Physiol 122:243–254

Köhler B, Wegner LH, Osipov V, Raschke K (2002) Loading of nitrate into the xylem: apoplastic nitrate controls the voltage dependence of X-QUAC, the main anion conductance in xylem-parenchyma cells of barley roots. Plant J 30:133–142

Konishi Y, Kobayashi S (2004) Microbial metabolites of ingested caffeic acid are absorbed by the monocarboxylic acid transporter (MCT) in intestinal caco-2 cell monolayers. J Agric Food Chem 52:6418–6424

Konishi Y, Kobayashi S, Shimizu M (2003) Transepithelial transport of p-coumaric acid and gallic acid in caco-2 cell monolayers. Biosci Biotechnol Biochem 67:2317–2324

Konishi Y, Shimizu M (2003) Transepithelial transport of ferulic acid by monocarboxylic acid transporter in Caco-2 cell monolayers. Biosci Biotechnol Biochem 67:856–862

Lacombe B, Pilot G, Gaymard F, Sentenac H, Thibaud JB (2000) pH control of the plant outwardly-rectifying potassium channel SKOR. FEBS Lett 466:351–354

Leigh RA (2001) Potassium homeostasis and membrane transport. J Plant Nutr Soil Sci 164:193–198

Lewis A, Hartness ME, Chapman CG, Fearon IM, Meadows HJ, Peers C, Kemp PJ (2001) Recombinant hTASK1 is an O_2-sensitive K^+ channel. Biochem Biophys Res Comm 285:1290–1294

Li HB, Vaillancourt R, Mendham N, Zhou MX (2008) Comparative mapping of quantitative trait loci associated with waterlogging tolerance in barley (*Hordeum vulgare* L.). BMC Genomics 9

Lopez-Barneo J, Lopez-Lopez JR, Urena J, Gonzalez C (1988) Chemotransduction in the carotid body: K^+ current modulated by PO_2 in type I chemoreceptor cells. Science 241:580–582

Lynch JM (1977) Phytotoxicity of acetic acid produced in the anaerobic decomposition of wheat straw. J Appl Bacteriol 42:81–87

Lynch JM (1978) Production and phytotoxicity of acetic acid in anaerobic soils containing plant residues. Soil Biol Biochem 10:131–135

Mancuso S, Boselli M (2002) Characterisation of the oxygen fluxes in the division, elongation and mature zones of *Vitis* roots: influence of oxygen availability. Planta 214:767–774

Marschner H (1995) Mineral nutrition of higher plants (2nd ed). Academic, London

Mertens E, Larondelle Y, Hers HG (1990) Induction of pyrophosphate: fructose 6-phosphate 1-phosphotransferase by anoxia in rice seedlings. Plant Physiol 93:584–587

Michelakis ED, Thebaud B, Weir EK, Archer SL (2004) Hypoxic pulmonary vasoconstriction: redox regulation of O_2-sensitive K^+ channels by a mitochondrial O_2-sensor in resistance artery smooth muscle cells. J Mol Cell Cardiol 37:1119–1136

Morard P, Lacoste L, Silvestre J (2004) Effect of oxygen deficiency on mineral nutrition of excised tomato roots. J Plant Nutr 27:613–626

O'Kelly I, Lewis A, Peers C, Kemp PJ (2000) O_2 sensing by airway chemoreceptor-derived cells: protein kinase C activation reveals functional evidence for involvement of NADPH oxidase. J Biol Chem 275:7684–7692

Osipenko ON, Tate RJ, Gurney AM (2000) Potential role for Kv3.1b channels as oxygen sensors. Circ Res 86:534–540

Palmqvist E, Hahn-Hagerdal B (2000) Fermentation of lignocellulosic hydrolysates. I: inhibition and detoxification. Bioresour Technol 74:17–24

Pang JY, Cuin T, Shabala L, Zhou MX, Mendham N, Shabala S (2007a) Effect of secondary metabolites associated with anaerobic soil conditions on ion fluxes and electrophysiology in barley roots. Plant Physiol 145:266–276

Pang JY, Newman I, Mendham N, Zhou MX, Shabala S (2006) Microelectrode ion and O_2 flux measurements reveal differential sensitivity of barley root tissues to hypoxia. Plant Cell Environ 29:1107–1121

Pang JY, Ross J, Zhou M, Mendham N, Shabala S (2007b) Amelioration of detrimental effects of waterlogging by foliar nutrient sprays in barley. Funct Plant Biol 34:221–227

Pang JY, Zhou MX, Mendham N, Shabala S (2004) Growth and physiological responses of six barley genotypes to waterlogging and subsequent recovery. Aust J Agric Res 55:895–906

Park WS, Han J, Kim N, Ko JH, Kim SJ, Earm YE (2005) Activation of inward rectifier K^+ channels by hypoxia in rabbit coronary arterial smooth muscle cells. Am J Physiol Heart Circ Physiol 289:H2461–H2467

Perez-Garcia MT, Colinas O, Miguel-Velado E, Moreno-Dominguez A, Lopez-Lopez JR (2004) Characterization of the Kv channels of mouse carotid body chemoreceptor cells and their role in oxygen sensing. J Physiol 557:457–471

Ponnamperuma FN (1972) The chemistry of submerged soil. Adv Agron 24:29–96

Pradet A, Bomsel JL (1978) Energy metabolism in plants under hypoxia and anoxia. In: Hook DD, Crawford RMM (eds) Plant Life in anaerobic environments. Ann Arbor Science, Ann Arbor, Michigan, pp 89–118

Qiu JD, Ke Y (1991) Study on determination of wet tolerance of 4572 barley germplasm resources (in Chinese). Acta Agricult Shanghai 7:27–32

Rao DN, Mikkelsen DS (1977) Effects of acetic, propionic, and butyric acids on rice seedling growth and nutrition. Plant Soil 47:323–334

Reggiani R (1997) Alteration of levels of cyclic nucleotides in response to anaerobiosis in rice seedlings. Plant Cell Physiol 38:740–742

Reyna N, Cornelious B, Shannon JG, Sneller CH (2003) Evaluation of a QTL for waterlogging tolerance in southern soybean germplasm. Crop Sci 43:2077–2082

Ricard B, Couee I, Raymond P, Saglio PH, Saintges V, Pradet A (1994) Plant metabolism under hypoxia and anoxia. Plant Physiol Biochem 32:1–10

Riesco-Fagundo AM, Perez-Garcia MT, Gonzalez C, Lopez-Lopez JR (2001) O_2 modulates large-conductance Ca^{2+}-dependent K^+ channels of rat chemoreceptor cells by a membrane-restricted and CO-sensitive mechanism. Circ Res 89:430–436

Roberts SK, Snowman BN (2000) The effects of ABA on channel-mediated K^+ transport across higher plant roots. J Exp Bot 51:1585–1594

Roberts SK, Tester M (1995) Inward and outward K^+-selective currents in the plasma membrane of protoplasts from maize root cortex and stele. Plant J 8:811–825

Russell JB (1992) Another explanation for the toxicity of fermentation acids at low pH – anion accumulation versus uncoupling. J Appl Bacteriol 73:363–370

Sanchez D, Lopez-Lopez JR, Perez-Garcia MT, Sanz-Alfayate G, Obeso A, Ganfornina MD, Gonzalez C (2002) Molecular identification of Kv alpha subunits that contribute to the oxygen-sensitive K^+ current of chemoreceptor cells of the rabbit carotid body. J Physiol 542:369–382

Setter TL, Waters I (2003) Review of prospects for germplasm improvement for waterlogging tolerance in wheat, barley and oats. Plant Soil 253:1–34

Setter TL, Waters I, Sharma SK, Singh KN, Kulshreshtha N, Yaduvanshi NPS, Ram PC, Singh BN, Rane J, McDonald G, Khabaz-Saberi H, Biddulph TB, Wilson R, Barclay I, McLean R, Cakir M (2009) Review of wheat improvement for waterlogging tolerance in Australia and India: the importance of anaerobiosis and element toxicities associated with different soils. Ann Bot 103:221–235

Shabala S (2003) Physiological implications of ultradian oscillations in plant roots. Plant Soil 255:217–226

Shabala S (2009) Salinity and programmed cell death: unravelling mechanisms for ion specific signalling. J Exp Bot 60:709–711

Shabala S, Demidchik V, Shabala L, Cuin TA, Smith SJ, Miller AJ, Davies JM, Newman IA (2006) Extracellular Ca^{2+} ameliorates NaCl-induced K^+ loss from *Arabidopsis* root and leaf cells by controlling plasma membrane K^+-permeable channels. Plant Physiol 141:1653–1665

Singh YV, Swarup A, Gupta SK (2002) Effect of short-term waterlogging on growth, yield and mineral composition of sorghum. Agrochimica 46:231–239

Smethurst CF, Garnett T, Shabala S (2005) Nutritional and chlorophyll fluorescence responses of lucerne (*Medicago sativa*) to waterlogging and subsequent recovery. Plant Soil 270:31–45

Stieger PA, Feller U (1994) Nutrient accumulation and translocation in maturing wheat plants grown on waterlogged soil. Plant Soil 160:87–95

Subbaiah CC, Bush DS, Sachs MM (1994a) Elevation of cytosolic calcium precedes anoxic gene expression in maize suspension cultured cells. Plant Cell 6:1747–1762

Subbaiah CC, Sachs MM (2000) Maize *cap1* encodes a novel SERCA-type calcium-ATPase with a calmodulin-binding domain. J Biol Chem 275:21678–21687

Subbaiah CC, Sachs MM (2003) Molecular and cellular adaptations of maize to flooding stress. Ann Bot 91:119–127

Subbaiah CC, Zhang JK, Sachs MM (1994b) Involvement of intracellular calcium in anaerobic gene expression and survival of maize seedlings. Plant Physiol 105:369–376

Szigligeti P, Neumeier L, Duke E, Chougnet C, Takimoto K, Lee SM, Filipovich AH, Conforti L (2006) Signalling during hypoxia in human T lymphocytes – critical role of the src protein tyrosine kinase p56Lck in the O2 sensitivity of Kv1.3 channels. J Physiol 573:357–370

Tamai I, Sai Y, Ono A, Kido Y, Yabuuchi H, Takanaga H, Satoh E, Ogihara T, Amano O, Izeki S, Tsuji A (1999) Immunohistochemical and functional characterization of pH-dependent intestinal absorption of weak organic acids by the monocarboxylic acid transporter MCT1. J Pharm Pharmacol 51:1113–1121

Tamura S, Kuramochi H, Ishizawa K (2001) Involvement of calcium ion in the stimulated shoot elongation of arrowhead tubers under anaerobic conditions. Plant Cell Physiol 42:717–722

Tanaka F, Ono S, Hayasaka T (1990) Identification and evaluation of toxicity of rice elongation inhibitors in flooded soils with added wheat straw. Soil Sci Plant Nutr 36:97–103

Trought MCT, Drew MC (1980) The development of waterlogging damage in wheat seedlings (*Triticum aestivum* L.). I. Shoot and root growth in relation to changes in the concentration of dissolved gases and solutes in the soil solution. Plant Soil 54:77–94

Verduyn C, Postma E, Scheffers WA, Vandijken JP (1992) Effect of benzoic acid on metabolic fluxes in yeasts – a continuous-culture study on the regulation of respiration and alcoholic fermentation. Yeast 8:501–517

Wang D, Youngson C, Wong V, Yeger H, Dinauer MC, Saenz Vega-, de Miera E, Rudy B, Cutz E (1996) NADPH-oxidase and hydrogen peroxide sensitive K^+ channel may function as an oxygen sensor complex in airway chemoreceptors and small cell lung carcinoma cell lines. Proc Natl Acad Sci USA 93:13182–13187

Wang S, Publicover S, Gu YC (2009) An oxygen-sensitive mechanism in regulation of epithelial sodium channel. Proc Natl Acad Sci USA 106:2957–2962

Wang TSC, Yang TK, Chuang TT (1967) Soil phenolic acids as plant growth inhibitors. Soil Sci 103:239–246

Watanabe H, Yashiro T, Tohjo Y, Konishi Y (2006) Non-involvement of human monocarboxylic acid transporter 1 (MCT1) in the transport of phenolic acid. Biosci Biotechnol Biochem 70:1928–1933

Wegner LH, De Boer AH (1997) Two inward K^+ channels in the xylem parenchyma cells of barley roots are regulated by G-protein modulators through a membrane-delimited pathway. Planta 203:506–516

Wegner LH, Raschke K (1994) Ion channels in the xylem parenchyma of barley roots – a procedure to isolate protoplasts from this tissue and a patch-clamp exploration of salt passageways into xylem vessels. Plant Physiol 105:799–813

White PJ, Broadley MR (2001) Chloride in soils and its uptake and movement within the plant: a review. Ann Bot 88:967–988

Williams SE, Wootton P, Mason HS, Bould J, Iles DE, Riccardi D, Peers C, Kemp PJ (2004) Hemoxygenase-2 is an oxygen sensor for a calcium-sensitive potassium channel. Science 306:2093–2097

Xia JH, Roberts JKM (1996) Regulation of H^+ extrusion and cytoplasmic pH in maize root tips acclimated to a low-oxygen environment. Plant Physiol 111:227–233

Ye SF, Zhou YH, Sun Y, Zou LY, Yu JQ (2006) Cinnamic acid causes oxidative stress in cucumber roots, and promotes incidence of Fusarium wilt. Environ Exp Bot 56:255–262

Yu JQ, Matsui Y (1997) Effects of root exudates of cucumber (*Cucumis sativus*) and allelochemicals on ion uptake by cucumber seedlings. J Chem Ecol 23:817–827

Zimmermann S, Ehrhardt T, Plesch G, Muller-Rober B (1999) Ion channels in plant signalling. Cell Mol Life Sci 55:183–203

Chapter 11
Ion Transport in Aquatic Plants

Olga Babourina and Zed Rengel

Abstract Aquatic higher plants have morphologically and physiologically adapted to their water environment. Similar to terrestrial plants that have been exposed to waterlogging and inundation, aquatic plants form aerenchyma that allows gas exchange in submerged organs. Additionally, to prevent gas and solute loss, aquatic plants have developed impermeable barriers in the cells and tissues, such as suberinisation of the cell walls. In addition to the roots, the major organ for nutrient acquisition in the terrestrial plants, aquatic plants have also developed the unique ability to take up ions via leaves or shoots. This chapter summarises findings on ion transport systems in aquatic plants. Most of the ion transport systems that are located in the plasma membranes share similarities with both aquatic and terrestrial plants. To date, only one system (Na^+/NO_3^- transporter) was found to be unique for a seagrass. Ion transport across the tonoplast of aquatic plants is poorly studied, with only two transporters described to date (an aquaporin and the SV channel). A molecular characterisation of the ion transporters of the aquatic plants is almost absent.

11.1 Introduction

The general term "aquatic plants" covers vascular plants living in a water environment, including ferns and angiosperms. The plants can be emergent [e.g., *Canna* spp, cattails (*Typha sp.*), reeds (*Phragmites sp.*), bulrushes *(Scirpus sp.)* and sedges (*Carex sp.*)], submergent (e.g., *Triglochin huegelii, Ludwigia repens*), rooted with floating leaves (e.g., *Potamogeton* spp.), and wholly floating plants (*Eichhornia*

O. Babourina (✉) and Z. Rengel
School of Earth and Environment, University of Western Australia, 35 Stirling Hwy, Crawley, WA 6009, Australia
e-mail: Olga.Babourina@uwa.edu.au

S. Mancuso and S. Shabala (eds.), *Waterlogging Signalling and Tolerance in Plants*,
DOI 10.1007/978-3-642-10305-6_11, © Springer-Verlag Berlin Heidelberg 2010

crassipes, Lemna spp, *Azolla* spp). This chapter covers ion transport only in aquatic higher plants or angiosperms.

Aquatic higher plants have been popular subjects of scientific research for decades. Since the 1930s, plants from the families *Lemnaceae* and *Hydrocharitaceae* have been the model plants in almost all areas of plant physiology, including fundamental studies on photosynthesis, growth, development, reproduction, nutrient uptake and circadian rhythms (Ashby 1929; Marx 1929; Clark and Fly 1930; White 1937; Clark and Frahm 1940; Steinberg 1946; Hillman 1959). In the 1980s, these species were favourite subjects for electrophysiological studies, primarily because they were easy to grow and perfect for lengthy exposure to bathing solutions, similar to their natural submerged environment (Jeschke 1970; Novacky et al. 1978; Bocher et al. 1980; Fischer and Luttge 1980; Hartung et al. 1980; Scheid et al. 1980; Ullrich-Eberius et al. 1981; Smith 1982). Given that aquatic plants are angiosperms, all of the data or conclusions derived from studies were assumed to be transferrable to other terrestrial angiosperms. Indeed, it is difficult to pinpoint features in ion transport that would be specific for aquatic plants in contrast to terrestrial plants, although it appears highly likely that these differences should exist.

11.2 Morphological and Physiological Adaptations of Aquatic Plants

In contrast to algae, aquatic higher plants have returned to a water environment, and therefore have adapted to this environment both morphologically and physiologically (Raven 1994).

Morphological adaptations include aerenchyma, the thickening of the cell wall of the epidermis, which is similar to adaptations observed in terrestrial plants exposed to waterlogging. Other specific adaptations include the expansion of leaf surfaces, a thinner cuticle in the aerial organs, and many others (for recent reviews see Bennett et al. 2009; Setter et al. 2009).

In seagrasses (e.g., *Ruppia maritima, Thalassia testudinum* and *Zostera marina*), the plasma membrane of the epidermal cells is invaginated, which increases the surface area of the plasma membrane and the capacity of these cells for ion and gas exchange (Jagels 1983). The plasma membranes of these cells have a high ATPase activity (Pak et al. 1995), and there are a large number of mitochondria located close to the plasma membrane to support energetic requirements for active ion transport across the plasma membrane (Iyer and Barnabas 1993; Fukuhara et al. 1996; Fernandez et al. 1999).

The functions of the aerenchyma vary from providing a flotation support for the leaves and stems, to sustaining gas exchange within the tissues and organs. The submerged tissues of aquatic plants are in a low oxygen environment, and require O_2 transport from the aerial parts, or gas diffusion in fully submerged plants.

The aquatic plants have developed aerenchyma with gas-filled lacunas (Sculthorpe 1967). The thickening and cutinising of the cell walls of external cells, and the hydrophobic surface of the lacunas, prevent oxygen loss to the water environment (Rascio et al. 1994; Raven 1996; Sorrell et al. 1997). The aerenchyma serves both O_2 and CO_2 exchange in submerged plant tissues. CO_2 has been suggested to be produced during photorespiration, or to be transported from the roots, as described for the isoetid species that live in an infertile environment (Raven et al. 1988). However, other species can also use CO_2 from sediments, as shown for *Vallisneria spiralis* (Kimber et al. 1999).

Some morphological adaptations are the result of the reduction of the original tissues of terrestrial plants, due to adaptation to a water environment, such as a thinner cuticle in the aerial organs compared to those of terrestrial plants, a smaller root biomass, and the lack of xylem in some species. These morphological changes are feasible because these plants do not need to prevent water loss. In addition, the leaves or fronds (plant tissues that are undifferentiated into stem and leaf) can take in nutrients or water straight from the surrounding environment, and the plants do not completely depend on nutrient uptake via the roots.

The expansion of the leaf surface and chloroplast-containing epidermal cells are considered to be adaptive features that are linked to the enrichment of photosynthetic tissues/cells with inorganic carbon that is dissolved in water. Generally, aquatic plants have adapted several mechanisms to enhance the CO_2 supply. First, in some plants HCO_3^- can be taken up by the leaf cells, with a subsequent reduction of HCO_3^- to CO_2. This reaction can be enhanced by increased activity of the plasma membrane H^+-ATPases, and an active co-transport of HCO_3^- with H^+ has been suggested in *Elodea nuttalii* (Eighmy et al. 1991; Fagerberg et al. 1991). All of the aquatic angiosperms that have been studied demonstrate the photoactivation of a proton pump at the plasmalemma in the cells of the lower epidermis (Marre et al. 1989a; Marre et al. 1989b; Prins and Elzenga 1989; Miedema et al. 1992; Krabel et al. 1995; Baur et al. 1996). Second, some aquatic plants have been suggested to be able to use HCO_3^- in addition to CO_2 during photosynthesis (Bodner 1994; Pokorny et al. 1985; Sand-Jensen et al. 1992). Third, the apoplastic conversion of HCO_3^- to CO_2 by carbonic anhydrases has also been suggested to occur in aquatic plants (Newman and Raven 1993; Sultemeyer et al. 1993; Badger and Price 1994; Rascio et al. 1999).

In addition to physiological adaptations at the cellular level, the whole leaf has been suggested to be a machine that supports carbon acquisition by the aquatic plants. The upper surface of *Elodea* leaves is negatively charged, whereas the lower surface is provided with H^+, which neutralises OH^- groups that are generated during the conversion of HCO_3^- to CO_2, and shifts the equilibrium of HCO_3^-/CO_2 towards CO_2, which enters the cell by passive diffusion (Prins et al. 1982).

Specific aquatic acid metabolism (AAM) has been found in some aquatic plants, especially the isoetid life forms (Raven and Spicer 1996). This type of metabolism is similar to the Crassulaceae acid metabolism (CAM)-like metabolism; however, instead of compartmentation within different tissues or cells, the AAM plants can compartmentalise organic acids within a cell (Raven and Spicer 1996).

11.3 Ion Transport

Despite the fact that aquatic plants are a popular model system among plant physiologists, not much is known about their ion transport mechanisms, neither electrophysiological nor biochemical nor molecular. At the same time, ion transport across membranes is linked to almost all vital processes in living organisms, such as nutrient uptake, growth, generation of energy, and signalling.

Despite the limited number of electrophysiological measurements that have been made on aquatic plants, it has always been thought that the ion transport systems in these plants are not significantly different from the transport systems of the terrestrial plants. Fig. 11.1 shows the major ion transport systems that are located at the plasma membrane and tonoplast that have been observed in terrestrial and aquatic plants. The grey colour indicates the transporters that are common for both types of plants. These transporters include K^+ outward and inward channels, non-selective cation channels (NSCC), anion channels, high affinity transporters for K^+ and NH_4^+, an anion/H^+ symporter, Na^+/H^+ antiporters, H^+- and Ca^{2+}-ATPases, a redox system located at the plasma membranes, and slow vacuolar channels at the tonoplast. Aquaporins have been found in both types of membranes in terrestrial and aquatic plants. Transporters that have not been confirmed in aquatic plants are not in colour, whereas the Na^+/NO_3^- transporter that has been found only in seagrasses is indicated by the black colour. This diagram is a modification of diagrams from recently published reviews on the ion transport systems in terrestrial plants and seagrasses (Touchette 2007; Ward et al. 2009).

Table 11.1 summarises findings on the ion transporters in aquatic plants. Most of the electrophysiological studies on aquatic plants are based on measurements of the

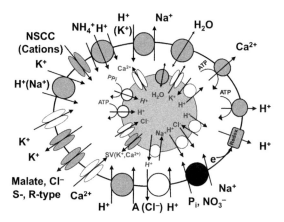

Fig. 11.1 The transport systems that have been observed in aquatic and terrestrial plants are indicated by a *grey* colour. The transport systems that have been found in higher plants, but have not been described in aquatic plants, are indicated by a *white* colour. The only transport system, which has been described for aquatic plants, but that has not been described in other terrestrial plants is indicated by a *black* colour

Table 11.1 Transport systems that have been discovered in aquatic higher plants

Type of transporters	Aquatic plant species	Habitat	Organ/Cells	Comments	References
Plasma membrane					
K^+ outward channel	*Posidonia oceanica*	Seawater	Sheath cells surrounding the vascular bundles of the leaves	Non-permeable to Na^+ and other monovalent cations	Carpaneto et al. (2004)
	Zostera muelleri	Seawater	Young leaves	Five times more permeable to K^+ than to Na^+	Garrill et al. (1994)
	Egeria densa	Freshwater	Leaves	Sensitive to pH; regulated by phosphorylation/ dephosphorylation by protein kinases and protein phosphatases	Natura and Dahse (1998)
K^+ inward channel	*Posidonia oceanica*	Seawater	Sheath cells surrounding the vascular bundles of the leaves	Non permeable to Na^+	Carpaneto et al. (2004)
	Zostera muelleri	Seawater	Young leaves	Permeable to Na^+	Garrill et al. (1994)
	Egeria densa	Freshwater	Leaves	regulated by phosphorylation/ dephosphorylation by protein kinases and protein phosphatases	Natura and Dahse (1998)
Non selective cation channel	*Posidonia oceanica*	Seawater	Sheath cells surrounding the vascular bundles of the leaves		Carpaneto et al. (2004)
Cl^- channel	*Zostera muelleri*	Seawater	Young leaves		Garrill et al. (1994)

(continued)

Table 11.1 (Continued)

Type of transporters	Aquatic plant species	Habitat	Organ/Cells	Comments	References
Ca^{2+}-ATPase	Egeria densa	Freshwater	Leaves		Beffagna et al. (2000)
H^+-ATPase	Elodea Egeria densa	Freshwater	Leaves		Marre et al. (1989b)
	Zostera marina	Seawater	Leaves		Pak et al. (1995) Fernandez et al. (1999)
Redox	Elodea canadensis	Freshwater	Leaves		Ivankina and Novak (1981)
	Egeria densa	Freshwater	Leaves		Elzenga and Prins (1989) Bernstein et al. (1989)
	Lemna gibba	Freshwater	Fronds		Lass and Ullrich-Eberius (1984)
	Limnobium stoloniferum	Freshwater	Roots		Grabov et al. (1993)
K^+/H^+ symporter	Elodea densa	Freshwater	Leaves		Bellando et al. (1995)
High affinity K^+ transporter	Cymodocea nodosa	Seawater	Leaves	CnHAK1	Garciadeblas et al. (2002)
K^+/Na^+ cotransporter	Egeria densa	Freshwater	Leaves and roots	High affinity	Maathuis et al. (1996)
	Elodea	Freshwater	Leaves and roots	High affinity	Maathuis et al. (1996) Walker (1994)
	Vallisneria	Freshwater	Leaves	High affinity	Maathuis et al. (1996) Walker (1994)
NH_4^+	Lemna gibba	Freshwater	Fronds		Ullrich et al. (1984)
HCO_3^-/Cation symporter	Elodea nuttallii	Freshwater	Leaves		Eighmy et al. (1991)
H^+/Na^+ antiporter	Zostera marina	Seawater	Leaves		Fernandez et al. (1999)
NH_4^+	Egeria densa	Freshwater	Leaves	One system is permeable to NH_4^+ and has lower affinity to $CH_3NH_3^+$, and is insensitive to Cs^+	Venegoni et al. (1997)

				The second system, operates at a higher H^+-ATPase activity and E_m hyperpolarisation, is inhibited by Cs^+ and is impermeable to $CH_3NH_3^+$	
Anion/H^+ symporter	*Lemna gibba*	Freshwater	Fronds	$H_2PO_4^-$: saturation at 50–100 μM	Ullrich-Eberius et al. (1981)
	Lemna gibba	Freshwater	Fronds	$H_2PO_4^-$: System I $K_m = 7.3$ μM	Ullrich-Eberius et al. (1989)
	Limnobium stoloniferum	Freshwater	Root hairs	System II $K_m = 88.2$ μM Cl^-, $H_2PO_4^-$, NO_3^- 1–1.8 mM for Pi	Ullrich and Novacky (1990)
P_i/Na^+	*Zostera marina*	Seawater	Leaves Roots	High affinity High affinity, $K_m = 1.5$ μM	Rubio et al. (2005)
NO_3^-/Na^+	*Zostera marina*	Seawater	Leaf mesophyll cells	High affinity, $K_m = 2.3$ μM	Garcia-Sanchez et al. (2000)
	Zostera marina	Seawater	Roots	High affinity, $K_m = 8.9$ μM	Rubio et al. (2005)
Aquaporin	*Posidonia oceanica*	Seawater	Leaves and roots	PoPIP; 1 was expressed in the meristematic region of the apical meristems shoot and root, and in the epidermal and sub-epidermal cells in the leaves, and also in the provascular and vascular tissues	Cozza and Pangaro (2009)
Tonoplast					
SV channel K^+ and Na^+	*Posidonia oceanica*	Seawater	Leaves		Carpaneto et al. (1997)
Aquaporin	*Posidonia oceanica*	Seawater	Leaves and roots	PoTIP; 1 was expressed with the tissues showing a well-differentiated vacuole compartment. Expression increased after hypersalinity treatments	Cozza and Pangaro (2009)

membrane potential. The only patch clamp study on a *freshwater* macrophyte was performed on *Egeria* (Natura and Dahse 1998), and two patch clamp studies on protoplasts derived from seagrasses were performed on *Zostera* and *Posidonia* (Garrill et al. 1994; Carpaneto et al. 2004). Several patch clamp studies have been performed on the vacuoles of *Posidonia* (Carpaneto et al. 1997; Carpaneto et al. 1998; Carpaneto et al. 1999). The small number of publications on patch clamp studies of aquatic plants is likely due to methodological difficulties in obtaining protoplasts from aquatic plants (Natura and Dahse 1998).

Many studies have focused on the capacity of aquatic submerged plants (*Egeria* or *Elodea*) to reach an extremely high electric cell potential in leaf cells under illumination (up to -300 mV), and the ability of these plants to make their leaves polarised. The latter ability is considered to be an adaptation in order to increase carbon uptake (Elzenga and Prins 1988). The measurements of E_m (membrane potential) under different external K^+ concentrations and pHs have suggested that this hyperpolarisation is mainly due to an increased activity of H^+-ATPases and K^+ channels in the plasma membrane (Miedema et al. 1992; Buschmann et al. 1996).

11.3.1 Cation Transport Systems

In the freshwater submerged plant *Egeria*, K^+ conductance was regulated by voltage, pH of the growth medium and phosphorylation/dephosphorylation. Also, it was found that regulation of K^+ uptake by an external pH was shifted to the more alkaline region in comparison to terrestrial plants (Natura and Dahse 1998).

Sheath protoplasts from the seagrass *Posidonia oceanica* contain both K^+ outward and K^+ inward channels, and a non-selective cation channel (Carpaneto et al. 2004). The K^+ outward channel was highly selective for K^+ and was impermeable to Na^+ and other monovalent cations, and resembled SKOR or GORK (Stelar K Outward Rectifier or Guard cell Outward Rectifying K channels identified in *A. thaliana* xylem parenchyma or guard cells). As in the GORK/SKOR channels, but unlike the outward K^+ channel identified in the halophytic angiosperm *Zostera muelleri* (Garrill et al. 1994), the voltage sensitivity of the *Posidonia* outward channel depends on the concentration of external K^+. It has been proposed that this outward channel is involved in K^+ release from the sheath cells to xylem vessels in *Posidonia* (Carpaneto et al. 2004).

The time-dependent component of the inward channel demonstrated a high selectivity of K^+, whereas the time-independent component was not selective. The highly selective time-dependent inward channel found in *Posidonia* is different from the homologous channel measured in *Zostera*, which was partially permeable to Na^+ (Garrill et al. 1994).

The non-selective channels were observed in *Posidonia* at hyperpolarised membrane potentials. These channels were inhibited by TEA^+, and were almost completely but reversibly blocked by La^{3+} (Carpaneto et al. 2004).

The ability of seagrasses to maintain a relatively low Na^+ concentration in tissues (e.g., Muramatsu et al. 2002) has sparked investigations on Na^+-ATPases that have previously been found in fungi and algae (Rodriguez-Navarro et al. 1994; Shono et al. 1996). However, several studies have failed to find such Na^+-ATPases in seagrasses, and therefore it was concluded that H^+-ATPases are the primary pumps at the plasma membrane of aquatic plants (Fernandez et al. 1999; Garciadeblas et al. 2001). Instead, a Na^+/H^+ antiporter has been proposed as a main transporter for Na^+ extrusion to the apoplast (Fernandez et al. 1999). To date, the only described Na^+/H^+ antiporter located at the plasma membrane is found in angiosperms (Shi et al. 2000). A search for SOS1 analogues resulted in the cloning of two *SOS1* genes from the seagrass *Cymodocea nodosa* (*CnSOS1A* and *CnSOS1B*), but the presence of additional *CnSOS1* genes was also detected (Garciadeblas et al. 2007). The expression of CnSOS1A and CnSOS1A-1 in a yeast mutant defective in Na^+ efflux demonstrated that CnSOS1A is a system very similar to AtSOS1, as was described by Shi et al (2001). However, unlike AtSOS1, CnSOS1A was not activated by the AtSOS2/SOS3 system (Garciadeblas et al. 2007).

The results that were found on K^+ and Rb^+ uptake in bacteria led the authors to conclude that in seagrasses, the SOS1 transport mechanism is more complex than electroneutral Na^+/H^+ exchange (Garciadeblas et al. 2007). A mechanism has therefore been proposed to exist that might simultaneously support the Na^+-efflux and K^+-uptake functions via a Na^+/H^+ antiporter, in which K^+ substitutes for H^+ in certain circumstances (Garciadeblas et al. 2007). Another type of Na^+/K^+ symporter, a Na^+-dependent K^+ high affinity uptake transporter, has been described in some freshwater angiosperms such as *Egeria*, *Elodea* and *Vallisneria* (Maathuis et al. 1996).

Two transport systems, a high affinity and low affinity system have been suggested to carry out NH_4^+ transport across the plasma membrane, as well as the transport of other cations. In terrestrial plants, the proton motive force has been proposed as a source of energy for NH_4^+ uptake against an electrochemical potential gradient (Wang et al. 1993; Wang et al. 1994). This means that NH_4^+ (when present in concentrations under/below 1 mM) is taken up by a NH_4^+/H^+ symporter. However, for aquatic plants the knowledge is scant, especially for the high affinity range under 1 mM NH_4^+. In nitrogen-starved plants of *Lemna gibba*, a combination of two transport systems has been observed: the first one was saturated and had a $K_m = 17\ \mu M$, and the second system had a linear relationship and was not saturated (Ullrich et al. 1984).

Two distinct transport systems for NH_4^+ uptake at concentrations under 1 mM were suggested for *Egeria densa* leaves, namely: one system that transports both NH_4^+ and $CH_3NH_3^+$ and that is not sensitive to Cs^+, and one system that is impermeable to $CH_3NH_3^+$, is strongly inhibited by Cs^+ and operates at a high H^+-ATPase activity and E_m hyperpolarisation (Venegoni et al. 1997).

Electrophysiological measurements in *Elodea densa* that were expressing cloned high affinity NH_4^+ transporters demonstrated that these transporters are in fact uniporters, which means they do not require H^+ as a co-transport ion (Smith 1982; Ludewig et al. 2002).

A hypothesis has recently been proposed to explain the apparent disagreement between the NH_4^+ uptake studies in intact plants (where a H^+ cotransporter appeared to be involved) and the electrophysiological studies of cloned NH_4^+ transporters (where properties of a uniporter were shown). For the lower NH_4^+ concentration range, two uptake systems have been suggested to be present in the water hyacinth (*E. crassipes*) (Fang, Babourina and Rengel, unpublished), namely: saturable system I (SSI) for concentrations under 100 μM NH_4^+ and saturable system II (SSII) for concentrations between 100 μM and 2 mM NH_4^+. SSI was found to be pH-independent; whereas SSII activity was non-competitively inhibited by H^+ (a lower pH decreased the V_{max} with no significant change in the K_m). In addition, pre-treatments with La^{3+} (a Ca^{2+} channel blocker) and a chelator (EDTA), which led to lower $[Ca^{2+}]_{cyt}$ (Fang, Babourina and Rengel, unpublished), resulted in higher NH_4^+ influx. Therefore, intracellular $[Ca^{2+}]_{cyt}$ might be involved in the regulation of the NH_4^+ transport system.

11.3.2 Anion Transport Systems

Anion uptake in all plants (including aquatic) has to be driven by active transport systems. Fresh water aquatic plants were among the first higher plants for which it was reported that symport of NO_3^- or $H_2PO_4^-$ with H^+ was required for high and low affinity anion transport systems (Ullrich and Novacky 1990; Ullrich-Eberius et al. 1981; Ullrich-Eberius et al. 1989) This mechanism has been proposed on the basis of NO_3^--induced depolarisation of the plasma membrane and concurrent NO_3^- and H^+ fluxes.

The seagrass *Z. marina* has been suggested to possess high-affinity Na^+-dependent NO_3^- and P_i (as $H_2PO_4^-$ and HPO_4^{-2}) transport systems. This seagrass uses a Na^+ gradient that is directed from the apoplast to the cytosol in a high saline environment to take NO_3^- with a stoichiometry of one NO_3^- for every two Na^+, one $H_2PO_4^-$ for every two Na^+, and one HPO_4^{-2} for every three Na^+ ions (Garcia-Sanchez et al. 2000; Rubio et al. 2005).

11.4 Root Versus Leaf Uptake

Both roots and fronds of *Lemnaceae* plants (duckweeds) are exposed to the surrounding media and, in theory, are able to take up nutrients. However, there is a long-standing controversy about root involvement in nutrient uptake in *Lemnaceae* species (Cedergreen and Madsen 2002). Only recent studies on *Lemna minor* have demonstrated that the roots are also involved in N uptake, and that the plants can regulate NO_3^- uptake via either fronds or roots (Cedergreen and Madsen 2002; Cedergreen and Madsen 2004; Fang et al. 2007a). The roots of floating macrophytes have been suggested to function primarily as anchors, whereas the fronds and leaves are the main organs involved in nutrient uptake (Hillman 1961). Subsequent studies, which concentrated on the fronds and roots of *Spirodela polyrrhiza* and *L. minor* (*Lemnaceae*) (using paraffin to cover the roots in one set of experiments, and physically removing the roots in another set of experiments) led to the conclusion that the duckweed roots had a small role in nutrient uptake (Muhonen et al. 1983; Ice and Couch 1987). However, the researchers based their conclusions on the observation that plants with excised roots multiplied more rapidly than the rooted plants (Muhonen et al. 1983). In earlier experiments, it has been demonstrated that covering the undersides of the fronds with lanolin decreased the growth of duckweed plants, even though the root length increased (Gorham 1941). Recently, it has been found that an increased ratio of the root surface to the frond surface led to increased NH_4^+ uptake rate in *L. minor* (Cedergreen and Madsen 2002). These studies indicate that duckweed plants can regulate their lifecycle, such as through an increased multiplication rate.

The conclusion that roots are of low importance for nutrient uptake in *Lemnaceae* was opposed by other studies (Oscarson et al. 1988; Cedergreen and Madsen 2002; Cedergreen and Madsen 2004). Moreover, in *L. minor*, the roots had a higher rate of uptake of both NH_4^+ and NO_3^- at a low external concentration when compared to the fronds, whereas a higher NH_4NO_3 supply reduced root uptake rates for both ions. This decreased uptake rate in the roots in the presence of a high NH_4NO_3 supply was compensated for by a higher uptake rate in the fronds (Cedergreen and Madsen 2002). However, in all of the studies mentioned above, the researchers used mechanical approaches to separate the organs; that is, covering the fronds or roots with paraffin or lanolin, and/or removing the roots. The involvement of roots in ion uptake was recently demonstrated in studies that directly measured ion fluxes along the roots of aquatic plants (Fang et al. 2007b; Fang et al. 2007a). In the floating plant *Landoltia punctata*, the roots contributed to N uptake as much as the fronds. Even though the rate of ion flux in the roots was lower than in the fronds, the root surface was twofold greater than the frond surface, and the ratio of the fronds/roots in N uptake ended up being close to 1:1.09 for NH_4^+ and 1:0.79 for NO_3^-. Therefore, the plants had an approximately equal capacity to use the fronds and roots for uptake of NO_3^- and NH_4^+ (Fang et al. 2007a).

In studies performed on emergent and floating aquatic plants (Fang et al. 2007b), uptake of NO_3^- and NH_4^+ was shown to vary spatially along the root tip (the first 6 mm), which is similar to many terrestrial plants (Colmer and Bloom 1998).

Also similar to terrestrial plants, H^+ fluxes showed a specific pattern that was dependent on the root zones in all four of the species tested [one floating (*Azolla* spp.) and three submergent (*Vallisneria natans*, *Bacopa monnieri* and *L. repens*)] (Fang et al. 2007b). The first few millimetres of a root are constantly undergoing cell division, expansion, and differentiation to produce zones with a distinct morphology and metabolic capacity. High H^+ excretion at the root apex, including the apical initials, the meristem and the distal elongation zones, has been shown for many terrestrial plants (Taylor and Bloom 1998; Newman 2001). *Lemnaceae* plants have a big root cap that covers the whole root apex, the full elongation zone and part of the mature zone, which makes them different from many other plants, including some wetland species (Landolt 1986). It has been shown that the root cap does not interfere with ion fluxes at the root surface; the H^+ flux profile of the root of *L. punctata* was similar to other terrestrial plants studied (Newman 2001).

Therefore, the submerged and floating plants can absorb ions by the leaf surfaces, in addition to their ability to take up ions from the sediment or the water table via their roots (Barko and Smart 1986; Eugelink 1998; Fang et al. 2007a; Thomaz et al. 2007).

The question whether submerged plants can take up nutrients by their roots is directly linked with their ability to deliver these nutrients to the leaves. There is a general perception that transpiration is a major driving force for the water and ion mass flow in higher plants (Campbell et al. 1999). However, submerged plants have adapted the strategy of water guttation from leaf hydatodes and the creation of impermeable barriers around the vascular tissues to avoid loss in solution. (Rascio 2002). In the submerged macrophyte *Sparganium emersum*, it has been shown that guttation is an energy-dependent process; the leaves ceased to guttate when the roots were cooled to $4°C$, and similarly, the addition of vanadate, an ATPase inhibitor, decreased guttation (Pedersen 1993). However, it has been noted that the estimated root pressure was smaller in aquatic plants than in terrestrial plants, and was not sufficient to drive water flow acropetally (Pedersen 1993). On the other hand, the separated roots and shoots of *Batrachium trichophyllum* demonstrated acropetal transport, and such transport was much higher in the roots, suggesting that the roots are particularly important for the flow of water (Pedersen and Sand-Jensen 1993).

11.5   Molecular Characterisation of Transporter Genes

Sequencing of the *Arabidopsis* and rice genome has promoted ion transport studies in these species, and has allowed researchers to be more specific in targeting genes of interest. It was found that membrane transport proteins comprise nearly 5% of the *Arabidopsis* genome. These proteins belong to 46 unique families with almost 880 members. It was estimated that several hundred putative transporters exist in *Arabidopsis* (Maser et al. 2001). In contrast, a molecular analysis of ion transport systems in aquatic plants is almost absent. Only three types of plasma-membrane-located transporters have been cloned in aquatic plants (so far: Ca^{2+}-ATPases, HAK and SOS1 transporters) (Garciadeblas et al. 2001, 2002, 2007). However, there are no observations of the physiological functions of these transporters in planta. Heterologous expression demonstrated that the functions of CnHAK1, CnHAK2, CnSOS1A and CnSOS1B after expression experiments in bacteria and yeast (Garciadeblas et al. 2002, 2007) were different from the functions of these genes' homologues in plants (Santa-Maria et al. 1997; Shi et al. 2000; Fulgenzi et al. 2008). Therefore, more detailed work on the characterisation of ion transporters in aquatic plants at the molecular level is urgently needed.

11.6 The Relevance of Aquatic Plants to Terrestrial Plants in Regards to Waterlogging and Inundation Stresses

Because aquatic plants have a superb evolutionary adaptation to the water environment, studying their mechanisms of adaptation might be useful in furthering our understanding of how terrestrial plants might tolerate waterlogging and inundation stresses (Bennett et al. 2009; Setter et al. 2009). Indeed, terrestrial plants that have been exposed to waterlogging and inundation develop morphological adaptations that are similar to aquatic plants, such as aerenchyma and barriers to radial oxygen loss. However, many research questions that have arisen from the knowledge of aquatic plant morphology and physiology still remain to be answered for terrestrial plant toleration of waterlogging and inundation stresses, such as the use of submerged shoot parts for ion and/or water uptake, or the use of bicarbonate ions as a carbon source etc. In saline environments, it is unknown whether terrestrial plants can begin to use Na^+ as a counter-ion for the active transport of other ions to mimic the adaptations that seagrasses have developed while growing in seawater. The answers to some of these interesting questions for terrestrial plants, stemming from the knowledge about adaptations in aquatic plants, may allow new insights into how tolerance to waterlogging and inundation stresses can be improved in terrestrial plants.

11.7 Conclusions

1. Aquatic higher plants have adapted to the water environment morphologically and physiologically. Similar to terrestrial plants that have been exposed to waterlogging and inundation, they form aerenchyma that allows gas exchange in submerged organs. Also, to prevent gas and solute loss, these plants have developed impermeable barriers in the cells and tissues, such as suberinisation of cell walls.
2. In addition to the common adaptations that aquatic plants share with waterlogged or inundated terrestrial plants, aquatic plants have also developed some unique features, such as invagination of the plasma membrane in leaf cells that border the water environment. In addition, aquatic plants have developed a unique ability for ion uptake via leaves or shoots.
3. Most of the ion transport systems that are located in the plasma membranes are similar in aquatic and terrestrial plants. However, the electrophysiological or biochemical analysis of these transporters in aquatic plants has been very limited. To date, only one system (Na^+/NO_3^- transporter) was found to be unique for a seagrass.
4. Ion transport across the tonoplast of aquatic plants is not well studied, and there have been only two transporters described to date (aquaporin and SV channel).
5. The molecular characterisation of ion transporters in aquatic plants is almost lacking, with only three transporters cloned from aquatic plants.

References

Ashby E (1929) The interaction of factors in the growth of Lemna. III. The interrelationship of duration and intensity of light. Ann Bot 43:333–354

Badger MR, Price GD (1994) The role of carbonic-anhydrase in photosynthesis. Ann Rev Plant Physiol Plant Mol Biol 45:369–392

Barko JW, Smart RM (1986) Sediment-related mechanisms of growth limitation in submersed macrophytes. Ecology 67:1328–1340

Baur M, Meyer AJ, Heumann HG, Lutzelschwab M, Michalke W (1996) Distribution of plasma membrane H^+-ATPase and polar current patterns in leaves and stems of *Elodea canadensis*. Bot Acta 109:382–387

Beffagna N, Romani G, Sforza MC (2000) H^+ fluxes at plasmalemma level: In vivo evidence for a significant contribution of the Ca^{2+}-ATPase and for the involvement of its activity in the abscisic acid-induced changes in *Egeria densa* leaves. Plant Biol 2:168–175

Bellando M, Marre MT, Sacco S, Talarico A, Venegoni A, Marre E (1995) Transmembrane potential-mediated coupling between H^+ pump operation and K^+ fluxes in *Elodea densa* leaves hyperpolarized by fusicoccin, light or acid load. Plant Cell Environ 18:963–976

Bennett SJ, Barrett-Lennard EG, Colmer TD (2009) Salinity and waterlogging as constraints to saltland pasture production: a review. Agric Ecosyst Environ 129:349–360

Bernstein M, Dahse I, Muller E, Petzold U (1989) The membrane potential as indicator for transport and energetic processes of leaf cells of the aquatic plant *Egeria densa*. 3. Evidence for electron transport through the plasmalemma. Biochem Physiol Pflanz 185:343–356

Bocher M, Fischer E, Ullrich-Eberius C, Novacky A (1980) Effect of fusicoccin on the membrane potential, on the uptake of glucose and glycine, and on the ATP level in *Lemna gibba* G1. Plant Sci Lett 18:215–220

Bodner M (1994) Inorganic carbon source for photosynthesis in the aquatic macrophytes *Potamogeton natans* and *Ranunculus fluitans*. Aquat Bot 48:109–120

Buschmann P, Sack H, Kohler AE, Dahse I (1996) Modeling plasmalemma ion transport of the aquatic plant *Egeria densa*. J Membr Biol 154:109–118

Campbell NA, Reece JB, Mitchell LG (1999) Biology, 5th edn. Addison Wesley Longman, Reading, MA

Carpaneto A, Cantu AM, Busch H, Gambale F (1997) Ion channels in the vacuoles of the seagrass *Posidonia oceanica*. FEBS Lett 412:236–240

Carpaneto A, Cantu AM, Gambale F (1998) Biophysical properties of ion channels in the vacuoles of *Posidonia oceanica*. Pflugers Archiv-Eur J Physiol 435:30

Carpaneto A, Cantu AM, Gambale F (1999) Redox agents regulate ion channel activity in vacuoles from higher plant cells. FEBS Lett 442:129–132

Carpaneto A, Naso A, Paganetto A, Cornara L, Pesce ER, Gambale F (2004) Properties of ion channels in the protoplasts of the Mediterranean seagrass *Posidonia oceanica*. Plant Cell Environ 27:279–292

Cedergreen N, Madsen TV (2002) Nitrogen uptake by the floating macrophyte *Lemna minor*. New Phytol 155:285–292

Cedergreen N, Madsen TV (2004) Light regulation of root and leaf NO_3^- uptake and reduction in the floating macrophyte *Lemna minor*. New Phytol 161:449–457

Clark NA, Fly CL (1930) The role of manganese in the nutrition of *Lemna*. Plant Physiol 5:241–248

Clark NA, Frahm EE (1940) Influence of auxins on reproduction of *Lemna major*. Plant Physiol 15:735–741

Colmer TD, Bloom AJ (1998) A comparison of NH_4^+ and NO_3^- net fluxes along roots of rice and maize. Plant Cell Environ 21:240–246

Cozza R, Pangaro T (2009) Tissue expression pattern of two aquaporin-encoding genes in different organs of the seagrass *Posidonia oceanica*. Aquat Bot 91:117–121

Eighmy TT, Jahnke LS, Fagerberg WR (1991) Studies of *Elodea nuttallii* grown under photo-respiratory conditions. 2. Evidence or bicarbonate active transport. Plant Cell Environ 14:157–165

Elzenga JTM, Prins HBA (1988) Adaptation of *Elodea* and *Potamogeton* to different inorganic carbon levels and the mechanism for photosynthetic bicarbonate utilization. Aust J Plant Physiol 15:727–735

Elzenga JTM, Prins HBA (1989) Light-induced polar pH changes in leaves of *Elodea canadensis*. 2. Effects of ferricyanide: evidence for modulation by the redox state of the cytoplasm. Plant Physiol 91:68–72

Eugelink AH (1998) Phosphorus uptake and active growth of *Elodea canadensis* Michx. and *Elodea nuttallii* (Planch.) St. John. Water Sci Tech 37:59–65

Fagerberg WR, Eighmy TT, Jahnke LS (1991) Studies of *Elodea nuttallii* grown under photo-respiratory conditions. 3. Quantitative cytological characteristics. Plant Cell Environ 14: 167–173

Fang YY, Babourina O, Rengel Z, Yang XE, Pu PM (2007a) Ammonium and nitrate uptake by the floating plant *Landoltia punctata*. Ann Bot 99:365–370

Fang YY, Babourina O, Rengel Z, Yang XE, Pu PM (2007b) Spatial distribution of ammonium and nitrate fluxes along roots of wetland plants. Plant Sci 173:240–246

Fernandez JA, Garcia-Sanchez MJ, Felle HH (1999) Physiological evidence for a proton pump and sodium exclusion mechanisms at the plasma membrane of the marine angiosperm *Zostera marina* L. J Exp Bot 50:1763–1768

Fischer E, Luttge U (1980) Mebrane potential changes related to active transport of glycine in *Lemna gibba* G1. Plant Physiol 65:1004–1008

Fukuhara T, Pak UY, Ohwaki Y, Tsujimura H, Nitta T (1996) Tissue-specific expression of the gene for a putative plasma membrane H^+-ATPase in a seagrass. Plant Physiol 110:35–42

Fulgenzi FR et al (2008) The ionic environment controls the contribution of the barley HvHAK1 transporter to potassium acquisition. Plant Physiol 147:252–262

Garcia-Sanchez MJ, Jaime MP, Ramos A, Sanders D, Fernandez JA (2000) Sodium-dependent nitrate transport at the plasma membrane of leaf cells of the marine higher plant *Zostera marina* L. Plant Physiol 122:879–885

Garciadeblas B, Benito B, Rodriguez-Navarro A (2001) Plant cells express several stress calcium ATPases but apparently no sodium ATPase. Plant Soil 235:181–192

Garciadeblas B, Benito B, Rodriguez-Navarro A (2002) Molecular cloning and functional expression in bacteria of the potassium transporters CnHAK1 and CnHAK2 of the seagrass *Cymodocea nodosa*. Plant Mol Biol 50:623–633

Garciadeblas B, Haro R, Benito B (2007) Cloning of two SOS1 transporters from the seagrass *Cymodocea nodosa*. SOS1 transporters from *Cymodocea* and *Arabidopsis* mediate potassium uptake in bacteria. Plant Mol Biol 63:479–490

Garrill A, Tyerman SD, Findlay GP (1994) Ion channels in the plasma membrane of protoplasts from the halophytic angiosperm *Zostera muelleri*. J Membr Biol 142:381–393

Gorham PR (1941) Measurement of the response of *Lemna* to growth promoting substances. Am J Bot 28:98–101

Grabov A, Felle H, Bottger M (1993) Modulation of the plasma membrane electron-transfer system in root cells of *Limnobium stoloniferum* by external pH. J Exp Bot 44:725–730

Hartung W, Ullrich-Eberius CI, Luttge U, Bocher M, Novacky A (1980) Effect of abscisic acid on membrane potential and transport of glucose and glycine in *Lemna gibba* G1. Planta 148:256–261

Hillman WS (1959) Experimental control of flowering in *Lemna*. 1. General methods. Photoperiodism in *L. pepusilla* 6746. Am J Bot 46:466–473

Hillman WS (1961) The Lemnaceae, or duckweeds. A review of the descriptive and experimental literature. Bot Rev 27:221–287

Ice J, Couch R (1987) Nutrient absorption by duckweed. J Aquat Plant Manage 25:30–31

Ivankina NG, Novak VA (1981) Localization of redox reactions in plasmalemma of *Elodea* leaf cells. Studia Biophys 83:197–206

Iyer V, Barnabas AD (1993) Effects of varying salinity on leaves of *Zostera capensis* Setchell. 1. Ultrastructural changes. Aquat Bot 46:141–153

Jagels R (1983) Further evidence for osmoregulation in epidermal leaf cells of seagrasses. Am J Botany 70:327–333

Jeschke WD (1970) Influx of K^+ ions in leaves of *Elodea densa*, dependence on light, potassium concentration, and temperature. Planta 91:111–128

Kimber A, Crumpton WG, Parkin TB, Spalding MH (1999) Sediment as a carbon source for the submersed macrophyte Vallisneria. Plant Cell Environ 22:1595–1600

Krabel D, Eschrich W, Gamalei YV, Fromm J, Ziegler H (1995) Acquisition of carbon in Elodea canadensis Michx J Plant Physiol 145:50–56

Landolt E (1986) The family of Lemnaceae – a monographic study. Veroffentlichungen des Geobotanischen Institues ETH, Stiftung Rubel, Zurich

Lass B, Ullrich-Eberius CI (1984) Evidence for proton sulfate cotransport and its kinetics in *Lemna gibba* G1. Planta 161:53–60

Ludewig U, von Wiren N, Frommer WB (2002) Uniport of NH_4^+ by the root hair plasma membrane ammonium transporter LeAMT1;1. J Biol Chem 277:13548–13555

Maathuis FJM, Verlin D, Smith FA, Sanders D, Fernandez JA, Walker NA (1996) The physiological relevance of Na^+-coupled K^+-transport. Plant Physiol 112:1609–1616

Marre MT, Albergoni FG, Moroni A, Marre E (1989a) Light-induced activation of electrogenic H^+ extrusion and K^+ uptake in *Elodea densa* depends on photosynthesis and is mediated by the plasma membrane H^+-ATPase. J Exp Bot 40:343–352

Marre MT, Albergoni FG, Moroni A, Pugliarello MC (1989b) Evidence that H^+ extrusion in *Elodea densa* leaves is mediated by an ATP-driven H^+-pump. Plant Sci 62:21–28

Marx D (1929) The effect of small electric currents ion the assimilation of *Elodea canadensis*. Ann Bot 43:163–172

Maser P et al (2001) Phylogenetic relationships within cation transporter families of *Arabidopsis*. Plant Physiol 126:1646–1667

Miedema H, Felle H, Prins HBA (1992) Effect of high pH on the plasma membrane potential and conductance in *Elodea densa*. J Membr Biol 128:63–69

Muhonen M, Showman J, Couch R (1983) Nutrient absorption by *Spirodela polyrrhiza*. J Aquat Plant Manage 21:107–109

Muramatsu Y, Harada A, Ohwaki Y, Kasahara Y, Takagi S, Fukuhara T (2002) Salt-tolerant ATPase activity in the plasma membrane of the marine angiosperm *Zostera marina* L. Plant Cell Physiol 43:1137–1145

Natura G, Dahse I (1998) Potassium conductance of Egeria leaf cell protoplasts: regulation by medium pH, phosphorylation and G-proteins. J Plant Physiol 153:363–370

Newman IA (2001) Ion transport in roots: measurement of fluxes using ion-selective microelectrodes to characterize transporter function. Plant Cell Environ 24:1–14

Newman JR, Raven JA (1993) Carbonic anhydrase in *Ranunculus penicillatus* spp. *pseudofluitans*: activity, location and implications for carbon assimilation. Plant Cell Environ 16:491–500

Novacky A, Ullrich-Eberius CI, Luttge U (1978) Membrane potential changes during transport of hexoses in *Lemna gibba* G1. Planta 138:263–270

Oscarson P, Ingemarsson B, Ugglas MA, Larsson CM (1988) Characteristics of NO_3^- uptake in *Lemna* and *Pisum*. Plant Soil 111:203–205

Pak JY, Fukuhara T, Nitta T (1995) Discrete subcellular localization of membrane-bound ATPase activity in marine angiosperms and marine algae. Planta 196:15–22

Pedersen O (1993) Long-distance water transport in aquatic plants. Plant Physiol 103:1369–1375

Pedersen O, Sand-Jensen K (1993) Water transport in submerged macrophytes. Aquat Bot 44:385–406

Pokorny J, Ondok JP, Koncalova H (1985) Photosynthetic response to inorganic carbon in *Elodea densa* (Planchon) Caspary. Photosynthesis 19:366–372

Prins HBA, Elzenga JTM (1989) Bicarbonate utilization: function and mechanism. Aquat Bot 34:59–83

Prins HBA, Snel JFH, Zanstra PE, Helder RJ (1982) The mechanism of bicarbonate assimilation by the polar leaves of *Potamogeton* and *Elodea*. CO_2 concentration at the leaf surface. Plant Cell Environ 5:207–214

Rascio N (2002) The underwater life of secondarily aquatic plants: some problems and solutions. Crit Rev Plant Sci 21:401–427

Rascio N, Cuccato F, Dalla Vecchia F, La Rocca N, Larcher W (1999) Structural and functional features of the leaves of *Ranunculus trichophyllus* Chaix., a freshwater submerged macrophophyte. Plant Cell Environ 22:205–212

Rascio N, Mariani P, Vecchia FD, Zanchin A, Pool A, Larcher W (1994) Ultrastructural and photosynthetic features of leaves and stems of *Elodea canadensis*. J Plant Physiol 144:314–323

Raven JA (1994) Photosynthesis in aquatic plants. In: Schulze ED, Caldwell M (eds) Ecology of photosynthesis. Berlin, Springer, pp 299–318

Raven JA (1996) Into the voids: the distribution, function, development and maintenance of gas spaces in plants. Ann Bot 78:137–142

Raven JA et al (1988) The role of CO_2 uptake by roots and CAM in acquisition of inorganic C by plants of the isoetid life-form: a review, with new data on *Eriocaulon decangulare* L. New Phytol 108:125–148

Raven JA, Spicer RA (1996) The evolution of CAM. In: Winter K, Smith JAC (eds) Crassulacean acid metabolism. Biochemistry, ecophysiology and evolution. Springer, Heidelberg, pp 360–385

Rodriguez-Navarro A, Quintero FJ, Garciadeblas B (1994) Na^+-ATPases and Na^+/H^+ antiporters in fungi. Biochim Biophys Acta 1187:203–205

Rubio L, Linares-Rueda A, Garcia-Sanchez MJ, Fernandez JA (2005) Physiological evidence for a sodium-dependent high-affinity phosphate and nitrate transport at the plasma membrane of leaf and root cells of *Zostera marina* L. J Exp Bot 56:613–622

Sand-Jensen K, Pedersen MF, Nielsen SL (1992) Photosynthetic use of inorganic carbon among primary and secondary water plants in streams. Freshwater Biol 27:283–293

Santa-Maria GE, Rubio F, Dubcovsky J, Rodriguez-Navarro A (1997) The HAK1 gene of barley is a member of a large gene family and encodes a high-affinity potassium transporter. Plant Cell 9:2281–2289

Scheid HW, Ehmke A, Hartmann T (1980) Plant NAD-dependent glutamate dehydrogenase. Purification, molecular properties and metal ion activation of the enzymes from *Lemna minor* and *Pisum sativum*. Zeitschr Naturforsch C 35:213–221

Sculthorpe CD (1967) The biology of aquatic vascular plants. Edward Arnold, London

Setter TL et al (2009) Review of wheat improvement for waterlogging tolerance in Australia and India: the importance of anaerobiosis and element toxicities associated with different soils. Ann Bot 103:221–235

Shi HZ, Ishitani M, Kim CS, Zhu JK (2000) The Arabidopsis thaliana salt tolerance gene SOS1 encodes a putative Na^+/H^+ antiporter. Proc Natl Acad Sci USA 97:6896–6901

Shono M, Hara Y, Wada M, Fujii T (1996) A sodium pump in the plasma membrane of the marine alga *Heterosigma akashiwo*. Plant Cell Physiol 37:385–388

Smith FA (1982) Transport of methylammonium and ammonium ions by *Elodea densa* J. Exp Bot 33:221–232

Sorrell BK, Brix H, Orr PT (1997) *Eleocharis sphacelata*: Internal gas transport pathways and modelling of aeration by pressurized flow and diffusion. New Phytol 136:433–442

Steinberg RA (1946) Mineral requirements of *Lemna minor*. Plant Physiol 21:42–48

Sultemeyer D, Schmidt C, Fock HP (1993) Carbonic anhydrases in higher plants and aquatic microorganisms. Physiol Plant 88:179–190

Taylor AR, Bloom AJ (1998) Ammonium, nitrate, and proton fluxes along the maize root. Plant Cell Environ 21:1255–1263

Thomaz SM, Chambers PA, Pierini SA, Pereira G (2007) Effects of phosphorus and nitrogen amendments on the growth of *Egeria najas*. Aquat Bot 86:191–196

Touchette BW (2007) Seagrass-salinity interactions: physiological mechanisms used by submersed marine angiosperms for a life at sea. J Exp Mar Biol Ecol 350:194–215

Ullrich-Eberius CI, Novacky A, Fischer E, Luttge U (1981) Relationship between energy-dependent phosphate uptake and the electrical membrane potential in *Lemna gibba* G1. Plant Physiol 67:797–801

Ullrich-Eberius CI, Sanz A, Novacky AJ (1989) Evaluation of arsenate-associated and vanadate-associated changes of electrical membrane potential and phosphate transport in *Lemna gibba* G1. J Exp Bot 40:119–128

Ullrich CI, Novacky AJ (1990) Extra- and intracellular pH and membrane potential changes induced by K^+, Cl^-, $H_2PO_4^-$, and NO_3^- uptake and fusicoccin in root hairs of *Limnobium stoloniferum*. Plant Physiol 94:1561–1567

Ullrich WR, Larsson M, Larsson CM, Lesch S, Novacky A (1984) Ammonium uptake in *Lemna gibba* G1, related membrane potential changes, and inhibition of anion uptake. Physiol Plant 61:369–376

Venegoni A, Moroni A, Gazzarrini S, Marre MT (1997) Ammonium and methylammonium transport in *Egeria densa* leaves in conditions of different H^+ pump activity. Bot Acta 110:369–377

Walker NA (1994) Sodium-coupled symports in the plasma membranes of plant cells. In: Blatt MR, Leigh RA, Sanders D (eds) Membrane transport in plants and fungi: molecular mechanisms and control society for experimental biology symposium XLVIII. The company of Biologists, Cambridge, UK, pp 179–192

Wang MY, Glass ADM, Shaff JE, Kochian LV (1994) Ammonium uptake by rice roots. III. Electrophysiology. Plant Physiol 104:899–906

Wang MY, Siddiqi MY, Ruth TJ, Glass ADM (1993) Ammonium uptake by rice roots. II. Kinetics of $^{13}NH_4^+$ influx across the plasmalemma. Plant Physiol 103:1259–1267

Ward JM, Maser P, Schroeder JI (2009) Plant ion channels: gene families, physiology, and functional genomics analyses. Ann Rev Physiol 71:59–82

White HL (1937) The interaction of factors in the growth of Lemna XI. The interaction of nitrogen and light intensity in relation to growth and assimilation. Ann Bot 1:623–647

Part IV
Agronomical and Environmental Aspects

Chapter 12
Genetic Variability and Determinism of Adaptation of Plants to Soil Waterlogging

Julien Parelle, Erwin Dreyer, and Oliver Brendel

Abstract Flooding or waterlogging, and associated soil hypoxia, affect severely the growth and fitness of plant species, from crops to forest ecosystems. An improved understanding of the intra-species genetic diversity of traits involved in hypoxia tolerance is a prerequisite for crop breeding programmes aimed at increasing the tolerance to waterlogging, as well as for assessing the adaptability of natural populations to waterlogging. Some genotypes within the species have developed adaptations to hypoxia, as shown by differences among populations in growth and fitness, and in traits conferring some degree of tolerance such as sequence, expression and activity of alcohol dehydrogenase, or the ability to develop adventitious roots, increased tissue porosity and hypertrophied lenticels. Genetic control has been estimated for a number of such traits. Overall, under waterlogging, specific tolerance traits show higher heritabilities compared to traits quantifying productivity, damage or overall performance. Genomic regions involved in the control of these traits (i.e., Quantitative Trait Loci QTL) have been detected for tolerance traits in a few species, and allow gaining some insight into the genetic basis of the observed natural diversity or may be a starting point for breeding purposes. However, only for submergence tolerance in rice (sub-1) has a successful gene candidate approach resulted in the detection of alleles that are directly involved in the tolerance process.

J. Parelle
University of Franche-Comté, UMR UFC/CNRS 6249 USC INRA "Chrono-Environnement",
Place Leclerc 25030 Besancon, France
e-mail: julien.parelle@univ-fcomte.fr

E. Dreyer and O. Brendel (✉)
INRA, UMR1137 "Ecologie et Ecophysiologie Forestières", F 54280 Champenoux, France

Nancy-Université, UMR1137 "Ecologie et Ecophysiologie Forestières", Faculté des Sciences,
54500 Vandoeuvre, France
e-mail: dreyer@nancy.inra.fr; brendel@nancy.inra.fr

S. Mancuso and S. Shabala (eds.), *Waterlogging Signalling and Tolerance in Plants*,
DOI 10.1007/978-3-642-10305-6_12, © Springer-Verlag Berlin Heidelberg 2010

Abbreviations

ADH	Alcohol dehydrogenase
LEI	Lowest elongated internode
PDC	Pyruvate decarboxylase complex
PEV	Per cent of explained variance
QTL	quantitative trait loci
RIL	Real isogenic lines
SNP	Single nucleotide polymorphism
Sub	Submergence tolerance locus

12.1 Introduction

Excess soil water due to flooding or temporary waterlogging can be a major constraint on growth and yield of crops (Tuberosa and Salvi 2004) and forest stands (Kozlowski 1997). It affects severely growth and probably also fitness and distribution of plant species in natural environments. Some species or genotypes within species have developed adaptive responses to flooding and waterlogging. In the case of crops, the occurrence of some genetic diversity in tolerance traits is a prerequisite for breeding programmes. In natural ecosystems, due to the local occurrence of temporarily waterlogged soils (often called hydromorphic soils, Lévy et al. 1999), the frequency and severity of episodes of waterlogging or flooding act as a selective pressure and differences in tolerance can develop among species, or populations within species. To gain insight into the degree of inter-specific variability, we need a careful quantification of the tolerance to waterlogging in individuals and methods to assess it as objectively as possible.

Two major situations of excess water can be identified (Colmer and Voesenek 2009). Flooding, the partial, or in some cases the complete submergence of the shoot, can be permanent, such as in mangrove ecosystems, or temporary, such as in floodplains or in rice paddies. Waterlogging, due to excess water in the soil, usually occurs temporarily with a water level below or not much above the soil surface that affects primarily the root system and can occur in natural as well as in cultivated ecosystems, depending on soil type and water table dynamics.

In both cases, a temporal sequence of chemical changes occurs in the soil following the onset of waterlogging or flooding (Setter and Waters 2003). Due to a reduced gas exchange between soil and atmosphere, changes in soil bacteria populations occur, oxygen concentration decreases rapidly (hypoxia), carbon dioxide and ethylene concentrations increase, reduced and toxic cations such as manganese (Mn^{++}) and iron (Fe^{++}) accumulate, and an intense de-nitrification occurs. In case of prolonged waterlogging, soils may be completely depleted of oxygen (anoxia) and hydrogen sulphide and methane are produced and diffuse into the atmosphere. Except the last, all of these steps occur usually within the first 20 days of waterlogging. In some soils, this sequence may even occur faster. In this review, we will concentrate on the hypoxia induced by waterlogging or total submergence with all the consequences it might have on respiration, metabolism and growth of affected plants.

To date, the processes conferring some degree of tolerance to waterlogging and hypoxia are still not fully understood despite accumulating information (e.g., Vartapetian 2006; Voesenek et al. 2006; Colmer and Voesenek 2009; Kawano et al. 2009; Jackson et al. 2009; Parolin 2009). The degree of tolerance to a given level of waterlogging may be assessed: (1) indirectly through damage indices or the observed, usually negative, impact on growth, productivity and survival, or (2) directly by evaluating the occurrence of traits contributing to the acclimation to hypoxia (adaptive traits). These traits can be constitutive (i.e., they occur already in individuals growing under optimal conditions and provide some advantage during waterlogging) or induced (i.e., they appear only during episodes of waterlogging in response to a signalling cascade). Induced traits can be roughly grouped into short-term responses (e.g., metabolic adjustments) and long-term acclimations (e.g., development of aerenchyma). A typical short-term response of roots is a decrease of respiration and an increase of glycolytic flux and alcoholic fermentation (Drew 1997). Some key enzymes in this process are alcohol dehydrogenase (McManmon and Crawford 1971; Chan and Burton 1992; Bailey-Serres and Voesenek 2008), sucrose synthase or hexokinases (Germain et al. 1997; Ricard et al. 1998). Long-term responses are mainly related to growth, either of existing or of newly formed structures. In rice, where total submergence clearly poses a major problem for productivity (Tuberosa and Salvi 2004), the elongation of internodes is an important adaptive trait, resulting either in quiescence or an escape strategy (Bailey-Serres and Voesenek 2008). The quiescence strategy consists in a lack of elongation (Xu and Mackill 1996), whereas the escape strategy consists in an enhanced growth rate that maintains the top of the shoot above the water level (Fukao et al. 2006). Adaptive morphological traits are slower to develop compared to purely physiological or metabolic adjustments. Assessment of such traits requires long-term experiments with the risk of an interaction between ontogenic development and stress response. A few anatomical traits, thought to allow transport of oxygen to roots and enable a partial maintenance of respiration, survival or even root growth, have commonly been measured in experiments on genetic variability. They include the development of hypertrophied lenticels (Parelle et al. 2007), of adventitious roots (Mano et al. 2005a, b) and of aerenchyma (porosity) in root or stem tissues (Zaidi et al. 2007; Mollard et al. 2008; Mano et al. 2007, 2008; Mano and Omori 2008; and see also Chap. 6 in this volume).

From an agronomic point of view, the maintenance of productivity, particularly yield, is of major importance. This can be evaluated by quantifying growth or biomass and also more indirectly by assessing, among others, leaf level gas exchange or photosynthetic capacity. Leaf gas exchange, for instance, has been used (Dreyer 1994; Wagner and Dreyer 1997) to characterise the overall performance under waterlogging. Such traits bring no information about the morphological and physiological mechanisms of tolerance; nevertheless, maintenance of productivity or photosynthesis contributes to fitness and survival of individuals. Survival rate under hypoxia is, together with shoot dieback and other fitness related traits (number of seeds produced, etc), an important means to assess the degree of tolerance of populations. Leaf epinasty, the downward growth of leaf petioles, is a specific response to root hypoxia in some species (Jackson and Campbell 1976) and a direct indicator of the level of hypoxia stress perceived by the individuals (Vartapetian and Jackson 1997). All these traits may respond to waterlogging with quite different intensities. Nevertheless, some traits obviously play a direct adaptive role, or at least are thought to do so. Such traits include the development of hypertrophied lenticels, of adventitious roots, of aerenchyma or the occurrence of physiological changes (switch from a respiratory to a fermentative metabolism). All these traits contribute to mitigate the impact of hypoxia in the soil, by maintaining a minimal supply of oxygen to roots.

The quantification of growth decline under stress provides a first indication about the level of tolerance of a genotype. Traits that have been measured include plant height, growth increment and shoot or root dry weight (see Table 12.1 for examples from quantitative genetic studies). Yield, or the reduction thereof, has also been quantified for crops under waterlogging (Vantoai et al. 2001; Githiri

Table 12.1 Traits tested during QTL experiments to identify hypoxia-tolerance related loci in different species

Stress type	Article	Genus	Trait types
Submergence	Xu and Mackill (1996)	Rice	Damage
	Toojinda et al. (2003)	Rice	Damage
	Nandi et al. (1997)	Rice	Survival
	Sripongpangkul et al. (2000)	Rice	Survival
	Toojinda et al. (2003)	Rice	Survival
	Toojinda et al. (2003)	Rice	Growth
	Ikeda et al. (2007)	Rice	Growth
	Nemoto et al. (2004)	Rice	Elongation
	Tang et al. (2005)	Rice	Elongation
	Hattori et al. (2007)	Rice	Elongation
	Sripongpangkul et al. (2000)	Rice	Elongation
	Toojinda et al. (2003)	Rice	Elongation
Waterlogging	Mano et al. (2006)	Maize	Damage
	Cornelious et al. (2005)	Soybean	Damage
	Martin et al. (2006)	Iris	Survival
	Vantoai et al. (2001)	Soybean	Growth
	Parelle et al. (2007)	Oak	Growth
	Qiu et al. (2007)	Maize	Growth
	Vantoai et al. (2001)	Soybean	Yield
	Githiri et al. (2006)	Soybean	Yield
	Mano et al. (2005a)	Maize	Adventitious roots
	Mano et al. (2005b)	Maize	Adventitious roots
	Zheng et al. (2003)	Rice	Adventitious roots
	Parelle et al. (2007)	Oak	Hypertrophied lenticels
Control	Mano et al. (2007, 2008)	Maize	Aerenchyma

et al. 2006). However, the use of these traits to detect genetic differences in waterlogging tolerance requires a careful interpretation to identify adaptive traits involved in the tolerance to hypoxia, in contrast to nonadaptive traits indicating merely a genetic difference, for instance in growth potential.

The quantification of damage induced by waterlogging may also provide an estimation of tolerance. Some authors use visual ordinal scales of damage (Xu and Mackill 1996; Sripongpangkul et al. 2000; Cornelious et al. 2005), others quantify leaf senescence (Toojinda et al. 2003), fraction of yellow leaves (Zhou et al. 2007a) or decline in leaf chlorophyll content (GuangHeng et al. 2006). Damage indices have been successfully used to study the genetic determinism of tolerance in crops. For example, the so-called *sub-1* locus (Xu and Mackill 1996) was identified in rice 10 years before the actual process controlled by the locus was understood (Xu et al. 2006). A large amount of damage eventually leads to mortality, which is a very simple approach to characterising tolerance (Nandi et al. 1997; Sripongpangkul et al. 2000; Toojinda et al. 2003; Martin et al. 2006). Despite the fact that survival is an ordinal trait, the approximately normal distribution within experiments allows using it for a genetic trait dissection in rice (Xu and Mackill 1996) and soybean (Cornelious et al. 2005). The advantage of survival and damage traits is that the variability tested is directly related to the stress tolerance, and can therefore be fully attributed to genetic diversity.

To assess the genetic variability of tolerance to waterlogging or flooding, the most obvious procedure is the quantitative analysis of traits conferring directly or indirectly some level of tolerance to waterlogging. Prerequisites for suitable traits include: (1) the relevance and specificity of the trait as an indicator of adaptation; (2) the repeatability of the measurement procedure and (3) the possibility of assessing a large number of individuals.

Genetic variability of tolerance traits can be studied *in situ* in natural populations only when detailed information on the environmental conditions and their spatial and temporal variability is available. Further, an already advanced knowledge of the genetic determinism of the trait studied is necessary, with a remaining risk of confusion between purely genetic differences and genotype and environment interactions. Therefore, all of the studies reviewed here were done using common conditions for all genotypes (vegetative copies or half-sib families), as provided by common-garden plantations (comparative plantations with a homogenized environment) or greenhouse experiments. However, even under such controlled conditions, statistical methods (such as complete or random blocks) should be used to minimise residual variations of environment or stress conditions. It is very difficult to control with large precision the level of soil hypoxia imposed to the different individuals, as the oxygen concentration in the soil depends also on rooting density, soil heterogeneity and the presence of soil microorganisms. One possibility to control more directly the amount of oxygen available to the root system is the use of hydroponic systems that are bubbled with a specific nitrogen/oxygen mixture (e.g. Ricard et al. 1998). However root growth; root anatomy and root system architecture differ widely between hydroponics and soils.

246   J. Parelle et al.

Some degree of intra-specific genetic diversity of traits induced by root hypoxia has been shown for a number of species and traits. In the first section of this chapter, we review common garden comparisons of natural populations which are exposed to different levels of waterlogging at their sites of origin. The second section concentrates on offspring from controlled crosses of specific genotypes, often preclassified as tolerant or sensitive to waterlogging, used for quantitative analyses of the genetic determinism, and ultimately for the detection of Quantitative Trait Loci (QTL).

12.2 Diversity Among Populations: Adaptation to Water-Logged Soils?

Common garden comparisons of individuals grown from seeds collected in diverse populations were mainly published for noncrop species. The detected diversity was interpreted, with some caution, as revealing differences in adaptation due to natural selection.

Interest in intra-species variation of tolerance to waterlogging or to flooding emerged in the 1970s. Some examples include among-family variation in *Veronica peregrina*, an annual dicotyledon found on moist sites (Linhart and Baker 1973); population differences in *Eucalyptus viminalis*, from dry or wet forests (Ladiges and Kelso 1977) or in *E. ovata*, Australian swamp gum (Clucas and Ladiges 1979). These studies showed already that phenoypic differences could be detected among populations from sites differing in susceptibility to waterlogging and hypoxia.

Growth is strongly affected by waterlogging, and is usually significantly depressed in plants originating from both well-drained and hydromorphic sites. This was the case for *Eucalyptus globulus* and *E. grandis* families (Marcar et al. 2002), where at least *E. globulus* is known to be sensitive to waterlogging. Similarly, during a hypoxia experiment on *Geum rivale* (wetland species), *Geum urbanum* (dry habitat species) and hybrid populations, root dry weight was reduced in all families (Waldren et al. 1988). However, as examples span a large range of plant species, from monocotyledon grasses to forest trees, responses are very diverse. In some species, growth increased during waterlogging for populations from wet environments, such as in *Panicum antidotale* (Ashraf 2003) or *Paspalum dilatatum* (Loreti and Oesterheld 1996) populations, whereas it decreased in other species. This differential growth response to waterlogging is an extreme case of environment × genotype interaction and can also be found to a lesser degree within species. Marcar et al. (2002) studied growth under waterlogging in different populations from two *Eucalyptus* species. A significant treatment × provenance effect was detected for shoot dry weight in *E. globulus*, a rather hypoxia-sensitive species, but none for *E. grandis*, a species growing on hydromorphic soils. Similarly, Waldren et al. (1988) found in *G. rivale* no population differentiation for growth during waterlogging, whereas *G. urbanum* and hybrid populations showed significant population × waterlogging interactions. Overall, these examples show that a diversity of adaptations to waterlogging can evolve in closely related species resulting in growth differences. This suggests a genetic differentiation among populations, and thus a genetic determinism of hypoxia tolerance. However, there are also examples of species where no specific adaptation has been detected for hypoxia-exposed populations that would result in growth differences. In *Acer rubrum* seedlings, no relationship was detected between population differences in response to controlled waterlogging and the conditions at the sites of origine (Will et al. 1995).

Adaptive differences among populations have also been detected using net CO_2 assimilation rate and stomatal conductance as indicators of fitness on lowland (wet) and upland (dry) populations of *P. dilatatum* (Mollard et al. 2008). Flooded plants displayed higher net CO_2 assimilation and stomatal conductance compared to controls in lowland populations, and stomatal closure and reduced net CO_2 assimilation in upland populations: under similar hypoxia, lowland populations were able to maintain water absorption by roots, while upland populations were not.

Variability in growth and leaf gas exchange among genotypes during waterlogging is the result of anatomical or physiological adaptations, such as the ability to develop hypertrophied lenticels, adventitious roots and aerenchyma in root or stem tissue. Under waterlogging, a significant increase in the number and height of hypertrophied lenticels was found in populations of *Luehea divaricata* from temporarily waterlogged soils versus those from well drained soils (De Carvalho et al. 2008, C.F. Ruas, pers. comm.). The development of hypertrophied lenticels or adventitious roots is typically an induced adaptive response to hypoxia with quite a large genetic diversity. A genetic basis for diversity in adventitious root growth was detected among *Carex flacca* populations (Heathcote et al. 1987). Continuous flooding increased adventitious root biomass in all populations to the same extent. Significant population differences and population × treatment interactions were detected during repeated transient episodes of flooding. This underlines that the modality of stress application may impact the degree of detected genetic diversity. Stress responses often differ between organs, as in *P. dilatatum*, where porosity did not increase in roots during flooding, while it did in the leaf sheath (Mollard et al. 2008). However, there were strong treatment × population interactions: root porosity was different between lowland (wet) and upland (dry) populations under control conditions but not under flooding, whereas leaf sheath porosity was different under flooding but not under control conditions. The hypoxia-adapted lowland populations had constitutively higher root porosity, with little increase during flooding, whereas upland populations showed a larger response to flooding for leaf sheath porosity. Overall, these examples, covering a range of different plant types, suggest that genetic differences seem to have evolved for morphological adaptations to root hypoxia among natural populations exposed to different levels of soil hypoxia and that an adaptation to different environments has taken place.

As described above, hypoxia induces changes in root metabolism. Genetic differences in the expression of alcohol dehydrogenase (ADH) have been studied intensively in a number of species. As early as the 1970s, ADH polymorphism has been shown to affect growth rate under waterlogging (Marshall et al. 1973; Brown et al. 1976) and population differences were detected (Torres et al. 1977; Brown 1978). A genetic variability was also detected for the gene coding for ADH (locus ADH-B) among five European populations of *Fraxinus excelsior,* but was not related to flooding frequency at the sites of origin (Ruedinger et al. 2008). Herzog and Krabel (1999) studied 17 isoenzyme loci, of which some are thought to be involved in waterlogging or hypoxia tolerance. They found no evidence for a selection on these loci when comparing a frequently flooded and a dry-land population of *Quercus robur.* Chan and Burton (1992) found for *Trifolium repens* a strong population × treatment interaction for ADH activity in roots, with higher activities in populations from frequently flooded sites. ADH activity under waterlogging was positively correlated with relative growth rate, suggesting that a higher ADH activity contributes to a higher tolerance to hypoxia. This contradicts inter-specific comparisons, where more tolerant species displayed lower ADH activity (McManmon and Crawford 1971). However, ADH activity varies with time during stress application: sensitive *Brassica rapa* L. plants displayed a higher ADH activity after 18 h of stress but not earlier or later (Daugherty and Musgrave 1994). Enzymes potentially involved in hypoxia tolerance have been studied in detail in diverse crop species, where genetically well-defined varieties or clones are available. Increased ADH activity was found for waterlogging tolerant compared to susceptible *Zea mays* genotypes (Zaidi et al. 2003). Similarly, more tolerant *Oryza sativa* cultivars with a higher internode elongation rate under hypoxic conditions (escape strategy) showed also higher ADH and pyruvate decarboxylase (PDC) activities and ATP concentration (Kato-Noguchi and Morokuma 2007). This difference in ADH activity seems specific for roots (Kato-Noguchi et al. 2003). Fukao et al. (2003) found with seeds of the weed *Echinochloa crus-galli* germinating under anoxic conditions, that aldolase, aldehyde dehydrogenase and PDC were more strongly induced in a tolerant compared to an intolerant variety, whereas sucrose synthase, enolase and ADH showed similar induction patterns for both. The occurrence of some genetic variation has been detected in the sequence, expression and activity of ADH, whereas less information is available for other enzymes involved in hypoxia responses. However, even for ADH, we still lack experimental support demonstrating that the genetic diversity that is observed results in variation in adaptation to waterlogging by natural populations. Such a demonstration could be provided, for example, by population genetic studies linking single nucleotide polymorphisms (SNP) within the ADH-gene or its promoter to survival and fitness in stressed environments. Further, population genetic models could then be applied (e.g., Beaumont and Nichols 1996), testing whether nucleotide differentiation patterns of SNP within the ADH-gene could depart from neutral patterns and result from natural selection. This has been done, for example, with candidate genes for drought tolerance in *Pinus pinaster* populations (Eveno et al. 2008).

12.3 Genetic Control of Traits Related to Hypoxia Tolerance

A more direct approach to estimate genetic control of traits related to hypoxia tolerance is the estimation of heritability, which is, in the simplest case, the ratio between genetic and total variance within a given experimental set up (Lynch and Walsh 1997). The calculation of the genetic variance requires not only controlled conditions for trait estimations, but also an assessment of the relatedness of individuals within the experimental set up, such as multi-parental crossings (diallels, half-dialleles, clonal repetitions, etc). This approach is rarely possible with wild populations, however, it has been frequently used for crops. Heritability is difficult to compare among experiments, as it depends on environmental variance induced by the specific experimental set up. However, it provides an indication of the importance of the genetic control on a trait in a given experiment and can be used to predict results of artificial and natural selection (Hartl and Clark 1997), where narrow-sense heritability (ratio of additive genetic variance to total variance) is more important for population responses to individual selection than broad sense heritability (ratio of total genetic variance to total variance).

Significant levels of heritability have been detected under waterlogging or flooding in a number of species for biomass and yield (Collaku and Harrison 2005, *Triticum aestivum;* Silva et al. 2007, *Z. mays*) as well as for traits assessing the sensitivity to hypoxia e.g., the percentage of yellow leaf (Zhou et al. 2007b, *Hordeum vulgare*). However, without an estimate of heritability under controlled conditions, it is impossible to infer whether the observed genetic control refers to a constitutive or an induced trait. Marcar et al. (2002) compared *E. globulus* (hypoxia sensitive) and *E. grandis* (tolerant) seedlings for shoot dry weight changes under water logging relative to control conditions. Narrow sense heritability for this trait was higher in the tolerant species and lower in the sensitive one. This species × environment interaction, suggests a larger genetic control of growth during hypoxia for the adapted plants. Kolodynska and Pigliucci (2003) observed during a three-generation selection experiment with *Arabidopsis thaliana* that heritability changed in response to selection, and that morphological traits displayed increasing heritabilities compared to life-history traits. Selection did not alter the overall shape of reaction norms but lowered the phenotypic means of some traits. Hybrid families of *G. rivale* × *urbanum* (wetland × dryland species) showed no significant heritability for the response of shoot biomass to waterlogging (Waldren et al. 1988). However, the response of root dry weight or shoot/root ratio was under significant genetic control in this experiment. Thus, integrative traits with no direct link to hypoxia tolerance such as above-ground biomass yield or growth might, in some situations seem not to be under genetic control even though some genetic diversity was detected in adaptive traits. In such cases, differences in fitness during stress might be explained better by survival rate than by aboveground growth and biomass production.

A typical adaptive trait studied in rice is internode elongation during submergence. Nemoto et al. (2004) studied the lowest elongated internode (LEI) for a diallel crossing of different *Oryza sativa* and *O. rufipogon* varieties. For this highly adaptive trait, high heritabilities were estimated (0.994 for broad sense and 0.962 for narrow sense) with a much larger additive than dominance variance, suggesting a high potential for individual selection. The heritability of adventitious root development was studied under flooding in *Cucumis sativus* (Yeboah et al. 2008). Narrow-sense heritability was higher for this trait (0.74) than for the overall tolerance score (0.60). The heritability of total root dry weight was higher under waterlogging than in controls. In a large test with 436 *Z. mays* inbred lines, a low broad-sense heritability was found for root porosity under normal conditions, which increased significantly during waterlogging (Zaidi et al. 2007). In contrast, heritability of biomass and yield declined during waterlogging compared to control. The tight correlation between root porosity and grain yield under stress, and its absence in controls, stresses the importance of root porosity for hypoxia tolerance in this species. We found no estimate of heritability for the development of hypertrophied lenticels in the literature. Nevertheless, in general, whenever heritability was estimated for morphological adaptive traits, a rather tight genetic control was shown, and it often increased under stress. There are few estimates of genetic control of enzyme activities related to hypoxia tolerance. Chan and Burton (1992) showed a strong genetic control for hypoxia-induced ADH activity in *T. repens* populations (broad sense heritability 0.55 ± 0.13). Overall, tolerance traits seem to show higher heritabilities in stressed conditions compared to productivity or traits quantifying damage or overall performance.

12.4 Genetic Determinism of Tolerance to Waterlogging and Identification of the Involved Genome Regions

Once the occurrence of a genetic control of a trait has been established, the next step is to identify the underlying genetic determinism, that is, how many genes control the expression of the trait and to what extent each gene controls its variability. The classical approach to this question is QTL (quantitative trait loci) mapping, the resolution of quantitative traits into discrete mendelian inherited components (Paterson et al. 1988). This requires a reference population screened for a high number ($\gg100$) of genetic markers. The recombination information produced by the progeny is then used to order the markers on a genetic map. The comparison of this genetic information across all individuals with their phenotype for a given trait allows identifying regions on the genetic map (QTL), that each determines a fraction of the observed phenotypic variability of the trait (called the phenotypically explained variance, PEV). The least likelihood (LOD), position on the genetic map, allelic substitution effect and PEV are estimated for each QTL. Depending on the statistical package used, the presence of a QTL is either determined by a LOD score threshold or a significance statistic calculated using permutation techniques. Further, bootstrap methods allow estimating standard deviations of all parameters.

However, most parameters estimated during QTL analyses, including the number of QTL detected, depend heavily on the number of genotypes in the reference population. Simulations showed that PEV values are overestimated and that the number of QTL detected does not correspond to the number of loci involved when the sample size of the mapping pedigree (N) is below 1,000 (Beavis 1994). QTL experiments for hypoxia-related traits never involved more that 300 genotypes (Table 12.2, range: 60–288), related to the large experimental set ups necessary, as for example, shown in Fig 12.1. Some caution is therefore needed when attempting to infer the actual genetic determinism of specific traits as the number of detected QTL is likely to be smaller than the actual number of genes involved. QTL experiments with a relatively small number of genotypes will mainly detect major QTL with high allelic substitution effects and PEV.

QTL detection encompassed the whole range of traits and conditions described above, focusing on short-term responses of physiological processes, or long-term acclimation with the objective to investigate morphological adaptations. The environmental conditions used for QTL detection, ranged from waterlogging and flooding to total submergence.

12.4.1 Methodology of the Detection of QTL for Hypoxia Tolerance: Caution and Strategies

12.4.1.1 Submergence Tolerance and Waterlogging Tolerance

To date, QTL detection for tolerance to total submergence concentrated on rice (Xu and Mackill 1996; Nandi et al. 1997; Sripongpangkul et al. 2000; Xu et al. 2000; Toojinda et al. 2003; GuangHeng et al. 2006; Hattori et al. 2007). All these experiments detected a major QTL on chromosome 9 (see Chen et al. 2002, for a detailed physical and genetic map of rice) and allowed the identification of the *sub-1* locus (Xu et al. 2006). QTL have also been detected for several species during partial submergence and root hypoxia (Vantoai et al. 2001; Zheng et al. 2003; Cornelious et al. 2005; Mano et al. 2005a, b; Cornelious et al. 2006; Githiri et al. 2006; Qiu et al. 2007; Parelle et al. 2007; Zhou et al. 2007a). Duration of water-logging as well as the height of the water table were highly variable, ranging from a few days (Qiu et al. 2007) to several weeks (Vantoai et al. 2001; Mano et al. 2005b), and from few centimetres (Mano et al. 2005b; Qiu et al. 2007) to 10 cm above soil surface (Vantoai et al. 2001; Cornelious et al. 2005). This diversity in experimental procedures may have contributed to the large variability in the number and locali-zation of the detected QTL (Table 12.2). Phenotyping after variable stress durations and intensities may detect different tolerance processes and thus result in a QTL detection that varies with environment. QTL detection is a statistical process, whereby minor QTL with low allelic substitution effects and thus low PEV will often be below the detection or significance limit. Parelle et al. (2007) detected a

Table 12.2 Number of QTL (N_Q) detected under hypoxic conditions (except Mano et al. 2007 and Mano et al. 2008) for different tolerance or growth traits for each experiment (where n_i is the number of genotypes of the tested family)

Article	Species	n_i	Trait	N_Q	LOD score	PEV
Martin et al. (2006)	Iris	120	Survival	2	–	0.11–0.14
Qiu et al. (2007)	Maize	288	Total dry weight	4	2.8–5.9	0.12–0.32
Qiu et al. (2007)	Maize	288	Shoot dry weight	4	2.6–7.0	0.05–0.21
Qiu et al. (2007)	Maize	288	Root length	7	2.5–3.5	0.04–0.07
Qiu et al. (2007)	Maize	288	Root dry weight	2	2.7–2.9	0.04–0.05
Qiu et al. (2007)	Maize	288	Plant height	3	2.7–3.2	0.05–0.07
Mano et al. (2008)	Maize	195	Aerenchyma	5	1.5–3.4	0.04–0.09
Mano et al. (2007)	Maize	141	Aerenchyma	6	1.7–4.9	0.07
Mano et al. (2006)	Maize	178	Degree of leaf injury	1	4.4	0.03–0.15
Mano et al. (2005a)	Maize	201	Adventitious root formation	2	3.9–5.1	0.09–0.1
Mano et al. (2005a)	Maize	94	Adventitious root formation	1	6.5	0.25
Mano et al. (2005b)	Maize	110	Adventitious root formation	3	3.2–5.1	0.1–0.21
Parelle et al. (2007)	Oak	119	Hypertrophied lenticel Nr.	1	3.4	0.15
Parelle et al. (2007)	Oak	100	Epinasty	5	7.2–15.2	0.08–0.12
Parelle et al. (2007)	Oak	100	Hypertrophied lenticel dev.	1	2.8	0.1
Zhou et al. (2007a)	Rice	282	Seedling height	1	2.3	0.04
Zhou et al. (2007a)	Rice	282	Seedling emergence	2	2.1–3.9	0.04–0.07
Zhou et al. (2007a)	Rice	282	Coleoptile emergence	2	2.3–4.9	0.04–0.08
Zheng et al. (2006)	Rice	96	Seminal root length	1	2.6	0.12
Zheng et al. (2006)	Rice	96	Total root dry weight	3	3.0–3.5	0.13–0.16
Zheng et al. (2006)	Rice	96	Adventitious root number	2	0.53–2.98	0?–0.133
Zheng et al. (2003)	Rice	96	Seminal root length	1	2.6	0.12
Zheng et al. (2003)	Rice	96	Lateral root number	1	2.8	0.13
Zheng et al. (2003)	Rice	96	Lateral root length	2	2.4	0.12
Zheng et al. (2003)	Rice	96	Adventitious root number	4	2.5–4.6	0.11–0.20
Xu and Mackill (1996)	Rice	169	Tolerance	1	36	0.69
Toojinda et al. (2003)	Rice	172	Total shoot elongation	4	3.9–28.1	0.01–0.52
Toojinda et al. (2003)	Rice	65	Total shoot elongation	3	6.2–27.3	0.24–0.74
Toojinda et al. (2003)	Rice	172	Tolerance score	9	3.2–52.6	0.07–0.63
Toojinda et al. (2003)	Rice	65	Tolerance score	2	12.3–49.1	0.28–0.72

Toojinda et al. (2003)	Rice	172	Relative shoot elongation	2	3.8–16.2	0.01–0.12
Toojinda et al. (2003)	Rice	172	Per cent plant survival	4	5.4–65.8	0.11–0.77
Toojinda et al. (2003)	Rice	65	Per cent plant survival	3	5.4–17.3	0.16–0.48
Toojinda et al. (2003)	Rice	172	Leaf senescence	4	4.0–29.3	0.1–0.53
Toojinda et al. (2003)	Rice	65	Leaf senescence	2	5.5–18.0	0.30–0.72
Tang et al. (2005)	Rice	192	Early elongation ability	2	3.8–18.2	0.09–0.41
Sripongpangkul et al. (2000)	Rice	165	Survival (qualitative)	13	–	–
Sripongpangkul et al. (2000)	Rice	165	Plant height increment	11	3.6–13.0	0.05–0.26
Sripongpangkul et al. (2000)	Rice	165	Leaf increment	4	2.5–10.8	0.09–0.25
Sripongpangkul et al. (2000)	Rice	165	Initial plant height	9	4.0–10.1	0.06–0.31
Sripongpangkul et al. (2000)	Rice	165	Total increment	5	4.0–10.8	0.06–0.36
Sripongpangkul et al. (2000)	Rice	165	Internode increment	3	7	0.09–0.37
Nemoto et al. (2004)	Rice	186	Early elongation ability	1	18.6	0.46
Nandi et al. (1997)	Rice	74	Survival (qualitative)	1	–	–
Nandi et al. (1997)	Rice	250	Survival	4	3.2–4.7	0.19–0.27
Ikeda et al. (2007)	Rice	98	No QTL specific for treatment			
Hattori et al. (2007)	Rice	94	Total internode elongation	5	3.4–6.2	0.17–0.36
Hattori et al. (2007)	Rice	94	Elongated internode Nr.	2	4.5–6.2	0.27
Hattori et al. (2007)	Rice	94	Lowest elongation internode	3	3.1–7.7	0.14–0.36
GuangHeng et al. (2006)	Rice	–	Relative damage	3	2.8–4.6	–
GuangHeng et al. (2006)	Rice	–	Plant height	3	2.8–3.9	–
GuangHeng et al. (2006)	Rice	–	Livability/survival	3	3.0–3.7	–
GuangHeng et al. (2006)	Rice	–	Lenghth of mesocotyl	4	2.3–3.8	–
GuangHeng et al. (2006)	Rice	–	Chlorophyll damage index	3	2.5–4.2	–
Vantoai et al. (2001)	Soybean	75	Seed yield	1	–	–
Vantoai et al. (2001)	Soybean	102	Seed yield	1	–	–
Vantoai et al. (2001)	Soybean	75	Growth	1	–	–
Vantoai et al. (2001)	Soybean	102	Growth	1	–	–
Githiri et al. (2006)	Soybean	60	Tolerance index	7	2.0–15.4	0.07–0.49
Cornelious et al. (2005)	Soybean	67	Tolerance	1	2	0.16
Cornelious et al. (2005)	Soybean	103	Tolerance	1	2.5	0.06
Finch-Savage et al. (2005)	Wild Mustard	95	Germination	1	2.8	0.13

Fig. 12.1 Example of the experimental procedure for QTL detection of waterlogging tolerance traits (Parelle et al. 2007). This photography shows 320 rooted cuttings of *Quercus robur* waterlogged for 4 weeks and phenotyped daily (photo Parelle)

QTL for epinasty of which the PEV varied from 2.8% to 11.6% depending on the observation date during permanent waterlogging. Furthermore, as discussed above, the estimation of PEV and allelic substitution effect of a QTL also depends on the number of genotypes and of vegetative copies within genotypes. This, together with the variations in environmental conditions, makes it difficult to compare QTL among different experiments. However, comparing the position of major QTL across experiments clarifies their significance across conditions and genotypes.

One major QTL detected in rice during total submergence is the *sub-1* locus (Xu and Mackill 1996), showing five to seven times higher LOD scores and at least two times higher PEV (Table 12.2) than QTL for any other adaptive trait despite the similar ($\pm 20\%$) number of genotypes involved in the different experiments, e.g.: Mano et al. (2007, 2008) and Mano and Omori (2008) for aerenchyma; Mano et al. (2005a, b) for adventitious roots; Mano et al. (2006) for leaf injury. Even when taking into account that the estimated LOD and PEV depend on the experiment, the *sub-1* locus stands out among all QTL detected for hypoxia tolerance. This might be due to the fact that it controls internode growth, which is a trait controlled probably by only few genes; whereas strategies of tolerance to waterlogging are more complex, probably relying on multiple traits, and thus depending on many, interacting genes. This would result in the detection of more QTL with lower PEV.

However, other than indicating effectively a stronger genetic determinism, the high LOD scores for *sub-1* might also be due to the experimental control of the stress intensity. The high water table present during total submergence experiments homogenizes the hypoxic stress among plants, thus reducing within-experiment environmental variability; whereas hypoxic stress during waterlogging experiments also depends on the homogeneity of the soil and the rooting density. As oxygen diffusion in water is slow, the actual oxygen deficiency in the soil depends on root and rhizospheric O_2 consumption, which creates a gradient from the soil surface to the root. This induces a large variability of the stress actually perceived among plants within an experiment, with evident effects on the phenotypes and on QTL detection. Indeed Parelle et al. (2007) detected QTL for dissolved oxygen content in water in the vicinity of the roots of oaks submitted to waterlogging, revealing a clear problem of stress control among plants, with genotypes influencing their environment.

12.4.1.2 QTL Detection for Constitutive Traits of Tolerance

The most recently published experiments on maize (Mano et al. 2007, 2008) detected QTL for aerenchyma formation, a key trait known to be highly related to hypoxia tolerance. This was achieved with an inter-specific cross between two species with different capacities of aerenchyma formation. They performed the analysis only under control conditions, which avoids stress heterogeneity among individuals (although not environmental variability). Aerenchyma formation is usually enhanced under stress, thereby showing genotype × environment interactions which could involve a different genetic determinism than under control conditions. This could change allelic effects and PEV of the detected QTL. Genotype × environment interaction could also be a cause for the small number of co-localised QTL between the two different inter-specific crosses in the same experimental set-up: only one out of seven detected QTL co-localized.

12.4.1.3 Comparison with a Control Environment

QTL detection for induced adaptive traits does not necessarily require a control treatment. Nevertheless, QTL detection for growth and productivity needs a comparison between stress and control. This is necessary to distinguish constitutive QTL that influence growth independently of the applied stress, and induced QTL that control growth specifically in response to stress. Qiu et al. (2007) and Githiri et al. (2006) compared the phenotypes expressed by the same genotypes growing in water-logged versus control conditions. Githiri et al. (2006) computed the ratio of seed production under waterlogging versus control, and used this ratio as a tolerance index. However, QTL for ratios between stress and control are difficult to interpret. This is mainly due to the fact that the condition that results in the highest among-genotype variability dominates the statistical analysis. Three situations might occur: (1) the variability is larger under stress, the QTL is then related to stress tolerance; (2) it is larger in controls (e.g. due to severe growth reduction under stress), and the QTL then describes the genetic variability of growth potential in the absence of stress, and not of tolerance; (3) the variability is similar in the two treatments and it is difficult to conclude whether the detected QTL is related to tolerance or not. Detecting QTL for each condition separately can provide support for the interpretation. This was for example used by Qiu et al. (2007), who detected QTL for the ratio of growth parameters between waterlogging and control as well as for each treatment separately and used the resulting co-localisations to interpret QTL as either nonadaptive, constitutive or induced. Other than using stressed/ control ratios as traits, both datasets can be used within the same statistical analysis, such as in multi-environment QTL detection models (Jansen et al. 1995) which allow a direct computation of QTL × environment interactions. This has been used for example by Jermstad et al. (2003) to detect QTL in a factorial experimental design using different winter chilling and spring flushing temperatures. Only one application of multi-environment QTL detection is known to the authors in the case of waterlogging. Parelle et al. (2007) recorded epinasty, root collar diameter and leaf chlorophyll content in a Q. *robur* full-sib progeny and showed that the allelic substitution effects of the detected QTL varied significantly during the 4 weeks of waterlogging. This method described QTL by the temporal pattern of the corresponding allelic substitution effect, and compared such patterns among different QTL and traits. Interestingly QTL with correlated effect patterns were dispersed over the whole genome, suggesting a polygenic determinism of tolerance to hypoxia.

12.4.2 Major Loci Detected for Hypoxia Tolerance

One of the main tools for breeding crops for agriculture in areas submitted to waterlogging or submergence is the detection of genomic regions or genes for marker aided selection (Vartapetian 2005). On the other hand, QTL studies also aim at elucidating the molecular mechanisms of hypoxia tolerance. However, only few QTL experiments were performed with the aim to detect loci for which the genetic diversity is effectively selected in natural populations. Table 12.1 summarizes the traits and species for which QTL related to hypoxia response were detected.

12.4.2.1 QTL for Traits Submitted to Natural Selection Pressure in Hypoxic Environments

Identifying candidate genes submitted to natural selection in water-logged or flooded soils would advance our understanding of speciation processes in such environments (Lexer et al. 2005). Martin et al. (2006) detected two QTL for survival in an Iris family. Their experiment was performed under water-logged

12 Genetic Variability and Determinism of Adaptation of Plants

conditions in a common garden for 4 years. Another experiment was performed by Parelle et al. (2007) who detected QTL for traits that are known to vary among natural population of two sympatric, hybridising oak species (*Q. robur* and *Q. petraea*). They detected two QTL for hypertrophied lenticel formation, five for the level of epinasty, but none for adventitious root development. Loci identified during these experiments could be a starting point in research strategies identifying candidate genes. Such genes could then be screened for genetic variability in natural populations.

12.4.2.2 QTL Detection for Breeding Purposes

Many QTL were detected for a large range of traits of interest for the maintenance of productivity and growth under waterlogging (see Table 12.2 for details). Shoot growth was the main indicator of productivity during hypoxia, and QTL were detected for shoot biomass by Qiu et al. (2007) or shoot height by GuangHeng et al. (2006) and Qiu et al. (2007). Vantoai et al. (2001) used shoot growth during stress as a tolerance index, considering that eliminating the growth before stress would detect induced rather than constitutive QTL. As the root system is directly affected by hypoxia, some authors use it as an indicator of tolerance, for example Qiu et al. (2007) detected QTL for root length under hypoxic condition, or Zheng et al. (2006) for total root biomass. QTL detected for these traits might be used for marker aided selection or to produce inbred lines to improve crop performance under hypoxia (Vantoai et al. 2001), without necessarily having detailed information on the tolerance strategy that is controlled by the QTL.

12.4.2.3 QTL Detection for Tolerance to Hypoxia

The most recent studies on QTL detection of hypoxia tolerance were performed on hypoxia-induced morphological traits: 11 QTL were identified for aerenchyma formation in maize (Mano et al. 2007, 2008) and two QTL for hypertrophied lenticel formation in oaks (Parelle et al. 2007). Six QTL for adventitious root development were detected in maize (Mano et al. 2005a, b), but none in *Q. robur* despite a visible development of such roots. This lack of genetic determinism for adventitious root development in oak might be due to heterogeneity of soil hypoxia, which dominated the phenotypic variance of this trait, and therefore diluted the genetic variance (Parelle et al. 2007).

To characterise the tolerance of rice to total submergence, internode elongation was used to quantify the capacity of quiescence or stress escape. In consequence, this trait allowed the detection of two types of loci related to those two strategies. This can be used in QTL detection studies when specific crosses are used, either combining two genotypes with different strategies or a tolerant and a non tolerant genotype. Indeed, the sign of the allelic substitution effect, combined with the knowledge of the strategy developed by the parents allowed to clearly attribute

Table 12.3 Traits for which a QTL related to an allele of the F13 A variety of *Oryza sativa* ssp *indica* was detected at the Sub1 Loci

Trait	Article	Methods of detection	LOD score	R^2
Tolerance scrore (visual scale)	Xu and Mackill 1996	Linear regression	36	0.69
Surviving or not	Nandi et al. 1997	Direct mapping of the qualitative trait	–	–
Per cent plant survival	Toojinda et al. 2003	Composite interval mapping	36.4–65.8	0.41–0.77
Relative shoot elongation compare to control			16.2	9.7
Tolerance score (visual scale)			26.6–38.3–52.6	0.38–0.61–0.63
Leaf senescence after submergence			29.3	0.53

For the LOD score and the R^2, values for repetitions of the QTL detection experiments are indicated

QTL to the two strategies. Several authors (Sripongpangkul et al. 2000; Toojinda et al. 2003; Nemoto et al. 2004; Tang et al. 2005; Hattori et al. 2007; Kawano et al. 2008) detected QTL related to the escape strategy of different rice varieties, and Xu and Mackill (1996) detected the *sub-1* locus related to the quiescence strategy by an inhibition of the internode elongation. In Table 12.3 all traits are listed for which QTL were detected at the *sub-1* locus. Sripongpangkul et al. (2000) and Hattori et al. (2007) performed multiple trait phenotyping of the early elongation ability, where the constitutive diversity included in each trait differed, whereas stress responses relating to the same tolerance mechanism would result in co-localisation of QTL. Both experiments resulted in the detection of the sub-1 locus, thus relating it clearly to low elongation ability (quiescence strategy). An important step for breeding was the successful introgression of the sub-1 locus into a rice variety of economic importance. Siangliw et al. (2003) crossed three tolerant varieties of rice, containing the *sub-1* allele conferring tolerance by quiescence strategy, with the hypoxia intolerant Thai jasmine rice. In the hybrid families, QTL were detected for hypoxia tolerance at the *sub-1* locus and the alleles related to quiescence were always from the tolerant parent. Introgression increased survival from 1.6% in Thai jasmine to 23%–31% in the hybrid families. This example shows how the detection of a QTL for tolerance can be used directly for breeding purposes.

The *sub-1* locus was the only QTL for which the underlying genes were clearly identified. It was reported for the first time by Xu and Mackill (1995, 1996), and further detected during all QTL detection experiments in rice, including the tolerant variety F13A, in which elongation is inhibited in order to decrease energy demand during hypoxia (quiescence strategy) (Nandi et al. 1997; Toojinda et al. 2003). Nandi et al. (1997) demonstrated that a cartography of the qualitative trait "surviving/not surviving to total submergence" was sufficient to detect the *sub-1* locus.

The *sub-1* locus was not only detected in F13A, but also in other tolerant varieties. For example Sripongpangkul et al. (2000) detected a QTL for elongation ability on the *sub-1* locus in a F8 RIL cross from two *indica* cultivars; the tolerant parent conferring the allele for a faster elongation ability. This locus therefore seems to be also involved in the escape strategy developed by some deep-water varieties of rice. However, the main QTL for the fast elongation capacity was detected on a different chromosome than *sub-1* (Tang et al. 2005; Hattori et al. 2007, 2008).

Candidate gene approaches were used to identify the gene(s) beneath the *sub-1* locus. Ruanjaichon et al. (2004) first mapped a small GTP-binding protein, belonging to a family known to be involved in signal transduction pathways and Kottapalli et al. (2006) identified 1,473 (putative) genes. However the large confidence interval for the position of the QTL for *sub-1* (6,4 cM for Kottapalli et al. 2006) did not allow the identification of the gene(s) responsible for submergence tolerance. It was finally the sequencing of the rice genome (International Rice Genome Sequencing Project 2005), the construction of a high resolution genetic map (Harushima et al. 1998; Xu et al. 2000) and the comparison of genetic and physical maps (Kamolsukyunyong et al. 2001; Chen et al. 2002) that allowed the identification of the *sub-1* cluster of genes (Xu et al. 2006). To date, the physical structure of this locus is well known (see Fukao et al. 2006 and Xu et al. 2006 for details). It contains 13 genes, including three ethylene response factors called Sub1-A, Sub1-B and Sub1-C. Sub1-A is present only in *O. sativa* ssp *indica*, including the tolerant variety F13A (Xu et al. 2006; Fukao et al. 2009). This gene originates probably from the duplication of the Sub1-B gene, as the two genes display a large sequence homology, and as the presence of the Sub1-A gene was correlated with variation of Sub1-B alleles (Fukao et al. 2009). Two alleles of Sub1-A, have been reported: Sub1-A-1 and Sub1-A-2. These alleles were correlated with variation of alleles of Sub1-C (Fukao et al. 2009). The tolerant variety F13A possesses the Sub1-A-1 allele and the corresponding alleles of the Sub1-C and Sub1-B. Recombinant crossing experiments among the three genes (Xu et al. 2006; Septiningsih et al. 2009) demonstrated that variation of the two alleles of the Sub1-A locus modify the submergence tolerance of rice, independently of the effect of the alleles present in Sub1-B and Sub1-C. This suggests that the QTL is controlled by the allelic effect of one single gene, of which only two states were detected: presence/absence of the Sub1-A-1 allele or presence/absence of the entire gene. Introgression of the Sub1-A-1 allele into other species than rice could improve productivity of crops under flooding. The Sub1-A-1 allele induces the quiescence strategy resulting from the inhibition of internode elongating (Toojinda et al. 2003). Actually the Sub1-A-1 allele inhibited the effects of the Sub1-C gene on elongation initiation in response to ethylene (Fukao et al. 2006). However, it is far from being clear whether the introgression of the Sub1-A-1 allele into other species actually confers a larger tolerance to total submergence, as interactions with other tolerance strategies need to be taken into account.

It is interesting to notice that the three *sub1* genes (A, B, and C) are ethylene response factors. Ethylene is known to be involved in a large number of hypoxia tolerance mechanisms, as for example the development of aerenchyma, adventitious roots and hypertrophied lenticels (Bailey-Serres and Chang 2005). To our knowledge no QTL experiment was performed directly for ethylene production or for other traits related to signalling of hypoxia stress.

12.5 Conclusions

Genetic diversity has been shown among populations or within mapping families, for indirect indicators of tolerance, such as growth and leaf level gas exchange, as well as for constitutive or induced adaptive traits. The genetic control was elucidated for only a small number of traits and even less gene candidates have actually been tested. At the enzyme level, some diversity was detected only for ADH. However, no signal transduction pathway has been put forward and related to the observed genetic differences. Further, the actual effect of genetic differences in ADH on survival and fitness in natural populations still lacks experimental support. The correlation between the observed genetic diversity of short term metabolic adjustments to hypoxia and long term morphological adaptations needs further investigation. The QTL detected for survival or traits known to vary among natural populations could be starting points for gene candidate approaches. Such gene candidates could then be screened for natural genetic variability, thereby generating knowledge on the adaptability of populations, especially with respect to environmental changes.

The major challenge for future QTL detection for traits conferring hypoxia tolerance is the definition of integrative traits (1) indicating different tolerance strategies and (2) well suited to high-throughput phenotyping required for quantitative genetic analyses. In addition to the *sub-1* locus, a large number of minor QTL have been detected. The combination of several favourable alleles will determine the tolerance of an individual. However to decompose these processes, future approaches should combine large scale QTL experiments using complex traits and detailed studies on selected genotypes to decompose overall tolerance into elementary components.

References

Ashraf M (2003) Relationships between leaf gas exchange characteristics and growth of differently adapted populations of blue panicgrass (*Panicum antidotale* retz.) under salinity or waterlogging. Plant Sci 165:69–75

Bailey-Serres J, Chang R (2005) Sensing and signalling in response to oxygen deprivation in plants and other organisms. Ann Bot 96:507–518

Bailey-Serres J, Voesenek L (2008) Flooding stress: acclimations and genetic diversity. Annu Rev Plant Biol 59:313–339

12 Genetic Variability and Determinism of Adaptation of Plants

Beaumont M, Nichols R (1996) Evaluating loci for use in the genetic analysis of population structure. Proc R Soc Lond B 263:1619–1626

Beavis W (1994) The power and deceit of QTL experiments: lessons from comparative QTL studies. In: Proceedings of the forty-ninth annual corn & sorghum industry research conference. pp 205–266

Brown AHD, Marshall DR, Munday J (1976) Adaptedness of variants at an alcohol dehydrogenase locus in *Bromus mollis*. Aust J Biol Sci 29:389–396

Brown A (1978) Isozymes, plant population genetic structure and genetic conservation. Theor Appl Genet 52:145–157

Chan J, Burton R (1992) Variation in alcohol-dehydrogenase activity and flood tolerance in white clover, *Trifolium repens*. Evolution 46:721–734

Chen M, Presting G, Barbazuk W, Goicoechea J, Blackmon B, Fang G, Kim H, Frisch D, Yu Y, Sun S, Higingbottom S, Phimphilai J, Phimphilai D, Thurmond S, Gaudette B, Li P, Liu J, Hatfield J, Main D, Farrar K, Henderson C, Barnett L, Costa R, Williams B, Walser S, Atkins M, Hall C, Budiman M, Tomkins J, Luo M, Bancroft I, Salse J, Regad F, Mohapatra T, Singh N, Tyagi A, Soderlund C, Dean R, Wing R (2002) An integrated physical and genetic map of the rice genome. Plant Cell 14:537–545

Clucas R, Ladiges P (1979) Variations in populations of *Eucalyptus ovata* Labill and the effects of waterlogging on seedling growth. Aust J Bot 27:301–315

Collaku A, Harrison S (2005) Heritability of waterlogging tolerance in wheat. Crop Sci 45:722–727

Colmer TD, Voesenek LACJ (2009) Flooding tolerance: suites of plant traits in variable environments. Funct Plant Biol 36:665–681

Cornelious B, Chen P, Chen Y, De Leon N, Shannon J, Wang D (2005) Identification of QTLs underlying waterlogging tolerance in soybean. Mol Breed 16:103–112

Cornelious B, Chen P, Hou A, Shi A, Shannon J (2006) Yield potential and waterlogging tolerance of selected near-isogenic lines and recombinant inbred lines from two southern soybean populations. J Crop Improv 16:97–111

Daugherty C, Musgrave M (1994) Characterization of populations of rapid-cycling *Brassica rapa* L selected for differential waterlogging tolerance. J Exp Bot 45:385–392

De Carvalho M, Da Silva D, Ruas P, Medri M, Ruas E, Ruas C (2008) Flooding tolerance and genetic diversity in populations of *Luehea divaricata*. Biol Plant 52:771–774

Drew M (1997) Oxygen deficiency and root metabolism: injury and acclimation under hypoxia and anoxia. Annu Rev Plant Physiol Plant Mol Biol 48:223–250

Dreyer E (1994) Compared sensitivity of seedlings from 3 woody species (*Quercus robur* L, *Quercus rubra* L and *Fagus silvatica* L) to waterlogging and associated root hypoxia: effects on water relations and photosynthesis. Ann For Sci 51:417–429

Eveno E, Collada C, Guevara M, Léger V, Soto A, Díaz L, Léger P, González-Martínez S, Cervera M, Plomion C, Garnier-Géré P (2008) Contrasting patterns of selection at *Pinus pinaster* Ait drought stress candidate genes as revealed by genetic differentiation analyses. Mol Biol Evol 25:417–437

Finch-Savage W, Come D, Lynn J, Corbineau F (2005) Sensitivity of *Brassica oleracea* seed germination to hypoxia: a QTL analysis. Plant Sci 169:753–759

Fukao T, Harris T, Bailey-Serres J (2009) Evolutionary analysis of the sub1 gene cluster that confers submergence tolerance to domesticated rice. Ann Bot 103:143–150

Fukao T, Kennedy R, Yamasue Y, Rumpho M (2003) Genetic and biochemical analysis of anaerobically-induced enzymes during seed germination of *Echinochloa crus-galli* varieties tolerant and intolerant of anoxia. J Exp Bot 54:1421–1429

Fukao T, Xu KN, Ronald PC, Bailey-Serres J (2006) A variable cluster of ethylene response factor-like genes regulates metabolic and developmental acclimation responses to submergence in rice. Plant Cell 18:2021–2034

Germain V, Ricard B, Raymond P, Saglio PH (1997) The role of sugars, hexokinase, and sucrose synthase in the determination of hypoxically induced tolerance to anoxia in tomato roots. Plant Physiol 114:167–175

Githiri S, Watanabe S, Harada K, Takahashi R (2006) QTL analysis of flooding tolerance in soybean at an early vegetative growth stage. Plant Breed 125:613–618

GuangHeng Z, DaLi Z, ShiKai H, Yan S, LaTie A, LongBiao G, Qian Q (2006) QTL analysis of traits concerned submergence tolerance at seedling stage in rice (*Oryza sativa* L). Acta Agron Sin 32:1280–1286

Hartl D, Clark A (1997) Principles of population genetics. Sinauer Associates, Sunderland

Harushima Y, Yano M, Shomura A, Sato M, Shimano T, Kuboki Y, Yamamoto T, Lin S, Antonio B, Parco A, Kajiya H, Huang N, Yamamoto K, Nagamura Y, Kurata N, Khush G, Sasaki T (1998) High-density rice genetic linkage map with 2275 markers using a single F2 population. Genetics 148:479–494

Hattori Y, Miura K, Asano K, Yamamoto E, Mori H, Kitano H, Matsuoka M, Ashikari M (2007) A major QTL confers rapid internode elongation in response to water rise in deepwater rice. Breed Sci 57:305–314

Hattori Y, Nagai K, Mori H, Kitano H, Matsuoka M, Ashikari M (2008) Mapping of three QTLs that regulate internode elongation in deepwater rice. Breed Sci 58:39–46

Heathcote C, Davies M, Etherington J (1987) Phenotypic flexibility of *Carex-flacca* Schreb – tolerance of soil flooding by populations from contrasting habitats. New Phytol 105:381–391

Herzog S, Krabel D (1999) Genetic structures of a flooded and a non-flooded oak (*Quercus robur*) population from the floodplains of the Rhein river. Ekológia (Bratisl) 18:160–163

Ikeda H, Kamoshita A, Manabe T (2007) Genetic analysis of rooting ability of transplanted rice (*Oryza sativa* L.) under different water conditions. J Exp Bot 58:309–318

International Rice Genome Sequencing Project (2005) The map-based sequence of the rice genome. Nature 436:793–800

Jackson M, Campbell D (1976) Waterlogging and petiole epinasty in tomato: the role of ethylene and low oxygen. New Phytol 76:21–29

Jackson M, Ishizawa K, Ito O (2009) Evolution and mechanisms of plant tolerance to flooding stress. Ann Bot 103:137–142

Jansen R, Van Oijen J, Stam P, Lister C, Dean C (1995) Genotype-by-environment interaction in genetic mapping of multiple quantitative trait loci. Theor Appl Genet 91:33–37

Jermstad K, Bassoni D, Jech K, Ritchie G, Wheeler N, Neale D (2003) Mapping of quantitative trait loci controlling adaptive traits in coastal Douglas fir. III. Quantitative trait loci-by-environment interactions. Genetics 165:1489–1506

Kamolsukyunyong W, Ruanjaichon V, Siangliw M, Kawasaki S, Sasaki T, Vanavichit A, Tragoonrung S (2001) Mapping of quantitative trait locus related to submergence tolerance in rice with aid of chromosome walking. DNA Res 8:163–171

Kato-Noguchi H, Morokuma M (2007) Ethanolic fermentation and anoxia tolerance in four rice cultivars. J Plant Physiol 164:168–173

Kato-Noguchi H, Ohashi C, Sasaki R (2003) Metabolic adaptation to flooding stress in upland and lowland rice seedlings. Acta Phys Plant 25:257–261

Kawano R, Doi K, Yasui H, Mochizuki T, Yoshimura A (2008) Mapping of QTLs for floating ability in rice. Breed Sci 58:47–53

Kawano N, Ito O, Sakagami J (2009) Morphological and physiological responses of rice seedlings to complete submergence (flash flooding). Ann Bot 103:161–169

Kolodynska A, Pigliucci M (2003) Multivariate responses to flooding in *Arabidopsis*: an experimental evolutionary investigation. Funct Ecol 17:131–140

Kottapalli KR, Sarla N, Kikuchi S (2006) In silico insight into two rice chromosomal regions associated with submergence tolerance and resistance to bacterial leaf blight and gall midge. Biotechnol Adv 24:561–589

Kozlowski T (1997) Responses of woody plants to flooding and salinity. Tree Phys Monogr 1:1–29

Ladiges P, Kelso A (1977) Comparative effects of waterlogging on 2 populations of *Eucalyptus viminalis* Labill and one population of *Eucalyptus ovata* Labill. Aust J Bot 25:159–169

Lévy G, Lefèvre Y, Becker M, Frochot H, Picard J, Wagner PA (1999) Excess water: effects on growth of oak trees. Rev For Fr 51:151–161

12 Genetic Variability and Determinism of Adaptation of Plants

Lexer C, Fay M, Joseph J, Nica M, Heinze B (2005) Barrier to gene flow between two ecologically divergent *Populus* species, *P. alba* (white poplar) and *P. tremula* (european aspen): the role of ecology and life history in gene introgression. Mol Ecol 14:1045–1057

Linhart Y, Baker I (1973) Intra-population differentiation of physiological response to flooding in a population of *Veronica peregrina* L. Nature 242:275–276

Loreti J, Oesterheld M (1996) Intraspecific variation in the resistance to flooding and drought in populations of *Paspalum dilatatum* from different topographic positions. Oecologia 108:279–284

Lynch M, Walsh B (1997) Genetics and analysis of quantitative traits. Sinauer Associates, Sunderland

Mano Y, Muraki M, Fujimori M, Takamizo T, Kindiger B (2005a) Identification of QTL controlling adventitious root formation during flooding conditions in teosinte (*Zea mays* ssp. *huehuetenangensis*) seedlings. Euphytica 142:33–42

Mano Y, Muraki M, Takamizo T (2006) Identification of QTL controlling flooding tolerance in reducing soil conditions in maize (*Zea mays* L) seedlings. Plant Prod Sci 9:176–181

Mano Y, Omori F (2008) Verification of QTL controlling root aerenchyma formation in a maize × teosinte "*Zea nicaraguensis*" advanced backcross population. Breed Sci 58:217–223

Mano Y, Omori F, Kindiger B, Takahashi H (2008) A linkage map of maize x teosinte *Zea luxurians* and identification of QTLs controlling root aerenchyma formation. Mol Breed 21:327–337

Mano Y, Omori F, Muraki M, Takamizo T (2005b) QTL mapping of adventitious root formation under flooding conditions in tropical maize (*Zea mays* l.) seedlings. Breed Sci 55:343–347

Mano Y, Omori F, Takamizo T, Kindiger B, Bird R, Loaisiga C, Takahashi H (2007) QTL mapping of root aerenchyma formation in seedlings of a maize x rare teosinte *Zea nicaraguensis* cross. Plant Soil 295:103–113

Marcar N, Crawford D, Saunders A, Matheson A, Arnold R (2002) Genetic variation among and within provenances and families of *Eucalyptus grandis* W Hill and *E. globulus* Labill. subsp *globulus* seedlings in response to salinity and waterlogging. For Ecol Manage 162:231–249

Marshall DR, Broué P, Pryor AJ (1973) Adaptive significance of alcohol dehydrogenase isozymes in maize. Nat New Biol 244:16–17

Martin N, Bouck A, Arnold M (2006) Detecting adaptive trait introgression between *Iris fulva* and *I. brevicaulis* in highly selective field conditions. Genetics 172:2481–2489

McManmon M, Crawford R (1971) A metabolic theory of flooding tolerance: the significance of enzyme distribution and behaviour. New Phytol 70:299–306

Mollard F, Striker G, Ploschuk E, Vega A, Insausti P (2008) Flooding tolerance of *Paspalum dilatatum* (Poaceae: Paniceae) from upland and lowland positions in a natural grassland. Flora 203:548–556

Nandi S, Subudhi PK, Senadhira D, Manigbas NL, SenMandi S, Huang N (1997) Mapping QTLs for submergence tolerance in rice by AFLP analysis and selective genotyping. Mol Gen Genet 255:1–8

Nemoto K, Ukai Y, Tang D, Kasai Y, Morita M (2004) Inheritance of early elongation ability in floating rice revealed by diallel and QTL analyses. Theor Appl Genet 109:42–47

Parelle J, Zapater M, Scotti-Saintagne C, Kremer A, Jolivet Y, Dreyer E, Brendel O (2007) Quantitative trait loci of tolerance to waterlogging in a european oak (*Quercus robur* L): physiological relevance and temporal effect patters. Plant Cell Environ 30:422–434

Parolin P (2009) Submerged in darkness: adaptations to prolonged submergence by woody species of the Amazonian floodplains. Ann Bot 103:359–376

Paterson A, Lander E, Hewitt J, Peterson S, Lincoln S, Tanksley S (1988) Resolution of quantitative traits into Mendelian factors by using a complete linkage map of restriction fragment length polymorphisms. Nature 335:721–726

Qiu FZ, Zheng YL, Zhang ZL, Xu SZ (2007) Mapping of QTL associated with waterlogging tolerance during the seedling stage in maize. Ann Bot 99:1067–1081

Ricard B, VanToai T, Chourey P, Saglio P (1998) Evidence for the critical role of sucrose synthase for anoxic tolerance of maize roots using a double mutant. Plant Physiol 116: 1323–1331

Ruanjaichon V, Sangsrakru D, Kamolsukyunyong W, Siangliw M, Toojinda T, Tragoonrung S, Vanavichit A (2004) Small GTP-binding protein gene is associated with QTL for submergence tolerance in rice. Russ J Plant Physiol 51:648–657

Ruedinger M, Glaeser J, Hebel I, Dounavi A (2008) Genetic structures of common ash (*Fraxinus excelsior*) populations in Germany at sites differing in water regimes. Can J For Res 38:1199–1210

Septiningsih E, Pamplona A, Sanchez D, Neeraja C, Vergara G, Heuer S, Ismail A, Mackill D (2009) Development of submergence-tolerant rice cultivars: the sub1 locus and beyond. Ann Bot 103:151–160

Setter T, Waters I (2003) Review of prospects for germplasm improvement for waterlogging tolerance in wheat, barley and oats. Plant Soil 253:1–34

Siangliw M, Toojinda T, Tragoonrung S, Vanavichit A (2003) Thai jasmine rice carrying QTL ch9 (subQTL) is submergence tolerant. Ann Bot 91:255–261

Silva S, Sereno M, Lemons eSilva C, de Oliveira A, Neto J (2007) Inheritance of tolerance to flooded soils in maize. Crop Breed Appl Biotechnol 7:165–172

Sripongpangkul K, Posa G, Senadhira D, Brar D, Huang N, Khush G, Li Z (2000) Genes/QTLs affecting flood tolerance in rice. Theor Appl Genet 101:1074–1081

Tang D, Kasai Y, Miyamoto N, Ukai Y, Nemoto K (2005) Comparison of QTLs for early elongation ability between two floating rice cultivars with a different phylogenetic origin. Breed Sci 55:1–5

Toojinda T, Siangliw M, Tragoonrung S, Vanavichit A (2003) Molecular genetics of submergence tolerance in rice: QTL analysis of key traits. Ann Bot 91:243–253

Torres AM, Diedenhoffen U, Johnstone IM (1977) The early allele of alcohol dehydrogenase in sunflower populations. J Hered 68:11–16

Tuberosa R, Salvi S (2004) Cereal genomics. In: Gupta P, Varshney R (eds) Springer, Netherlands, pp 253–315

Vantoai T, St Martin S, Chase K, Boru G, Schnipke V, Schmitthenner A, Lark K (2001) Identification of a QTL associated with tolerance of soybean to soil waterlogging. Crop Sci 41:1247–1252

Vartapetian B (2005) Plant anaerobic stress as a novel trend in ecological physiology, biochemistry, and molecular biology: 1. Establishment of a new scientific discipline. Russ J Plant Physiol 52:826–844

Vartapetian B (2006) Plant anaerobic stress as a novel trend in ecological physiology, biochemistry, and molecular biology. 2. Further development of the problem. Russ J Plant Physiol 53:711–738

Vartapetian B, Jackson M (1997) Plant adaptations to anaerobic stress. Ann Bot 79:3–20

Voesenek LACJ, Colmer TD, Pierik R, Millenaar FF, Peeters AJM (2006) How plants cope with complete submergence. New Phytol 170:213–226

Wagner P, Dreyer E (1997) Interactive effects of waterlogging and irradiance on the photosynthetic performance of seedlings from three oak species displaying different sensitivities (*Quercus robur*, *Q petraea* and *Q rubra*). Ann Sci For 54:409–429

Waldren S, Etherington J, Davies M (1988) Comparative studies of plant-growth and distribution in relation to waterlogging. 15. The effect of waterlogging on growth of various populations of and hybrids between geum-rivale 1 and geum-urbanum 1. New Phytol 109:97–106

Will R, Seiler J, Feret P, Aust W (1995) Effects of rhizosphere inundation on the growth and physiology of wet and dry-site *Acer rubrum* (red maple) populations. Am Midl Nat 134:127–139

Xu K, Mackill D (1995) RAPD and RFLP mapping of a submergence tolerance locus in rice. Rice Genet Newsl 12:244–245

Xu K, Xu X, Fukao T, Canlas P, Maghirang-Rodriguez R, Heuer S, Ismail A, Bailey-Serres J, Ronald P, Mackill D (2006) Sub1a is an ethylene-response-factor-like gene that confers submergence tolerance to rice. Nature 442:705–708

Xu K, Xu X, Ronald PC, Mackill DJ (2000) A high-resolution linkage map of the vicinity of the rice submergence tolerance locus sub1. Mol Gen Genet 263:681–689

Xu KN, Mackill DJ (1996) A major locus for submergence tolerance mapped on rice chromosome 9. Mol Breed 2:219–224

Yeboah M, Chen X, Liang G, Gu M, Xu C (2008) Inheritance of waterlogging tolerance in cucumber (*Cucumis sativus* L). Euphytica 162:145–154

Zaidi P, Maniselvan P, Sultana R, Yadav M, Singh R, Singh S, Dass S, Srinivasan G (2007) Importance of secondary traits in improvement of maize (*Zea mays* L) for enhancing tolerance to excessive soil moisture stress. Cereal Res Commun 35:1427–1435

Zaidi P, Rafique S, Singh N (2003) Response of maize (*Zea mays* L) genotypes to excess soil moisture stress: morpho-physiological effects and basis of tolerance. Eur J Agron 19:383–399

Zheng B, Yang L, Mao C, Zhang W, Wu P (2006) QTLs and candidate genes for rice root growth under flooding and upland conditions. Acta Genet Sin 33:141–151

Zheng B, Yang L, Zhang W, Mao C, Wu Y, Yi K, Liu F, Wu P (2003) Mapping QTLs and candidate genes for rice root traits under different water-supply conditions and comparative analysis across three populations. Theor Appl Genet 107:1505–1515

Zhou L, Wang J, Yi Q, Wang Y, Zhu Y, Zhang Z (2007a) Quantitative trait loci for seedling vigor in rice under field conditions. Field Crops Res 100:294–301

Zhou M, Li H, Mendham N (2007b) Combining ability of waterlogging tolerance in barley. Crop Sci 47:278–284

Chapter 13
Improvement of Plant Waterlogging Tolerance

Meixue Zhou

Abstract Sources of tolerance and a reliable trait evaluation method are crucial in breeding for abiotic stress tolerance. Waterlogging is one of the most important abiotic stresses in high rainfall areas. Waterlogging tolerances have been reported in different plant species. However, the complexity of the trait makes it very difficult to evaluate, thus hard to breed for. A reliable screening method can make the breeding programme more successful. This chapter will summarize: genetic resources and genetic behaviour of waterlogging tolerance; different selection criteria; and QTLs controlling the tolerance. The importance of accurate phenotyping in screening for QTLs controlling the tolerance is also discussed.

13.1 Introduction

Waterlogging is one of the most important constraint factors for crop production. Nearly 16% of the total territory of the United States suffers from waterlogging (Boyer 1982). In South-East Asia, 15% of all maize growing areas are affected by waterlogging, causing 25–30% of yield losses every year (Rathore et al. 1998). The yield losses in soybean due to waterlogging can be 17–43% if waterlogged at the vegetative growth stage or 50–56% if waterlogged at reproductive growth stages (Oosterhuis et al. 1990; Scott et al. 1990). In barley, Bandyopadhyay and Sen (1992) reported more than 50% loss in yield after 2 days and 80% loss in yield after 3 days of super-saturation treatment after 6 weeks normal growth in a coastal saline soil. Even for the relatively tolerant wheat (Wang et al. 1996; Ikeda et al. 1955), the average yield losses of 39–44% were also found under waterlogging conditions (Musgrave and Ding 1998; Collaku and Harrison 2002). The inhibition

M. Zhou
Tasmanian Institute of Agricultural Research, University of Tasmania, P.O. Box 46, Kings Meadows, TAS 7249, Australia
e-mail: mzhou@utas.edu.au

S. Mancuso and S. Shabala (eds.), *Waterlogging Signalling and Tolerance in Plants*, 267
DOI 10.1007/978-3-642-10305-6_13, © Springer-Verlag Berlin Heidelberg 2010

of nitrogen uptake and the consequent redistribution of nitrogen within the shoot are important contributory factors in the early senescence of leaves and the retarded growth of shoots in flooded plants (Drew and Sisworo 1977). A decrease in the nitrogen concentration in shoots of seedlings can occur rapidly after the onset of flooding and precede leaf chlorosis (Drew and Sisworo 1977; Wang et al. 1996) and consequently reduces shoot and root growth, dry matter accumulation and final yield (Kozlowski 1984; Drew 1991; Huang et al. 1994a, b; Malik et al. 2001). Roots are also injured by O_2 deficiency and metabolism changes during acclimation to low concentrations of O_2 (Drew 1997).

Waterlogging tolerance is defined in physiological studies as the survival or the maintenance of growth rates under waterlogging at different stages of development relative to nonwaterlogged conditions, whereas the agronomic definition of waterlogging tolerance is the maintenance of relatively high grain yields under waterlogging relative to nonwaterlogged conditions (Setter and Waters 2003). The agronomic definition based on grain yields alludes to the possibility that a waterlogging tolerant variety may possess a mechanism of tolerance associated with escaping from anaerobic conditions through dormancy or slow growth during a stress period, and have a rapid recovery following stress (Setter and Waters 2003). Therefore, evaluation of crop varieties should consider both the physiological performance during waterlogging and their recovery ability after waterlogging. Germplasm evaluation based on grain yield may be confounded because of the possibility that tolerance and recovery mechanisms only partly contributed to the grain yield after the waterlogging stress was terminated. This is especially the case in environments where waterlogging is for a short time, and other environmental factors or stress may also affect the grain yield. Sometimes other stress may even contribute more to the final grain yield than waterlogging stress, unless the waterlogging events are during or close to the grain-filling period.

The most economic way to reduce the damage caused by waterlogging is to introduce waterlogging tolerance into current varieties. To achieve this target, both sources of tolerance and a reliable trait evaluation method are crucial. The understanding of the genetic behaviour of waterlogging tolerance is also needed to make the selection more efficient. This chapter will briefly review genetic resources and genetic studies on waterlogging tolerance, selection criteria for waterlogging tolerance and QTLs controlling waterlogging tolerance related traits.

13.2   Genetic Resources of the Tolerance

Extensive screening of barley germplasm for waterlogging or wet tolerance has occurred in China and Japan. Work by Qiu and Ke (1991) involved screening 4,572 varieties in Shanghai province, China. Waterlogging was imposed at three stages (leaf 3 stage, stem elongation and heading) for 10–15 days each. Calculation of a "damage index" was based on yield of plants in waterlogging treatments expressed as a percentage of yield under nonwaterlogged conditions. Varieties were classified into five groups according to waterlogging induced damage: 0.4% of varieties had 1% damage; 5% had 1–10% damage; 30% had 10–20% damage; 32% had 20–40% damage; the remaining 33% had >40% damage. The majority of the 16 varieties identified with the highest waterlogging tolerance also had very early or medium maturity, indicating that recovery was not the mechanism of tolerance (Qiu and Ke 1991). Recently, a germplasm screening project was conducted in several Chinese universities/institutes to search barley germplasm for waterlogging tolerance (Zhou, Final report on the project of Australia China collaboration on barley germplasm research, unpublished data). According to the loss of yield per plot after waterlogging treatment, the varieties were classified into three groups: tolerant varieties (yield loss less than 25%); medium (yield loss between 25 and 75%) and susceptible varieties (yield loss more than 75%). Landrace barley had a higher percentage of tolerant varieties (64%) than bred varieties (11%). Six-row barley showed a higher percentage (50%) of waterlogging-tolerant varieties than two-row varieties (13%), which was partly because most of the landrace varieties were six-rowed. Naked barleys had a higher proportion of tolerant (42%) than the hulled (18%). Xiao et al. (2007) also found a Chinese landrace variety, Yongjiahong Liuleng, showed the least yield loss after 12 d waterlogging at tillering stage. The tolerance of some varieties was also confirmed in Australia under controlled waterlogging conditions (Fig. 13.1) (Pang et al. 2004; Zhou et al. 2007). Takeda and

Fig. 13.1 A Chinese variety showed much better waterlogging tolerance than an Australian variety (Zhou et al. 2007)

Fukuyama (1986) tested 3,457 cultivars (preserved at the Barley Germplasm Center, Okayama University) by submerging 50 sterilized grains of each in deionized water in a test tube for 4 days at 25°C and subsequently determining their germination percentage after 4 days on moistened filter paper at 25°C. The germination percentage ranged from 0 to 100. The collections from China, Japan and Korea contained many tolerant cultivars (average indices 71.6, 66.3, and 60.5, respectively) while those from North Africa, Ethiopia and southwest Asia showed few tolerant cultivars (19.6, 13.8, and 13.2, respectively). The most tolerant cultivars retained complete germinability after 8 days of soaking at 25°C. In a glasshouse experiment, 20 different barley varieties (lines) showed significant differences in waterlogging tolerance based on grain yield and yield components with some varieties including Weisubuzi, Su5078, Tong83-11, Tong88-58 being significantly better than others (Xu et al. 2005). Setter et al. (1999) demonstrated a genetic diversity of waterlogging tolerance in barley exposed to intermittent waterlogging over 4 weeks, and waterlogging tolerance was assessed using leaf chlorosis following waterlogging. According to their results, grain yield of barley was reduced by 51–84% of nonwaterlogged plants, but the order of yield reduction did not coincide with that of leaf chlorosis. Yang et al. (1999) compared the waterlogging tolerance of eight barley dwarf-mutants. The results showed that physiological and biochemical characters such as green leaf number of main stem, fresh weight of plant and activity of superoxide dismutase (SOD) in the flag leaf were greatly changed by waterlogging stress, which also resulted in a decrease in grain yield. The results of that experiment also showed that there is a significant difference of tolerance to waterlogging among the mutants, and 95-39, 95-31 and 95-53 were better than others in waterlogging tolerance.

Genetic differences exist for tolerance to waterlogging in wheat (Davies and Hillman 1988; Thomson et al. 1992; Ding and Musgrave 1995; Huang et al. 1994a, b; McKersie and Hunt 1987; Gardner and Flood 1993). For example, Huang et al. (1994b) showed that there is good genetic diversity for tolerance of wheat to hypoxic solution cultures. In a glasshouse experiment with 14 wheat varieties and several doubled haploid wheat lines, Setter et al. (1999) showed that there was good diversity for waterlogging tolerance based on shoot growth during continuous waterlogging for 4 weeks, and after waterlogging during 3 weeks recovery period following drainage. Davies and Hillman (1988) demonstrated variation in vegetative growth and yield under continuous flooding of 4-week-old plants of various wheat species, with the hexaploid *Triticum macha* and the tetraploid *T. dicoccum* being the most tolerant. Inter-variety differences in wheat seedling survival after 7 days flooding with cold treatments have also been reported by McKersie and Hunt (1987).

Genetic variation was also reported in many other plant species, including oats (Lemons e Silva et al. 2003), cucumber (Yeboah et al. 2008a, b), Soybean (VanToai et al. 1994; VanToai and Nurjani 1996; Sayama et al. 2009; Hou et al. 1995) and maize (Anjos e Silva et al. 2006; Mano et al. 2005c). Mano et al. (2002) screened 46 maize inbred lines collected from Japan and the United States for pregermination flooding tolerance (germination rates of the seeds soaked for 8 days at 25°C) and found wide variations among the lines. The waterlogging tolerance of the other 223 inbred lines also showed wide variation for the tolerance at the seedling stage. Hou et al. (1995) tested 730 soybean varieties from different sources for seed germination after a 4 day soaking at 25°C. A large variation in seed flooding tolerance existed in the soybean germplasm as reflected in the germination rate which ranged from 0 to 100%. Most of the varieties tested were sensitive to seed flooding, and only 4% of the tested varieties had high tolerance (germination rate >90%).

13.3 Selection Criteria

Waterlogging tolerance is likely to be a complex trait affected by several mechanisms and complicated by confounding factors such as temperature, plant development stage, nutrient availability, soil type and sub-soil topography. Direct selection on grain yield has low effectiveness since the heritability of yield after waterlogging has been reported to be very low (Collaku and Harrison 2005). While the ability to produce high seed yield in flooded fields is the ultimate criterion of flooding tolerance, other traits, including leaf color, plant height, root, and shoot biomass, have been used frequently as determinants of flooding tolerance. Burgos et al. (2001) found that the lines derived from a cross between wheat spelt, which survived flooding well, germinated early, emerged fast, preserved their membrane integrity, and that the biggest seed can suffer more from flooding.

In soybean, Sayama et al. (2009) found that pigmented seed coat and small seed weight tended to give a positive effect on seed-flooding tolerance. Githiri et al. (2006) used relative seed weight and 100-grain weight as indications of early vegetative growth stage waterlogging tolerance for soybean in a pot experiment in a vinyl plastic greenhouse. They found that one of the major QTLs for the tolerance was at the similar position to a large QTL for days to the time of flowering, with that the late maturity may have conferred a longer growth period for recovery from flooding stress. A comparison of cross section area of the hilum revealed that the tolerant cultivars tended to have a larger area than the susceptible cultivars, suggesting that an inner space of the hilum can act as a reservoir at the initial stage of inundation, thereby reducing water absorption speed in tolerant cultivars (Muramatsu et al. 2008). Injury ratings (0 being no damage and 9 being 90% or more of the plants dead, Fig. 13.2) after waterlogging were used by Cornelious et al. (2005).

Different traits have been used as indirect selection indices for waterlogging tolerance in maize. Mano et al. (2005a, b, 2006) used adventitious root formation (Fig. 13.3), leaf injury and dry matter production as indications of waterlogging tolerance of maize. Adventitious root formation was also suggested to provide an alternative for some teosinte to address soil flooding or waterlogging (Bird 2000). Mano et al. (2002) studied pre-germination flooding tolerance and waterlogging tolerance at the seedling stage of maize and found no correlations between them,

Fig. 13.2 Visual ratings for waterlogging injury of RIL mapping populations of soybean in the field. Injury ratings range from 0 to 9 with 0 being no damage and 9 being 90% or more of the plants dead (Cornelious et al. 2005)

Fig. 13.3 Two-week-old flooded maize inbred lines. *Left*: tolerant with adventitious roots; *right*: susceptible without adventitious roots. Shoots were removed from the plants (Mano et al. 2005b)

i.e. pre-germination flooding tolerance was independent of waterlogging tolerance at the seedling stage.

Among all the different criteria, leaf chlorosis after waterlogging has been one of the major indices used by researchers in different crops such as wheat (Boru et al. 2001; Cai et al. 1996; Cao et al. 1995; Ikeda et al. 1954), soybean (Reyna et al. 2003) and barley (Hamachi et al. 1990). van Ginkel et al. (1992) demonstrated that there is a high negative correlation between leaf chlorosis (or death) and grain yield in wheat. Dead leaf percentage under excess soil moisture was thought to be the best criterion for selection for flooding tolerance in early generations because its heritability values are relatively constant and it is easy to measure (Hamachi et al. 1990) and was correlated with reduction of grain yield/plant and culm length (Hamachi et al. 1989). Wang et al. (2007), using principal components analysis, identified three principal components, two for spike-grain factor and one for the number of green leaves that can be used as selecting indices for waterlogging tolerance. Other indices, such as plant survival and reduction in dry matter accumulation, were also used in barley (Li et al. 2008).

Close relationships have been reported between some physiological traits and waterlogging tolerance. Waterlogging tolerant varieties showed better ability to develop more adventitious roots and larger percentage of aerenchyma (Pang et al. 2004), to uptake K^+ in root mature zone in WL conditions, to maintain larger O_2 uptake in root mature zone in WL conditions (Pang et al. 2006) and to tolerate secondary metabolites associated with WL soil conditions (Pang et al. 2007). These physiological traits cannot be easily used in routine screening program but they are useful criteria for further waterlogging related QTL identification which leads to marker assisted selection.

13.4 Genetic Studies on Waterlogging Tolerance

Waterlogging tolerance is considered to be a quantitative trait, even though some reports have found that the tolerance has been found to be controlled by one dominant gene in common wheat (Cao et al. 1995), Makha wheat (Fang et al. 1997) and maize (Sachs 1993). Most of the early published research in genetic studies on waterlogging tolerance was done in wheat, and almost all of this work measured waterlogging tolerance using leaf chlorosis or leaf/plant death and some other traits (Cao et al. 1992, 1994, 1995; Cai et al. 1996). These researchers indicated that waterlogging tolerance is under genetic control, and is heritable, with a broad sense heritability estimated to be over 70%. They concluded that it is possible to improve waterlogging tolerance in wheat by selecting progeny in early generations based on related traits. Cao et al. (1992, 1995) found that waterlogging tolerance based on leaf chlorosis was controlled by one dominant gene, but tolerance based on traits such as green leaves/main stem, plant height, grains per ear and 1000-grain weight could be controlled by multiple genes in the varieties involved in their study (Cao et al. 1994).

Boru (1996) extended the research of van Ginkel et al. (1992)'s work at CIM-MYT by continuing genetic studies involving several of the tolerant wheat varieties. In three waterlogging tolerant wheat genotypes, tolerance was conditioned by four major genes. The three tolerant wheat genotypes used in his study carried different genes, although they all possessed one tolerant gene (*Wt1*) in common. It was proposed that these different genes could control different mechanisms of tolerance to waterlogging, therefore waterlogging tolerance could be substantially improved by combining all tolerance genes into one genotype (Boru 1996). This may not be so where genes are related to the presence of different strategies of growth versus nongrowth during waterlogging. Some of the work in China (Cao et al. 1994) also indicated that additive gene action is the major determinant of the inheritance of waterlogging tolerance. Boru et al. (2001) further studied the inheritance of water-logging tolerance in wheat by using three tolerant (Prl/Sara, Ducula and Vee/Myna) and two sensitive (Seri-82 and Kite/Glen) spring bread-wheat lines. Leaf chlorosis was used as a measure of waterlogging tolerance. The sensitive by sensitive cross, seri-82 Kite/glen, showed the highest mean values for percentage leaf chlorosis and area under chlorosis progress curve, and the lowest mean values for plant height, biomass, grain yield, and kernel weight. The expression of waterlogging tolerance was not influenced by a maternal effect. The F1 hybrids were intermediate for leaf chlorosis, indicating that tolerance was additive. Quantitative analysis also indicated that additive gene effects mainly controlled waterlogging tolerance in these crosses.

The only early work to evaluate the heritability of waterlogging tolerance based on plant grain yield was conducted by Bao (1997) using 20 wheat varieties. He found that heritability for tolerance to 15 days waterlogging in the field at the tillering stage and the booting stage was 74.7 and 80.2%, respectively. However, Collaku and Harrison (2005) found that grain yield had the lowest heritability ($h^2 = 0.25$) while relatively higher heritabilities were found for kernel weight (0.47), chlorophyll content (0.37) and tiller number (0.31). They suggested that selecting waterlogging tolerance in early generations using relatively highly inheritable traits, such as kernel weight, would be an efficient way as grain yield has a low heritability.

Both additive and nonadditive effects were important in the determination of the inheritance of flooding tolerance in maize (Anjos e Silva et al. 2006). A maize F_2 population developed from a waterlogging tolerant variety and a sensitive variety showed transgressive segregation in both directions for most traits under waterlogging conditions, indicating that both parents transmitted favourable alleles for each trait. Broad-sense heritabilities were from 0.28 for root length to 0.82 for total dry weight under waterlogging conditions. Root length was more easily affected by waterlogging stress (Qiu et al. 2007). Hou et al. (1995) found that seed flooding tolerance in soybean was controlled by both additive and dominant genes. A small number of effective factors and high narrow sense heritability in the diallel analysis indicated that selection for tolerance would be effective in early generations. Wang et al. (2008) reported three major genes with the heritability of 0.42 dominating submergence tolerance of soybean. A simple additive model explained the variations of tolerance score, adventitious root formation and water-logged root dry weight in cucumber. Nonallelic interactions were detected for waterlogged vine length and root length. Complementary epistasis occurred in waterlogged vine length while additive × additive, additive × dominance and dominance × dominance epistastic effects were significant for waterlogged root length. Transgressive segregation was also observed in most of the traits in the F_2

generation. The estimates of narrow-sense heritabilities for tolerance score and adventitious root formation were moderately high ($h_N^2 = 0.54$–0.74) (Yeboah et al. 2008a). In another experiment, they found the broad sense heritability was from 0.43 for adventitious root formation to 0.88 for vine length (Yeboah et al., 2008b).

Hamachi et al. (1989) reported that heterosis for tolerance expressed as reduction in damage was observed in F_1s, and frequency distributions of damage in F_2s showed continuous variation. A 6×6 half diallel analysis was conducted in barley from crosses of three waterlogging tolerant Chinese cultivars and three susceptible Australian or Japanese cultivars (Zhou et al. 2007). The waterlogging treatment was imposed starting from the 3-leaf stage. The percentage of yellow leaf was recorded after waterlogging treatment. Three Chinese cultivars showed significantly higher general combining ability for waterlogging tolerance while the variance of specific combining ability was not significant, indicating that the tolerance was mainly controlled by additive effects. High heritability ($h_B^2 = h_N^2 = 0.73$) of waterlogging tolerance indicated that selection in early generations could be very efficient. They concluded that when selections are made in a segregating populatin, the most effective selection strategy is to discard the plants with severe leaf chlorosis (Zhou et al. 2007).

13.5 Marker-Assisted Selection

Even though the heritability was relatively high for leaf chlorosis (Zhou et al. 2007) and early generation selection could be efficient, well-controlled waterlogging conditions are still crucial for the precise evaluation of this trait. In practice, it is very difficult for breeders to control the multiple confounding environmental factors in a field experiment over thousands of barley genotypes. Development of molecular markers associated with barley waterlogging tolerance and marker-assisted selection (MAS) could effectively avoid environmental effects. QTL analysis has proven to be very useful in identifying the genetic components of the variation for important economic traits (Mazur and Tingey 1995). A molecular marker closely linked to the target gene or QTL can act as a "tag" which can be used for indirect selection of the gene(s) in a breeding program (Babu et al. 2004).

13.5.1 QTL Controlling Waterlogging Tolerance

Earlier efforts involved trying to locate the waterlogging tolerance genes onto different chromosomes. Poysa (1984) used substitution lines to study the flooding tolerance of wheat and found that all three substitution lines survived severe flooding stress (7 days) and showed significantly better tolerance than Chinese Spring, but under moderate flooding stress (5 days) only substitution line 5D was better than Chinese Spring. The author suggested that genes controlling resistance to flooding stress are present on all three chromosomes. Taeb et al. (1993) reported that the related species *Thinopyrum elongatum* and *Elytrigia repens* had better waterlogging tolerance than wheat when comparing a number of *Triticeae* species for tiller production, shoot dry matter production and root penetration in water-logged soil. Tests of a number of wheat-alien amphiploids showed that there was at least partial expression of this exotic genetic variation in a wheat genetic background. The presence of chromosome 2E and 4E of *Th. Elongatum* was associated with a positive effect on root growth in waterlogged conditions. The positive effect of the 4E chromosome addition was mimicked by tetrasomic lines carrying extra doses of wheat homoeologous 4B and 4D, and it was concluded that the beneficial effect contributed by the presence of 4E was probably due to an increased dosage of group 4 chromosomes. However, the positive effect of adding chromosome 2E to wheat could not be reproduced by added doses of chromosomes 2A, 2B, or 2D, suggesting that this alien chromosome carries genes for tolerance not present on its wheat homoeologues. This gene was further located to the long arm of chromosome 2E by testing ditelosomic addition lines (Taeb et al. 1993).

Great effort has been put in identifying quantitative trait loci (QTL) controlling the tolerance. In most instances, waterlogging tolerance related traits were used as indications of the tolerance. In a cross between spelt and wheat (relatively suscep-tible to waterlogging), five QTLs explaining 41% of the phenotypic variance were found for survival to flooding, which were localized on the chromosomes 2B, 3B, 5A and 7S. Ten QTLs were found for seedling growth index after flooding and they were localized on 2A, 2B, 2D, 3A, 4B, 5A, 5B, 6A and 7S (Burgos et al. 2001).

Six QTLs were found for early vegetative growth stage waterlogging tolerance in soybean (Githiri et al. 2006) in 2 years greenhouse experiments. The major QTL was found in both years and accounted for 30–49% of the variation. Some minor QTLs were also identified but most of them were different from different experi-ments. Three QTLs on linkage groups A1, D1a and G were reported by Wang et al. (2008), explaining 4.4–7.6% of the total phenotypic variation. Sayama et al. (2009) identified four QTLs for germination rate and normal seedling rate of soybean under waterlogging conditions. Among them, Sft1 on the linkage group H exhibited a large effect on germination rate after a 24-h treatment; Sft2 exhibited the largest effect on seed-flooding tolerance which was involved in seed coat pigmentation. Sft1, Sft3 and Sft4 were independent of seed coat colour and seed weight. From two different populations of soybean, VanToai et al. (2001) identified a single QTL from the tolerant parent which was associated with improved plant growth and grain yield. This QTL was not associated with maturity, normal plant height and grain yield. Lines with the tolerant allele showed 95% higher yielding and 16% taller on average. The QTL was also validated in another population later (Reyna et al. 2003). Further studies were conducted using two populations with 103 and 67 recombinant inbred lines to investigate QTLs controlling waterlogging tolerance in soybean (Cornelious et al. 2005). These two populations contained a common tolerant variety. In each population, one significant QTL was found, explaining

10 and 16% of the phenotypic variation, respectively. The QTL was at different position in two populations but the alleles provide the waterlogging tolerance in both populations were from the tolerant variety.

Adventitious root formation under waterlogging conditions was used to study the waterlogging tolerance of maize and teosinte. QTLs associated with waterlogging tolerance (adventitious root formation) were found on chromosome 4 and 8 from a cross between maize and teosinte with teosinte contributing all the tolerance. In an F_2 population of maize, QTLs controlling adventitious root formation on the soil surface under flooding conditions were identified on chromosomes 3, 7 and 8. The one on chromosome 8 was also identified from a different cross (Mano et al. 2005a). Later, Mano et al. (2006) identified a QTL for flooding tolerance in maize evaluated by either leaf injury or dry weight production after flooding treatment. However, this QTL only explained 10% of phenotypic variation of dry weight production and 14% of leaf injury. A major QTL controlling trait associated with relative shoot dry weight and relative total dry weight of maize was mapped to the same region of chromosome 9 which could be consistently identified in different experiments (Qiu et al. 2007). This QTL was located near a known anaerobic response gene. They also identified many other minor QTLs on chromosomes 1, 2, 3, 4, 6, 7 and 10, explaining 3.9–14.3% of the variation. These minor QTLs were specific to particular traits or environments (Qiu et al. 2007). The major QTL on chromosome 9 for relative shoot dry weight was not found in Anjos e Silva et al. (2005)'s study. Instead, by using a single marker analysis they detected three markers for shoot dry matter under waterlogging conditions. These three markers were located on chromosomes 3, 4 and 5. The markers on chromosomes 3 and 4 were also associated to root dry matter.

In cucumber, Yeboah et al. (2008b) used a set of 112 $F_{2:3}$ lines derived from the cross between two inbred lines PW0832 (tolerant) to PW0801 (susceptible) to evaluate waterlogging tolerance traits: tolerance score, adventitious root formation (Fig. 13.4), waterlogged shoot dry weight and waterlogged vine length. A total of 14 QTLs were detected for the different waterlogging traits. The QTL for the waterlogged traits accounted for 7.9–33.2% of the phenotypic variations.

There are few reports on waterlogging tolerance related QTL research in barley. Li et al. (2008) selected two double haploid populations (crosses between tolerant and susceptible varieties) to investigate the QTLs for waterlogging tolerance. Leaf chlorosis was chosen as the main indicator for waterlogging tolerance, and

Fig. 13.4 Variations among flooded F_2 plants of cucumber in adventitious root formation (ARF). Numbers 0, 1, 2 and 3 are the score ratings (Yeboah et al. 2008a, b)

plant biomass reduction and plant survival were also recorded. Twenty QTLs for waterlogging tolerance related traits were found in the two barley double haploid (DH) populations. Several of these QTLs were validated through replication of experiments across seasons or by co-location across populations. Some of these QTLs affected multiple waterlogging tolerance related traits. A consensus map (Wenzl et al. 2006) was used to compare QTLs from two different populations and summarized seven QTLs for waterlogging tolerance. These seven QTLs were located on all the different chromosomes except 6H. Among them, the QTL on 4H (Q_{wt}4-1) contributed not only to reducing barley leaf chlorosis, but also increasing plant biomass under waterlogging stress, whereas other QTLs controlled both leaf chlorosis and plant survival.

13.5.2 Accurate Phenotyping is Crucial in Identifying QTLs for Waterlogging Tolerance

Breeding for stress tolerance such as waterlogging tolerance controlled by multiple genes is difficult because of low heritability, variability among stress treatments, and the difficulty of screening a large number of lines in the field or under controlled conditions. These factors make it difficult for breeders to manipulate quantitative traits. Marker assisted selection could be very effective. Molecular markers give unambiguous, single sit genetic differences that can easily be scored and mapped in most segregating populations (Kearsey 1998). However, QTL analysis depends on the fact that the linkage between markers and QTL is such that the marker locus and the QTL will not segregate independently, and differences in the marker genotypes will be associated with different trait phenotypes (Kearsey 1998). The success of MAS depends on the development of reliable markers (accurate QTL location). Accurate phenotyping is imperative to the success of the QTL "genetic dissection" approach.

Genotyping and phenotyping errors are the two major reasons that reduce the accuracy of QTL results. As the development of new techniques, for example DArT technology (Wenzl et al. 2004), and the construction of consensus maps (Varshney et al. 2007; Wenzl et al. 2006), genotyping in barley have become more and more accurate. This leaves accurate phenotyping the major barrier to accurately locate QTLs controlling quantitative traits which are easily affected by environment. To increase phenotyping accuracy, we need to use highly reliable screening systems which are known to differentiate resistant from susceptible lines; to conduct analysis on the means of repeated screens rather than single trials and to ensure that repeatability of the screen is as high as possible. This section will use barley as an example to discuss the importance of accurate phenotyping in QTL analysis.

A same DH population from the cross between Yerong (waterlogging tolerant) and Franklin (waterlogging sensitive) was screened for waterlogging tolerance in glasshouse pot experiments (Li et al. 2008) and big tank experiments (Fig. 13.5) outside, during the normal barley growing season. In pot experiments, leaf

Fig. 13.5 Pot experiments (*left*) and tank experiments used to screen barley DH population for waterlogging tolerance. Much greater differences between tolerant varieties and susceptible varieties were shown in tank experiments

chlorosis, biomass reduction and plant survival were scored after waterlogging treatments. In tank experiments, one combined score system (plant healthiness which is a combined score of leaf chlorosis, plant survival after waterlogging, 0 = no affected and 10 = all died, Fig. 13.5) was used.

Pot experiments revealed six QTLs controlling waterlogging tolerance related traits. These QTLs explained 5–22% of the phenotypic variation with the biggest contribution from the QTL on 4H (Li et al. 2008). In contrast, only four significant QTLs were identified in the tank experiments. These three QTLs explained a total of more than 45% of the phenotypic variation (Fig. 13.6). In both pot and tank experiments, the QTL on 4H explained the greatest phenotypic variation. Figure 13.7 shows the effectiveness of using closely linked markers to select for waterlogging tolerance. As can be seen from the Figure, when selecting only one single major marker on 4H, the average score of the lines with this marker was 3.5 (more tolerant) which was significantly lower than the average score of 5.4 (more susceptible) from the lines without this marker. When all three markers can be selected, the score of all the lines showed tolerance or medium tolerance to waterlogging with the average score of 2.9. In contrast, when none of the three markers was selected, most of the lines were very susceptible with the average score of 7.4.

As mentioned above, the evaluation of waterlogging tolerance can be affected by many environmental factors, which include soil properties, temperature, water level, time of waterlogging treatment and barley development stage when waterlogging treatment starts. The difference between a pot experiment and a tank experiment is that the tank environment is closer to actual field conditions and the environment can be better controlled while variation among pots was unavoidable. Relatively longer times of waterlogging treatment resulted in greater differences between tolerant (quite healthy), medium tolerant (survived but not as healthy as the tolerant ones) and susceptible (dead) ones (Fig. 13.5). It is not surprising that QTLs identified from the tank experiment would be more reliable than those from pot experiment.

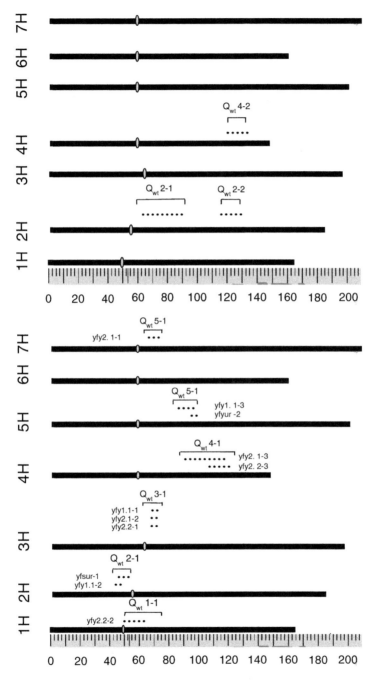

Fig. 13.6 The Franklin/Yerong chromosomes showing the locations of QTLs for waterlogging related trait(s) analyzed in pot experiments (Li et al. 2008) (*left*) and big tank experiments (*right*). A general name (such as $Q_{wt}1$-1) was given to each chromosome region associated with waterlogging tolerance, the first number was the chromosome number and the second number was the serial number of regions identified on that chromosome. yfy = leaf chlorosis; yfsur = survival rate

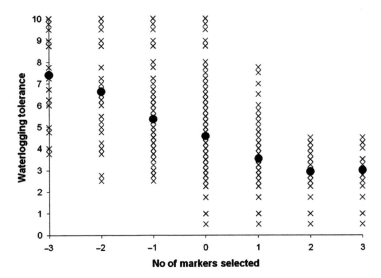

Fig. 13.7 The effectiveness of selecting molecular marker(s) to improve waterlogging tolerance as evaluated in the tank experiments. Waterlogging damage: 0 = very tolerant; 10 = very susceptible. No of markers selected: 3 and −3: three most important markers were selected and unselected, respectively; 2 and −2: three most important markers were selected and unselected, respectively; 1 and −1: only one most important marker was selected and unselected, respectively; 0: no marker was selected. *times*: original data; *filled circle*: average value

In conclusion, waterlogging tolerance exists in different plant species and is genetically inherited which is most likely controlled by several QTLs. Since the heritability of waterlogging tolerance was low and the evaluation of the tolerance can be very hard and easily affected by environmental conditions, the use of MAS could be very effective in selecting tolerance related traits. However, the effectiveness of MAS relies on the accurate location of the QTLs and closely linked markers. Among all the factors that affect the accuracy of QTL locations, phenotyping is the most important one, which needs very reliable screening facilities and selection indices.

References

Anjos e Silva S D dos, Sereno MJC de M, Lemons e Silva CF, Oliveira AC de, Barbosa Neto JF (2005) Genetic parameters and QTL for tolerance to flooded soils in maize. Crop Breed Appl Biotechnol 5:287–293

Anjos e Silva SD dos, Sereno MJC de M, Lemons e Silva CF, Barbosa Neto JF (2006) Combining ability of maize genotypes for flooding tolerance. Ciencia Rural 36:391–396

Babu R, Nair SK, Prasanna BM, Gupta HS (2004) Integrating marker-assisted selection in crop breeding – prospects and challenges. Curr Sci 87:607–619

Bandyopadhyay BK, Sen HS (1992) Effect of excess soil water conditions for a short period on growth and nutrition of crops on coastal saline soil. J Indian Soc Soil Sci 40:823–827

Bao XM (1997) Study on identification stage and index of waterlogging tolerance in various wheat genotypes (*Triticum aestivum* L.). Acta Agric Shanghai 13:32–38

Bird RMcK (2000) A remarkable new teosinte from Nicaragua: growth and treatment of progeny. Maize Gen Coop Newsl 74:58–59

Boru G (1996) Expression and inheritance of tolerance to waterlogging stresses in wheat (*Triticum aestivum* L.). PhD thesis, Oregon State University

Boru G, van Ginkel M, Kronstad WE, Boersma L (2001) Expression and inheritance of tolerance to waterlogging stress in wheat. Euphytica 117:91–98

Boyer JS (1982) Plant productivity and environment. Science 218:443–448

Burgos MS, Messmer MM, Stamp P, Schmid JE (2001) Flooding tolerance of spelt (*Triticum spelta* L.) compared to wheat (*Triticum aestivum* L.) – a physiological and genetic approach. Euphytica 122:287–295

Cai SB, Cao Y, Fang XW (1996) Studies on the variability and combining ability of waterlogging tolerance in common wheat. Jiangsu J Agric Sci 12:1–5

Cao Y, Cai SB, Wu ZS, Zhu W, Fang XW, Xiong EH (1995) Studies on genetic features of waterlogging tolerance in wheat. Jiangsu J Agric Sci 11:11–15

Cao Y, Cai SB, Zhu W, Fang XW (1992) Genetic evaluation of waterlogging resistance in the wheat variety Nonglin 46. Crop Genet Res 4:31–32

Cao Y, Cai SB, Zhu W, Xiong EH, Fang XW (1994) Combining ability analysis of waterlogging tolerance and main agronomic traits in common wheat. Scientia Agric Sinica 27:50–55

Collaku A, Harrison SA (2002) Losses in wheat due to waterlogging. Crop Sci 42:444–450

Collaku A, Harrison SA (2005) Heritability of waterlogging tolerance in wheat. Crop Sci 45:722–727

Cornelious B, Chen P, Chen Y, de Leon N, Shannon JG, Wang D (2005) Identification of QTLs underlying water-logging tolerance in soybean. Mol Breed 16:103–112

Davies MS, Hillman GC (1988) Effects of soil flooding on growth and grain yield of populations of tetraploid and hexaploid species of wheat. Ann Bot 62:597–604

Ding N, Musgrave ME (1995) Relationship between mineral coating on roots and yield performance of wheat under waterlogging stress. J Exp Bot 46:939–945

Drew MC (1991) Oxygen deficiency in the root environment and plant mineral nutrition. In: Jackson MB et al (eds) Plant life under oxygen deprivation. Academic, The Hague, pp 301–316

Drew MC (1997) Oxygen deficiency and root metabolism: Injury and acclimation under hypoxia and anoxia. Annu Rev Plant Physiol Plant Mol Biol 48:223–250

Drew MC, Sisworo EJ (1977) Early effects of flooding on nitrogen deficiency and leaf chlorosis in barley. New Phytol 79:567–571

Fang XW, Cao Y, Cai SB, Xiong EH, Zhu W (1997) Genetic evaluation of waterlogging tolerance in *Triticum macha*. Jiangsu J Agric Sci 13:73–75

Gardner WK, Flood RG (1993) Less waterlogging damage with long season wheats. Cereal Res Comm 21:337–343

Githiri SM, Watanabe S, Harada K, Takahashi R (2006) QTL analysis of flooding tolerance in soybean at an early vegetative growth stage. Plant Breed 125:613–618

Hamachi Y, Furusho M, Yoshida T (1989) Heritability of wet endurance in malting barley. Jpn J Breed 39:195–202

Hamachi Y, Yoshino M, Furusho M, Yoshida T (1990) Index of screening for wet endurance in malting barley. Jpn J Breed 40:361–366

Hou FF, Thseng FS, Wu ST, Takeda K (1995) Varietal differences and diallel analysis of pre-germination flooding tolerance in soybean seed. Bull Res Inst Bioresour (Okayama University) 3:35–41

Huang BR, Johnson JW, Nesmith S, Bridges DC (1994a) Growth, physiological and anatomical responses of two wheat genotypes to waterlogging and nutrient supply. J Exp Bot 45:193–202

Huang BR, Johnson JW, Nesmith S, Bridges DC (1994b) Root and shoot growth of wheat genotypes in response to hypoxia and subsequent resumption of aeration. Crop Sci 34:1538–1544

Ikeda T, Higashi S, Kawaide T (1955) Studies on the wet-injury resistance of wheat and barley varieties. (II) Varietal difference of wet-injury resistance of wheat and barley. Bull Division Plant Breed Cultivation, Tokai-Kinki, National Agricultural Experiment Station 2:11–16

Ikeda T, Higashi S, Kawaide T, Saigo S (1954) Studies on the wet-injury resistance of wheat and barley varieties. (I) Studies on the method of testing wet-injury resistance of wheat and barley varieties. Bull Division Plant Breed Cultivation, Tokai-Kinki, National Agricultural Experiment Station 1:21–26

Kearsey MJ (1998) The principles of QTL analysis (a minimal mathematics approach). J Exp Bot 49:1619–1623

Kozlowski TT (1984) Extent, causes, and impact of flooding. In: Kozlowski TT (ed) Flooding and plant growth. Academic, London, pp 9–45

Lemons e Silva CF, Mattos LAT de, Oliveira AC de, Carvalho FIF de, Freitas FA de, Anjos e Silva SD dos (2003) Flooding tolerance in oats. Crop Breed Appl Biotechnol 5:29–42

Li HB, Vaillancourt R, Mendham NJ, Zhou MX (2008) Comparative mapping of quantitative trait loci associated with waterlogging tolerance in barley (*Hordeum vulgare* L.). BMC Genomics 9:401

Malik AI, Colmer TD, Lambers H, Schortemeyer M (2001) Changes in physiological and morphological traits of roots and shoots of wheat in response to different depths of waterlogging. Aust J Plant Physiol 28:1121–1131

Mano Y, Muraki M, Fujimori M, Takamizo T, Kindiger B (2005a) Identification of QTL controlling adventitious root formation during flooding conditions in teosinte (*Zea mays ssp. huehuetenangensis*) seedlings. Euphytica 142:33–42

Mano Y, Muraki M, Fujimori M, Takamizo T (2005b) Varietal difference and genetic analysis of adventitious root formation at the soil surface during flooding in maize and teosinte seedlings. Jpn J Crop Sci 74:41–46

Mano Y, Muraki M, Komatsu T, Fujimori M, Akiyama F, Takamizo T (2002) Varietal difference in pre-germination flooding tolerance and waterlogging tolerance at the seedling stage in maize inbred lines. Jpn J Crop Sci 71:361–367

Mano Y, Muraki M, Takamizo T (2006) Identification of QTL controlling flooding tolerance in reducing soil conditions in maize (*Zea mays* L.) seedlings. Plant Prod Sci 9:176–181

Mano Y, Omori F, Muraki M, Takamizo T (2005c) QTL mapping of adventitious root formation under flooding conditions in tropical maize (*Zea mays* L.) seedlings. Breed Sci 55:343–347

Mazur BJ, Tingey SV (1995) Genetic mapping and introgression of genes of agronomic importance. Curr Opin Biotechnol 6:175–182

McKersie BD, Hunt LA (1987) Genotypic differences in tolerance of ice encasement, low temperature flooding, and freezing in winter wheat. Crop Sci 27:860–863

Muramatsu N, Kokubun M, Horigane A (2008) Relation of seed structures to soybean cultivar difference in pre-germination flooding tolerance. Plant Prod Sci 11:434–439

Musgrave ME, Ding N (1998) Evaluating wheat cultivars for waterlogging tolerance. Crop Sci 38:90–97

Oosterhuis DM, Scott HD, Hampton RE, Wullschleger SD (1990) Physiological responses of two soybean [*Glycine max* (L.) *Merr*] cultivars to short-term flooding. Environ Exp Bot 30:85–92

Pang JY, Cuin T, Shabala L, Zhou MX, Mendham NJ, Shabala S (2007) Effect of secondary metabolites associated with anaerobic soil conditions on ion fluxes and electrophysiology in barley roots. Plant Physiol 145:266–276

Pang JY, Mendham N, Zhou MX, Newman I, Shabala S (2006) Microelectrode ion and O_2 flux measurements reveal differential sensitivity of barley root tissues to hypoxia. Plant Cell Environ 29:1107–1121

Pang JY, Zhou MX, Mendham NJ, Shabala S (2004) Growth and physiological responses of six barley genotypes to waterlogging and subsequent recovery. Aust J Agric Res 55:895–906

Poysa VW (1984) The genetic control of low temperature, ice-encasement, and flooding tolerances by chromosomes 5A, 5B, and 5D in wheat. Cereal Res Comm 12:135–141

Qiu FZ, Zheng YL, Zhang ZL, Xu SZ (2007) Mapping of QTL associated with waterlogging tolerance during the seedling stage in maize. Ann Bot 99:1067–1081

Qiu JD, Ke YA (1991) Study of determination of wet tolerance of 4572 barley germplasm resources. Acta Agric Shanghai 7:27–32

Rathore TR, Warsi MZK, Singh NN, Vasal SK (1998) Production of Maize under excess soil moisture (Waterlogging) conditions. 2nd Asian regional maize workshop PACARD, Laos Banos, Phillipines, 23–27 February 1998, p 23

Reyna N, Cornelious B, Shannon JG, Sneller CH (2003) Evaluation of a QTL for waterlogging tolerance in Southern soybean germplasm. Crop Sci 43:2077–2082

Sachs MM (1993) Molecular genetic basis of metabolic adaptation to anoxia in maize and its possible utility for improving tolerance of crops to soil waterlogging. In: Jackson MB, Black CR (eds) Interacting stresses on plants in a changing climate. Springer-Verlag GmbH, Berlin, pp 375–393

Sayama T, Nakazaki T, Ishikawa G, Yagasaki K, Yamada N, Hirota N, Hirata K, Yoshikawa T, Saito H, Teraishi M, Okumoto Y, Tsukiyama T, Tanisaka T (2009) QTL analysis of seed-flooding tolerance in soybean (*Glycine max* [L.] *Merr.*). Plant Sci 176:514–521

Scott HD, DeAngulo J, Wood LS, Pitts DJ (1990) Influence of temporary flooding at three growth stages on soybeans grown on a clayey soil. J Plant Nutr 13:1045–1071

Setter TL, Burgess P, Waters I, Kuo J (1999) Genetic Diversity of barley and wheat for waterlogging tolerance in Western Australia. In 9th Australian Barley technical symposium, Melbourne, Australia. pp 2.17.1–2.17.7

Setter TL, Waters I (2003) Review of prospects for germplasm improvement for waterlogging tolerance in wheat, barley and oats. Plant soil 253:1–34

Taeb M, Koebner RMD, Forster BP (1993) Genetic variation for waterlogging tolerance in the *Triticeae* and the chromosomal location of genes conferring waterlogging tolerance in *Thinopyrum elongatum*. Genome 36:825–830

Takeda K, Fukuyama T (1986) Variation and geographical distribution of varieties for flooding tolerance in barley seeds. Barley Genet Newsl 16:28–29

Thomson CJ, Colmer TD, Watkin ELJ, Greenway H (1992) Tolerance of wheat (*Triticum aestivum* cvs. Gamenya and Kite) and triticale (*Triticosecale* cv. Muir) to waterlogging. New Phytol 120:335–344

van Ginkel M, Rajaram S, Thijssen M (1992) Waterlogging in wheat: germplasm evaluation and methodology development. Seventh regional wheat workshop for eastern, central and southern Africa, Nakuru, Kenya, 16–19 September 1991

VanToai TT, Beuerlein JE, Schmitthenner AF, Martin SK St (1994) Genetic variability for flooding tolerance in soybeans. Crop Sci 34:1112–1115

VanToai TT, Martin SK St, Chase K, Boru G, Schnipke V, Schmitthenner AF, Lark KG (2001) Identification of a QTL associated with tolerance of soybean to soil waterlogging. Crop Sci 41:1247–1252

VanToai TT, Nurjani N (1996) Screening for flooding tolerance of soybean. Soyb Genet Newsl 23:210–213

Varshney RK, Marcel TC, Ramsay L, Russell J, Röder MS, Stein N, Waugh R, Langridge P, Nike RE, Graner A (2007) A high density barley microsatellite consensus map with 775 SSR loci. Theor Appl Genet 114:1091–103

Wang F, Zhao TJ, Yu DY, Chen SY, Gai JY (2008) Inheritance and QTL analysis of submergence tolerance at seedling stage in soybean [*Glycine max* (L.) *Merr.*]. Acta Agron Sinica 34: 748–753

Wang SG, He LR, Li ZW, Zeng JG, Chai YR, Hou L (1996) A comparative study on the resistance of barley and wheat to waterlogging. Acta Agron Sinica 22:228–232

Wang J, Zhou MX, Xu RG, Lu C, Huang ZL (2007) Studies on selecting indices and evaluation methods for waterlogging tolerance in barley (*Hordeum vulgare* L.). Scientia Agric Sinica 40:2145–2152

Wenzl P, Carling J, Kudrna D, Jaccoud D, Huttner E, Kleinhofs A, Kilian A (2004) Diversity arrays technology (DArT) for whole-genome profiling of barley. Proc Nat Acad Sci 101:9915–9920

Wenzl P, Li HB, Carling J, Zhou MX, Raman H, Paul E, Hearnden P, Maier C, Xia L, Caig V, Ovesná J, Cakir M, Poulsen D, Wang JP, Raman R, Smith PK, Muehlbauer GJ, Chalmers KJ, Kleinhofs A, Huttner E, Kilian A (2006) A high-density consensus map of barley linking DArT markers to SSR, RFLP and STS loci and agricultural traits. BMC Genomics 7:206–227

Xiao YP, Wei K, Chen JX, Zhou MX, Zhang GP (2007) Genotypic difference in growth inhibition and yield loss in barley under waterlogging stress. J Zhejiang University (Agriculture and Life Science). 33:525–532

Xu RG, Lu C, Huang ZL, Huang ZR, Xu JH, Gong ZS (2005) Identification on Waterlogging Tolerance of Barley. Barley Sci 2:11–15

Yang J, Shen Q, Wang N, Li X, Yang W (1999) Wet endurance of Barley dwarf mutants. Acta Agriculturae Nucleatae Sinica 13:147–151

Yeboah MA, Chen XH, Liang GH, Gu MH, Xu CW (2008a) Inheritance of waterlogging tolerance in cucumber (*Cucumis sativus* L.). Euphytica 1620:145–154

Yeboah MT, Chen XH, Chen RF, Alfandi M, Liang GH, Gu MH (2008b) Mapping Quantitative Trait Loci for Waterlogging Tolerance in Cucumber Using SRAP and ISSR Markers. Biotechnology 7:157–167

Zhou MX, Li HB, Mendham NJ (2007) Combining ability of waterlogging tolerance in barley (*Hordeum vulgare* L.). Crop Sci 47:278–284

Index

A

ABA, 162
Acclimation, 80
ACC oxidase, 107
Acetaldheyde emission
    ALDH, 187
    ALDH inhibitor, 187
Acetic acid, 26
Acid, 171
    (elongation) growth, 90
    load, 165
Acidosis, 32, 79
Adenosine triphosphate (ATP), 85
Adventitious roots, 43, 154, 241
Aerenchyma, 7, 99–113, 154, 223, 243
Aeschynomene fluminensis, 48
Affinity, 25, 31
Agave, 152
Alanine, 162, 171
Alanine aminotransferase, 132
Alcohol dehydrogenase (ADH), 132,
    168, 243
    acetaldheyde reduction, 185
    ethanol oxidation, 185
Alcoholic fermentation, 122
Allele, 249
Alternative oxidase (AOX), 130, 132
Amino acids, 171
Aminocyclopropane-1-carboxylate
    (ACC), 99, 107
Ammonia permeation, 167
Ammonium, 29
Anaerobic conditions, 29

Anaerobic metabolism, 191
    alcohol dehydrogenase, 183
    fermentation, 183
    product
        acetate, 186
        ethanol, 186
    pyruvate decarboxylase, 183
Anaerobiosis, 4, 80
Anisohydric, 171
Anoxia, 27, 31
    acidification, 80
    cytoplasmic acidification, 83
    flooding-sensitive, 189
    flooding-tolerant, 189
    intolerance, 80
    tolerance, 91
Anoxic conditions, 29
Antioxidants, 133
Apoplastic acidification, 90
Apoplastic alkalinization, 83
Apoplast pathway, 167
AQP1, 161
Aquaporins, 42, 151, 158–171
    gating, 162
    genes, 159
Arabidopsis, 152
    A. thaliana, 166, 169
Arbuscular mycorrhizal (AM), 30
Aromatic-arginine region (ar/R), 161, 167
Ascorbate/dehydroascorbate, 169
Ascorbate peroxidases, 137–138
Ascorbic acid, 134–135
Asparagine-proline-alanine (NPA), 161

AtNIP1;1, 169
AtNIP1;2, 169
AtNIP2;1, 164, 170
ATP, 24, 26–28, 31, 32, 161
  decay, 87
  formation, 204
  hydrolytic activity, 208
AtPIP1;1, 169
AtPIP1;2, 166
AtPIP2;1, 166, 169
AtPIP2;2, 166, 167
AtPIP2;3, 166
AtPIP2;4, 169
AtTIP1:2, 164
AtTIP2;3, 169
AtTIP4:1, 164
AtUCP1, 129
Axial diffusion, 67
Azorhizobium caulinodans, 51

B

Bacopa monnieri, 231
Barriers to reduce oxygen, 156
Barrier to radial oxygen loss, 11
Batrachium trichophyllum, 232
Beta vulgaris, 160, 166, 168
Biochemical pH-stat, 80
Breeding, 260
Buffering, 25
Buffering strength, 26
Butyric acid, 26

C

Ca^{2+}-ATPase, 205, 232
Ca^{2+} channels, 206
Ca^{2+} concentration, 160
Ca^{2+}-dependent protein kinase, 168
Ca^{2+} effects on aquaporins, 168
Ca^{2+} influx, 168
Calmodulin, 100, 107, 112
Calmodulin-like Ca-binding protein, 132
Caproic acid, 26
Catalases, 132, 137–138
Cell energy, 204
Cell-to-cell pathway, 167
Cellular elimination process (CEP), 110
Cellular hydraulic conductivity, 165

Cellulase, 111
Cell walls, 170
Central (fifth) pore, 167
CH_4, 158
Channel activation, 84
Chemiosmotic principle, 89
Cl^-, 164
Class 1 hemoglobin, 128
CO_2, 63, 158, 161, 171
Coleoptiles, 90
Conjugated dienes, 124, 125
Convection, 13
Cotransport, 85
Critical oxygen pressure for respiration
  (COPR), 131
CTR1, 107
Cyanide, 85
Cymodocea nodosa, 229
Cytoplasmic acidosis, 122
Cytoplasmic pH, 31
Cytoplasmic pH change, 79
Cytosolic free Ca^{2+}, 168
Cytosolic pH, 165–168

D

Darcy's law, 67
Daucus carota, 170
Davis-Roberts hypothesis, 79
D_e, 28, 29
Desaturation, 170
DHA/ascorbate exchanger, 169
Diffusion, 24, 32
Diffusion coefficient, 25, 26, 28–29, 32, 67
Diffusivity, 67
Discolobium pulchellum, 47
Disrupt the membrane pH gradient, 167
Dithiothreitol, 170
Diversion of flow, 167
Divert flows, 167

E

Echinochloa phyllopogon, 124
EDS1, 107
Egeria, 228
Electrical signals, 82
Electrochemical gradient, 168
Electron transport chain (ETC), 122

Index 289

Eleocharis, 17
Elodea nuttalii, 228
Elongation rate, 29
Energy crisis, 79
Environment genotype interaction, 246
Epinasty, 254
Epoxy fatty acids, 124
ERF1, 107
Error signal, 79
Escape, 257
Ethane, 169
Ethanol, 171
Ethanolic, 162
Ethanolic fermentation, 85, 166
Ethylene, 30, 52, 105–110, 158
Ethylene binding to the receptor, 107
Ethylene glycol-*bis* -aminoethyl ether
  (EGTA), 106
Ethylene insensitive 2 (EIN2), 107
Ethylene insensitive 3 (EIN3), 107
ETR2, 132
Evaporation, 30–31
Expansins, 90, 111
Extensibility, 170
External calcium, 162

F
Fe, 28, 30, 32
Fe^{3+}, 28
Fenton reaction, 127, 169
Fermentable sugars, 191
Fermentation, 4
  ANPs, 183
Fermentative metabolism
  acetaldehyde, 184
  ethanol, 184
  lactate, 184
  pyruvate, 184
Fick's first law, 67
Fitness, 260
Flavonoids, 136
Flooded plants
  Suc, 190
  sugars, 190
Flooded trees
  acetaldehyde, 188

ADH activity, 188
  leaves, 186
Flooding, 262
Flooding stress
  acetaldehyde concentration, 187
  anoxia-related injuries, 187
Fractional porosity, 10
Free calcium, 105
Free fatty acids (FFA), 122
Fructokinase, 132

G
Gapped xylem mutant, 104
Gas diffusion, 67
Gas diffusion barrier, 39
Gas flow, 67
Gene activation, 93
Gene candidate, 260
Genetic diversity, 242
Genetic resources, 268–271
Genetic studies, 273–275
Genomic regions, 256
Germplasm, 284
Gibberellin, 53
Glutamic acid, 167
Glutathione, 134
Glyceraldehyde-3-phosphate
  dehydrogenase, 132
Glycerol, 161
Glycine decarboxylase complex
  (GDC), 99, 109
Glycolysis, 24, 27, 31
  metabolic depression, 190
  Pasteur effect, 190
  strategy, 190
Glycoprotein, 45
Guanylate cyclase, 128

H
HaberWeiss reaction, 127
Haemoglobin, 109
HAK, 232
H^+-ATPase activity, 205
H^+ circulation, 85
H^+ driving force (pmf), 89
Heartwood, 62
Heat stress, 105

Helianthus, 152
  H. annuus, 102
Henry's law, 69
Heritability, 241
High water permeabilities, 166
His 197, 167
Histidine kinases, 107
H^+ leak, 89
H_2O_2, 53, 106, 161, 168, 169
Hordeum marinum, 156
Hordeum vulgare, 102, 164, 168
Hormonal imbalance, 171
H^+ pump deactivation, 83
H^+ pumps, 205
H_2S, 29
H^+-translocating ATPase, 3
Humidity-induced pressurization/
  convection, 14–15
Hydraulic conductance, 151
Hydraulic redistribution, 167
Hydraulic signalling, 172
Hydrocharitaceae, 222
Hydrolic conductance, 31
Hydrolic conductivity, 31
Hydronium, 167
Hydrostatic gradient, 157
Hydroxycinnamate esters, 136
Hydroxyl, 167
Hypertrophied lenticel, 247
Hypoxia, 27, 29–32, 90, 247
  acetaldehyde, 185
  ethanol, 185

I

Infected zone, 44–45
Infection pocket, 51
Infection thread (IT), 50
Inhibitory subunit IF1, 130
Inner cortex, 44–45
Inositol phosphatides, 112
Ion toxicity, 162
Ion transport, 199
Isohydric, 171

J

Jasmonic acid, 53
Juncus effusus, 101, 102, 108, 118

K

K-525a, 112
K^+ channels, 210
K^+ efflux, 212
K^+ leak, 208
k_m, 26
K_m for oxygen of cytochrome oxidase, 109
K^+-outward rectifier channels, 164
K^+ transporters, 164
K^+ uptake, 206

L

Lactate dehydrogenase, 85, 132
Lactate formation, 79
Lactic acid, 27, 161, 162, 165, 170
Landoltia punctata, 231
Larix laricina, 157
Lateral root boundary (LRB), 50
Leaf, 182
  ADH, 183
  biochemistry, 191
  chlorosis, 211
  wilting, 171
Lemna
  L. gibba, 229
  L. minor, 231
Lemnaceae, 222
Lenticel, 43
Lignification, 156
Lignin, 136
Lipid
  hydroperoxides, 124
  nitration, 126
  peroxidation, 125–126
  peroxidation products, 169
  structure, 170
Loop D, 167
Lotus
  L. corniculatus, 49
  L. uliginosus, 49
Lowest elongated internode (LEI), 249
Ludwigia repens, 231
Lupinus, 153, 171
LVDTs, 68
Lycopersicon, 152
  L. esculentum, 102, 171
Lysigenous aerenchyma, 11, 100,
  104–110

Index

M
MAC236-glycoprotein, 45
MAC265-glycoprotein, 49
Maize, 168
Major intrinsic proteins (MIPs), 159
MAPKs, 107
Marker assisted selection, 275
Mechanical impedance, 105
Medicago, 108
Mehler reaction, 127
Membrane
 depolaris, 32
 fluidity, 170
 lipids, 124
 transporters, 198, 211–212
Microelectrode, 68
Microelectrode ion flux measurement, 208
MIPs, 170
Mitochondrial transition pore (MTP), 103
MMK3, 107
Monocarboxylic acid transporter, 209
MPK6, 107
Mycorrhizal, 29, 30

N
N, 32
NADPH oxidase (NOX), 100, 106, 109, 169
Na^+/H^+ antiporter, 229
Na^+/K^+ symporter, 229
Neptunia, 48
New set-point, 79
NH_4^+, 28, 29, 210
Nick-end labeling, 100, 104
Nitrate, 28, 165
 reductase, 122
 respiration, 28
Nitric oxide (NO), 109, 127, 161, 164, 171
Nitric oxide synthase (NOS), 109
Nitrite-NO reductase, 122
Nitrogenase, 46
Nitrogen starvation, 108
Nitrosothiols, 109
Nitrosylhemoglobin, 129
NO_2, 128
N_2O_2, 128
NO_3^-, 28, 29, 164, 165, 230
Nod factors, 50
NOD26-intrinsic proteins (NIPs), 159

Nodulation, 50–53
Nodule cortex, 42
NO_x, 93
NtAQP1, 167
NtPIP2:1, 167
Nucleotide triphosphates, 79, 86
Nutrient
 delivery, 199
 uptake, 23–32

O
OH^-, 169
$ONOO^-$, 109, 128
Organic acid anions, 85
Organic acids, 166
Oryza sativa, 101, 102, 114, 116, 156, 159
Osmotic gradients, 157, 161
Oxidative phosphorylation, 85
Oxygen (O_2), 63, 171, 198
 concentration, 8
 concentration profile, 7
 deficiency, 181
  biochemical/molecular level, 182
  energy availability, 188
 deprivation stress, 120
 diffusion barrier, 38–43
 microelectrodes, 9
 sensing, 161, 94
 sensors, 201–202
 stress, 80, 92
 transport, 6
 uptake, 212

P
PAD4, 107
Parenchyma, 62
Pasteur effect, 91, 162
Pectinase, 111
Pentatricopeptide repeat, 130
Permeability transition pore (PTP), 103
Peroxidases, 137–138
Peroxynitrite, 109
pH, 161
 gradients, 168
 for half-maximum inhibition, 166
 sensitivity, 167
 signals, 82

transmission, 83, 84
Phaseolus vulgaris, 102
Phenolic compounds, 136
Phenolics, 207
Phenotyping, 278
Phosphatidic acid, 124
Phospholipase D, 124
Phospholipid hydroperoxide glutathione
  peroxidase, 138
Phosphor, 30
Phosphorus (P), 29, 32, 108
Phosphorylation, 160, 165
Phragmites, 17
Physcomitrella patens, 159
Physiological drought, 151
P*i*, 230
PIP1, 160, 165, 167
PIP2, 160, 165, 167
Pisum sativum, 170
pKa, 170
Plasma membrane intrinsic protein (PIP),
  42, 159
Polarographic techniques, 7
Polysome
  regulation in protein, 189
  synthesis, 189
  translational level, 189
Populus trichocarpa, 171
Posidonia, 228
Preferential water uptake, 157
Pressure gradients, 161
Programmed cell death (PCD), 99–113
Propionic acid, 26
Protein kinase, 165
Protein kinase inhibitor, 112
Proton gradients, 166
Proton-motive-force (PMF), 90
Proton permeation, 167
Pyrophosphate, 205
Pyrophosphate-dependent glycolysis, 121
Pyruvate decarboxylase, 85, 132
Pyruvate phosphate dikinase, 122

Q

Quantitative trait loci (QTL), 248,
  275–278
Quiescence, 257

R

Radial diffusion, 68
Radial oxygen loss (ROL), 8, 107, 110
Reactive nitrogen species (RNS), 110, 120,
  127–129
Reactive oxygen, 161
Reactive oxygen species (ROS), 48, 53, 109,
  168–169
Recovery period, 186
Redox potential, 170
Reflection coefficient, 165
Respiration, 4
Rhizosphere, 18
Ricinus, 152
  R. communis, 131, 171
Root, 181
  cortical cells, 162
  death, 153–154
  diameter, 25, 32
  elongation, 25–28, 32
  morphology and anatomy, 153–158
  region, 157–158
  respiration, 103
  to shoot, 83
  tip, 153
Root hair curling (RHC), 50
Root hydraulic conductance (L_r),
  152–153
ROP (RHO-like small G protein), 169
  family proteins, 109
  signalling, 106
RopGAP4, 169
Rumex palustris, 108
Ruthenium red, 112

S

Sagittaria lancifolia, 101
Salicylic acid, 169
Salinity, 162
Sapwood, 62
Sapwood/heartwood boundary, 63
Sapwood respiration, 70
Schizogenous aerenchyma, 11, 100, 154
Scirpus, 100
Secondary
  aerenchyma, 43–44
  metabolites, 206–207
Selection, 256

Index 293

Selection criteria, 271
Sesbania rostrata, 48
SH groups, 170
Shoot elongation, 90
Signalling, 171–172
Slow vacuolar (SV), 170
Small basic intrinsic proteins (SIPs), 159
SO_4^{2-}, 28
Soil ecology, 18
Solanum tuberosum, 170
SoPIP2:1, 167
SoPIP2;1, 168
SOS1, 229
Soybean, 39
Sparganium emersum, 232
Spartina patens, 101, 102
Species, 171
Spinacia, 165
    S. oleracea, 167
Spirodela polyrrhiza, 231
Stomata, 29, 32, 171
Strong-ion-difference, 81
Sub-1, 241
Suberin, 156
Suberisation, 156
Submergence, 254
Succinate, 92
Sucrose synthase, 132
Sulphide, 26, 29
Sulphur (S), 32
Superoxide dismutases (SODs), 109, 127, 136
SUT1, 170

T
Tannins, 136
Thermal osmosis, 15
Thiol groups, 109
TIP1;1, 169
TIP1;2, 169, 171
TIP4;1, 171
Tocopherol (Vitamin E), 126, 135–136
Tonoplast, 89
Tonoplast intrinsic protein (TIP), 42, 159
Toxic secondary metabolites, 198
Tracheary element development, 104
Transmembrane gradients, 79

Transpiration, 30
Trans-tonoplast H^+ gradient, 89
Tricarboxylic acid cycle (TCA), 162
Trifolium subterraneum, 102, 111
Tripsacum dactyloides, 101
Triticum, 103
    T. aestivum, 102, 157, 165, 166, 168, 169
Tropical legume, 47
Tulipa gesnerina, 168
TUNEL, 104, 112
Turgor pressure, 153, 165
Typha, 17

U
Ubiquinone cytochrome b, 129
Uncoupling, 124
Ureide, 45

V
Vacuolar pH, 89
Vacuolar processing enzyme (VPE), 110
Vallisneria
    V. natans, 231
    V. spiralis, 223
Venturi-induced convection, 15
Viscoelastic hysteresis loops, 170
Vitis vinifera, 159
V_{max}, 31
Volatile and toxic compounds, 158
Volatile fatty acids, 207
Volumetric elasticity modulus, 170

W
W-7, 112
Water
    permeability, 158–171
    potential, 162
Waterlogging, 191, 198, 241
Waterlogging tolerance, 212
Wetland legumes, 46–47
Whole-genome response to oxygen deprivation, 132
WRKY transcription factors, 132

X

Xanthine oxidase, 127
Xanthine oxidoreductase, 128
Xenopus, 166
Xenopus oocytes, 166
Xylem
    loading, 200
    vessels, 153
Xyloglucan endotransglycosidases
    (XET), 112

Y

Yeast, 166

Z

Zea mays, 100, 102, 104, 114, 115,
    152, 170
Zinc finger protein ZAT12, 132
Zn, 32
Zostera marina, 222